This book is dedicated to Bill Characklis, the founder and former director of the Center for Interfacial Microbial Process Engineering at Montana State University. Bill was a man of courage, passion, and vision. We look back with gratitude at the challenges Bill offered us and the moments we shared.

«« **William G. Characklis 1941–1992** »»

BIOFOULING and BIOCORROSION in Industrial Water Systems

Edited by
Gill G. Geesey
Zbigniew Lewandowski
Hans-Curt Flemming

LEWIS PUBLISHERS
Boca Raton Ann Arbor London Tokyo

Library of Congress Cataloging-in-Publication Data

Biofouling and biocorrosion in industrial water systems/edited by Gill G. Geesey,
Zbigniew Lewandowski, Hans-Curt Flemming.
p. cm.
Includes bibliographical references and index.
ISBN 0-87371-928-X
1. Water-supply, industrial. 2. Fouling. I. Geesey, G. G.
II. Lewandowski, Zbigniew. III. Flemming, Hans-C. (Hans-Curt), 1947–
TD353.B54 1993
628.1—dc20 93-929625
CIP

Lewis Publishers is an imprint of CRC Press

No claim to original U.S. Government works
International Standard Book Number 0-87371-928-X
Library of Congress Card Number 93-29625
Printed in the United States of America 1 2 3 4 5 6 7 8 9 0
Printed on acid-free paper

Acknowledgment

We wish to acknowledge the American Chemical Society for providing the forum for the symposium on ''Biofouling and Biocorrosion in Water Systems'', and, for providing a supplement to cooperative agreement ECD8907039 between the National Science Foundation and Montana State University for supporting travel and registration costs of the symposium organizers and speakers. We also thank Alma Roe and Susan Cooper for organizational and editorial support, respectively.

Contributors

Peter Angell
Department of Microbiology
University of Surrey
Guildford, England

Mark L. Angles
School of Microbiology and Immunology
University of New South Wales
Kensington, NSW, Australia

Philip J. Bremer
Seafood Research Laboratory
Port Nelson, New Zealand

Anne K. Camper
Center for Biofilm Engineering
Montana State University
Bozeman, Montana

Anthony H. L. Chamberlain
Department of Microbiology
University of Surrey
Guildford, England

William G. Characklis
(Deceased)
Center for Biofilm Engineering
Montana State University
Bozeman, Montana

Ching-I Chen
Biology and Biotechnology Research Program
Lawrence Livermore National Laboratory
Livermore, California

J. W. Costerton
Center for Biofilm Engineering
Montana State University
Bozeman, Montana

Wolf R. Fischer
Korrosionsschutztechnik
Markische Fachhochschule
Iserlohn, Germany

Hans-Curt Flemming
Institute for Sanitary Engineering, Water Quality and Solid Waste
University of Stuttgart
Stuttgart, Germany

Gill G. Geesey
Department of Microbiology and Center for Biofilm Engineering
Montana State University
Bozeman, Montana

Amanda E. Goodman
School of Microbiology and Immunology
University of New South Wales
Kensington, NSW, Australia

Thomas Griebe
Center for Biofilm Engineering
Montana State University
Bozeman, Montana

W. Allan Hamilton
Department of Molecular and Cell Biology
Marischal College
University of Aberdeen
Aberdeen, Scotland

Slawomir W. Hermanowicz
Department of Civil Engineering
University of California
Berkeley, California

Joanne Jones-Meehan
Naval Surface Warfare Center
Biotechnology Laboratory
Silver Spring, Maryland

W. L. Jones
Civil Engineering Department and Center for Biofilm Engineering
Montana State University
Bozeman, Montana

C. William Keevil
Pathology Division
PHLS Centre for Applied Microbiology and Research
Porton Down, Salisbury, England

Frederico Lage-Filho
Department of Civil Engineering
University of California
Berkeley, California

WhonChee Lee
Center for Biofilm Engineering
Montana State University
Bozeman, Montana

Zbigniew Lewandowski
Civil Engineering Department and Center for Biofilm Engineering
Montana State University
Bozeman, Montana

Brenda J. Little
Naval Research Laboratory
Stennis Space Center, Mississippi

Florian B. Mansfield
Department of Materials Science and Engineering
University of Southern California
Los Angeles, California

Kevin C. Marshall
School of Microbiology and Immunology
University of New South Wales
Kensington, NSW, Australia

Richard McDonogh
Institute for Chemische Verfahren
University of Stuttgart
Stuttgart, Germany

Per H. Nielson
Environmental Engineering Laboratory
University of Aalborg
Aalborg, Denmark

Satoshi Okabe
Department of Civil and Environmental Engineering
Miyazaki University
Miyazaki, Japan

Kaye N. Power
School of Microbiology and Immunology
University of New South Wales
Kensington, NSW, Australia

Richard I. Ray
Naval Research Laboratory
Stennis Space Center, Mississippi

Harry F. Ridgeway
Orange County Water District
Fountain Valley, California

Frank L. Roe
Center for Biofilm Engineering
Montana State University
Bozeman, Montana

Gabriela Schaule
Institute for Sanitary Engineering, Water Quality and Solid Waste
University of Stuttgart
Stuttgart, Germany

Rene P. Schneider
School of Microbiology and Immunology
University of New South Wales
Kensington, NSW, Australia

Rohini Srinivasan
Center for Biofilm Engineering
Montana State University
Bozeman, Montana

Philip S. Stewart
Chemical Engineering Department and Center for Biofilm Engineering
Montana State University
Bozeman, Montana

Hector A. Videla
Bioelectrochemistry Section
INIFTA, University of La Plata
La Plata, Argentina

Dirk Wagner
Korrosionsschutztechnik
Markische Fachhochschule
Iserlohn, Germany

Patricia Wagner
Naval Research Laboratory
Stennis Space Center, Mississippi

Marianne Walch
Naval Surface Warfare Center
Biotechnology Laboratory
Silver Spring, Maryland

Jimmy Walker
Pathology Division
PHLS Centre for Applied Microbiology and Research
Porton Down, Salisbury, England

Oskar Wanner
Department of Computer and System Sciences
Swiss Federal Institute for Water Resources and Water Pollution Control (EAWAG)
Duebendorf, Switzerland

Bryan Warwood
Center for Biofilm Engineering
Montana State University
Bozeman, Montana

Ralf Waters
Center for Biofilm Engineering
Montana State University
Bozeman, Montana

Eric Wentland
Center for Biofilm Engineering
Montana State University
Bozeman, Montana

H. Xiao
Department of Materials Science and Engineering
University of Southern California
Los Angeles, California

Nick Zelver
Center for Biofilm Engineering
Montana State University
Bozeman, Montana

Table of Contents

Introduction: Biofouling/Biocorrosion in Industrial Water Systems

Gill G. Geesey

The accumulation of microorganisms and related products at equipment surfaces contributes to a variety of phenomena that reduce performance and/or lifetime of industrial materials. Biofilm buildup on pipe surfaces increases fluid frictional resistance, reduces heat transfer efficiency on heat exchange surfaces, and, under certain conditions, causes pitting corrosion of a variety of industrially important metal alloys. The costs to industry for treatment and repair of these and other biofilm-related processes increases yearly, adding significantly to operating costs that cut into profits and competitiveness.

Biofilms consist of a variety of microorganisms, the products they excrete, and other particles that become trapped in the biofilm matrix or slime. Processes leading to biofilm accumulation on surfaces vary according to system geometry, hydrodynamics, and environmental conditions. Our present understanding of biofilm processes limits the degree to which we can control biomass accumulation on industrial surfaces. For example, the application of nonspecific biocides typically involves a trial-and-error approach until acceptable protection to the system is achieved. Stricter requirements and higher costs associated with the use, handling, and disposal of biocides leave fewer alternatives for biofouling control, thereby providing the motivation for advancing our understanding of biofilm processes. Certain environmental conditions appear to control matrix polysaccharide production in some biofilm-forming bacteria. By establishing conditions that minimize matrix polysaccharide production, biofilm accumulation may also be minimized. In many cases, attachment of a layer of cells to a substratum will not cause problems. Rather it is the accumulation of cells mediated by the matrix polysaccharide that leads to unacceptable biofilm accumulation.

Efficient control of undesirable biofilm reactions on industrial surfaces requires a fundamental understanding of biofilm ecology, identification of the critical biological reactions responsible for fouling-related problems, knowledge of the rate-limiting steps and substrates for the reactions, and the development of novel, practical approaches to reduce the rates of these reactions to acceptable levels. Cooperation between scientists, engineers, and water system operators is required to achieve these goals.

Integration of efforts to achieve the goals described above may be realized through a process analysis approach. This approach has been employed by Characklis et al. to address biofouling problems in industry.[1] It begins with an understanding of the influence of substratum effects on initial colonization events at the microscale.[2] Subtle heterogeneities in surface topography and chemistry may be responsible for the patchiness in cell distribution and activity in mature biofilms, which are responsible for or contribute to localized attack on the substratum.[3] It includes an understanding of the hydrodynamics at the bulk liquid-biofilm interface and its influence on mesoscale phenomena such as mass, heat, and momentum transfer.[4] These mesoscale phenomena are the symptoms of a number of biofilm-related industrial problems. Mesoscale test-bed facilities may be used to simulate and model interfacial microbial processes likely to be important in a full-scale industrial process. For example, biomass accumulation, detachment, and downstream plugging in mesoscale systems can be evaluated using design and operational features characteristic of industrial-scale systems, yet can be controlled, monitored, and modeled more conveniently and economically than in a full-scale system. Understanding how operating parameters influence key interfacial microbial processes through well-designed monitoring programs will ultimately lead to more effective control of biofouling and biocorrosion problems at an industrially relevant scale.

The process analysis approach requires knowledge of the stoichiometry and rates of biofilm processes.[1] This type of information has been obtained primarily at the mesoscale through quantitative measurements of metabolites in the bulk fluid that enter and exit a fixed-film bioreactor. While such an approach may be acceptable for describing biofilm phenomena such as heat transfer resistance and fluid frictional resistance, it offers little opportunity to characterize localized reactions within the biofilm that contribute to phenomena such as detachment (sloughing) and pitting corrosion.

In order to characterize localized reactions within biofilms and their consequent effects in the system, it is necessary to use sensitive, nondestructive, real-time, analytical techniques that are capable of defining the chemical properties and reactions at or near a surface. Studies that combine microelectrodes (see Lewandowski, Chapter 11) and attenuated total reflectance spectroscopy[5] with the powerful, albeit destructive, surface-sensitive techniques (such as secondary ion mass spectrometry and Auger and X-ray photoelectron spectroscopy[6]) provide the opportunity to relate chemical changes at a substratum to biological distributions and interactions based on microscopic observations. From these types of studies evolve the conceptual models that become the objects of scrutiny at the different scales of evaluation.

Model development and verification are also important aspects of a process analysis approach to understand and control biofouling and biocorrosion. A tripartite modeling strategy that involves the formulation of conceptual, mathematical, and physical models through collaborations among laboratory experimentalists, mathematicians, and computer scientists provides an efficient way to evaluate the relevance of specific biofilm processes. Conceptual models based on laboratory-scale observations are transformed into mathematical expressions which, in turn, are assembled into computer models that predict interfacial microbial processes, such as biofilm accumulation in a pipeline or porous media,[4] oxygen removal rates at a biofilm-containing surface (see Chapter 11), or pit penetration times and frequencies over biofilm-coated metal surfaces susceptible to microbially influenced corrosion (see Mansfeld, Chapter 7). The computer models can be used to perform sensitivity analyses of potentially important interfacial microbial process parameters, for predicting biofilm behavior under untested conditions, or in conjunction with biofilm monitoring equipment and biocide applications to control fouling, corrosion, or other undesirable surface-associated biological phenomena.

A symposium entitled "Biofouling/Biocorrosion in Water Systems", organized by W. G. Characklis — founder and first director of the Center for Interfacial Microbial Process Engineering* at Montana State University — was held in August, 1992 at the Annual National Meeting of the American Chemical Society in Washington, D.C. The symposium brought together researchers with expertise in biofilm processes to discuss the state of knowledge of microbial activities on surfaces that impact industrial systems. This forum provided the opportunity to introduce analytical chemists and surface scientists to biologically mediated surface reactions and to encourage them to focus their powerful analytical capabilities and expertise on these industrially important, microbial problems.

This text summarizes the information that was discussed in the symposium, expanding the knowledge of biofilm processes that has been

* In March, 1993 the name of the National Science Foundation Engineering Research Center was changed to the Center for Biofilm Engineering.

published previously.[7–11] The papers prepared by the speakers and their co-investigators are arranged in three sections: the first section introduces the reader to the biofilm mode of growth, the second section describes the consequences of biofouling on various industrial surfaces and how biocides may be applied to control the fouling phenomenon, and the third section focuses on biofilm activities that promote localized corrosion of various types of industrial surfaces.

REFERENCES

1. **Bryers, J.D. and Characklis, W.G.,** Biofilm accumulation and activity: a process engineering analysis, in *Biofilms — Science and Technology,* Melo, L. et al., Eds., NATO Advanced Studies Institute Series, Kluwer Academic Publishers, Doordricht, 1992, 221.
2. **Mueller, R.F., Characklis, W.G., Jones, W.L., and Sears, J.T.,** Characterization of initial events in bacterial surface colonization by two *pseudomonas* species using image analysis, *Biotechnol. Bioeng.,* 39, 1161, 1992.
3. **Costerton, J.W., Geesey, G.G., and Jones, P.A.,** Bacterial biofilms in relation to internal corrosion monitoring and biocide strategies, *Mater. Perform.,* 27 (4), 49, 1988.
4. **Cunningham, A.B., Characklis, W.G., Abedeen, F., and Crawford, D.,** Influence of biofilm accumulation on porous media hydrodynamics, *Environ. Sci. Technol.,* 25 (7), 1305, 1991.
5. **Bremer, P.J. and Geesey, G.G.,** Laboratory-based model of microbially induced corrosion of copper, *Applied Environ. Microbiol.,* 57, 1956, 1991.
6. **Jolley, J.G., Geesey, G.G., Hankins, M.R., Wright, R.B., and Wichlacz, P.L.,** Auger electron spectroscopy and X-ray photoelectron spectroscopy of the biocorrosion of copper by gum arabic, BCS and *Pseudomonas atlantica* exopolymer, *J. Surface Interface Anal.,* 11, 371, 1988.
7. **Mittelman, M.W. and Geesey, G.G., Eds.,** *Biological Fouling of Industrial Water Systems: A Problem Solving Approach,* Water Micro Associates, San Diego, 1987.
8. **Characklis, W.G. and Wilderer, P.A., Eds.,** *Structure and Function of Biofilms,* John Wiley & Sons, Chichester, 1989.
9. **Characklis, W.G. and Marshall, K.C., Eds.,** *Biofilms,* John Wiley & Sons, New York, 1990.
10. **Dowling, N.J., Mittelman, M.W., and Danko, J.C., Eds.,** *Microbially Influenced Corrosion and Biodeterioration,* University of Tennessee, Knoxville, 1990.
11. **Flemming, H.-C. and Geesey, G.G., Eds.,** *Biofouling and Biocorrosion in Industrial Water Systems,* Springer-Verlag, Heidelberg, 1991.

Section I
BIOFILM STRUCTURE AND FUNCTION

1

Structure of Biofilms

J. W. Costerton

I. INTRODUCTION

When natural, industrial, and medical ecosystems were examined by direct microscopic methods, it soon became apparent that most of the bacteria in aquatic systems grow in biofilms attached to available surfaces. Some of these biofilms were seen to be thick and highly structured, as in algal "mats" and dental plaque, and some were seen to be simple accretions of microcolonies of bacterial cells embedded in a hydrated exopolysaccharide matrix. This initial realization[1] that biofilms were the predominant mode of bacterial growth in most aquatic systems, including many of great interest and importance to man, came as a profound shock to microbiologists who had, collectively, been studying cells in the planktonic mode of growth for more than 200 years. The importance of the recent American Chemical Society Symposium on Biofouling/Biocorrosion in Water Systems depends, in large measure, on the extent to which bacterial cells in biofilms differ from planktonic cells of the same species floating in the bulk fluid. If biofilm cells and planktonic cells are structurally and functionally identical, the study of biofilms becomes a somewhat esoteric and scholarly exercise. If biofilm bacteria are substantially different from planktonic bacteria, most of the fundamental concepts of microbiology must now be reexamined because these concepts have been based entirely on the study of planktonic cells. If the differences between planktonic cells and biofilm cells are profound, we must expect to be able to explain many microbial phenomena that

0-87371-928-X/94/$0.00+$.50

have not yet been rationalized because the old microbiological model — the planktonic cell growing in monospecies culture in a rich medium — does not reflect reality. Because microbiology involves the study of very small organisms, it has necessarily been extrapolative, and extrapolation from a model to reality is only accurate if the model approximates reality.

It is entirely fitting to highlight the importance of choosing models that approximate reality, which was the contribution made by Bill Characklis to the study of biofilms. As an engineer, Bill was inherently free of the preconceptions and conventions of the traditional microbiologist, and he had a natural instinct for direct examination of real systems; therefore, it was to Bill and his precise engineering methods that founding members of the biofilm group turned for quantitation and confirmation of direct observations.

II. THE STRUCTURE OF BIOFILMS

While thin accretions of bacteria on glass surfaces can readily be examined by light microscopy, fully formed biofilms were too complex and too refractile to yield useful images by this method.

For this reason, we relied on electron microscopy for our initial explorations of biofilm structure. Electron microscopy was informative only if the viewer could mentally rehydrate the images, because both scanning electron microscopy (SEM) and transmission electron microscopy (TEM) require the rigorous removal of water from the specimen, and the biofilm matrix is 99% water.[2] However, TEM images of bacterial biofilms (Figure 1) showed condensed residue of the biofilm matrix that invoked the living hydrated structure, and SEM images showed relatively rigid bacterial cells among the condensed amorphous remnants of a biofilm matrix (Figure 2) that could be mentally rehydrated. These images of biofilms, which are collected, described, and interpreted in a 1979 review,[3] allowed one to deduce the presence of bacterial cells and a hydrated matrix, and most biofilm diagrams of that era show these cells randomly distributed through a homogenous hydrated matrix. We guessed that bacteria and algae formed microcolonies within the homogenous matrix (Figure 3), but hard spatial data were lacking.

The modern use of the confocal scanning laser (CSL) microscope has allowed the examination of hydrated biofilms by a totally nonintrusive technique that yields[4] clean, three-dimensional images (Figure 4) of living biofilms in real time. This revolutionary new technique has shown[5] that more of the volume of bacterial biofilms is occupied by matrix (±75 to 95%) than by bacterial cells (±5 to 25%) and that these cells may be concentrated either in the lower or upper regions of these biofilms. Detailed spatial examinations[6] have shown that cells of many species

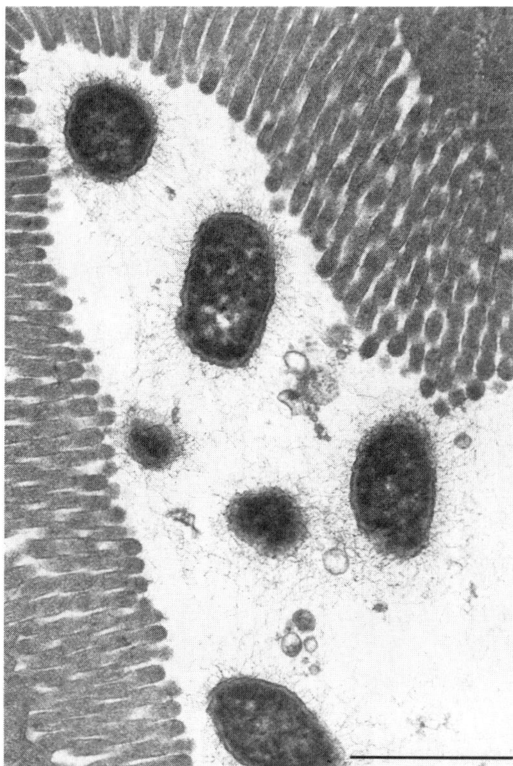

FIGURE 1. Transmission electron micrograph of a ruthenium red-stained preparation of cells of *Escherichia coli* that had formed a biofilm on the microvillar surface of the gut of a calf. Note the dehydration-condensation of glycocalyces to produce radial patterns of fibrils in regions that would be fully occupied by the polysaccharide matrix in its living hydrated state. The bar indicates 1.0 μm.

tend to form microcolonies within the biofilm and that the areas of the matrix between these aggregates of sister cells may be occupied by "water channels" within which the matrix is less dense (Figure 5). Because CSL microscopy allows the continuous observation of living biofilms, Lawrence and colleagues[7] have been able to monitor the initial stages of biofilm formation and to describe how the different post-adhesion "behaviors" of cells of various species produce local aggregates of cells that may become well-defined microcolonies within mature biofilms.

While the study of monospecies biofilms allows us to determine the surface colonization strategies of a particular species, we must remember that single-species surface populations are exceedingly rare in nature. When we examined a wide variety of natural biofilms, we noticed that the microcolonies of particular morphotypes (e.g., curved rods) were

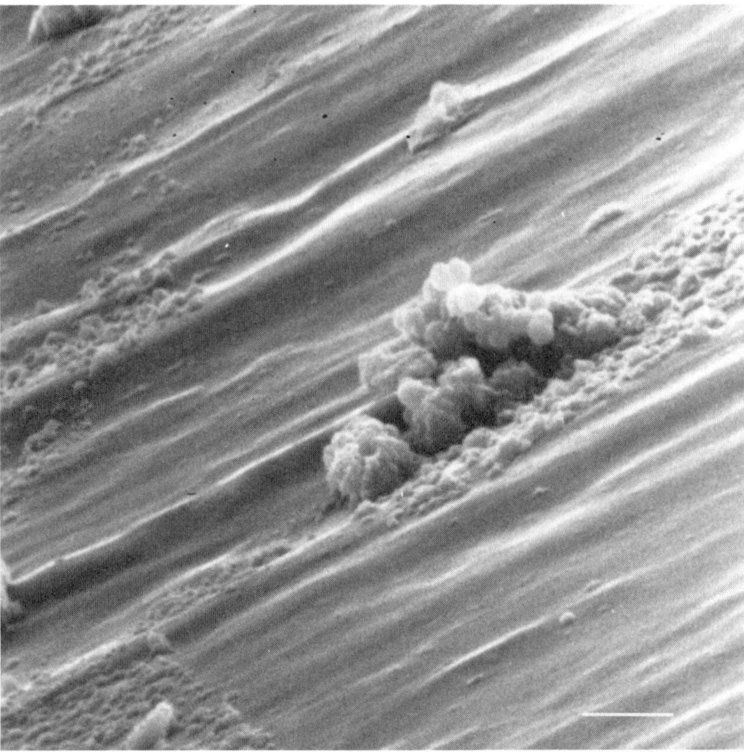

FIGURE 2. Scanning electron micrograph of a critical-point, dried preparation in which a microcolony of cells of *Staphylococcus aureus* had formed on the battery pack of a cardiac pacemaker. Note the dehydration-condensation of the matrix material that would have entirely covered the rigid bacterial cells in their living hydrated state. The bar indicates 5 μm.

often juxtaposed to microcolonies of another morphotype (e.g., a filamentous cyanobacterium) within the mature biofilm. The frequency and intimacy of these associations led us to suggest[8] that cells of one species may associate with those of another species for mutual physiological benefit. Since that time, the concept of microbial consortia has emerged, and this exciting perception is intimately bound up with biofilms because it is only in stable, multispecies aggregates that the spatial relationships necessary for physiological cooperation can be established and maintained. It is obvious that physiological cooperation between planktonic cells of different species floating past each other in the bulk fluid is difficult to sustain, but we can readily visualize how similar cells immobilized in adjacent, or even mixed, microcolonies within a biofilm could initiate and sustain the cooperative exchange of metabolic substrates and products. Two properties of the bacterial cell equip it perfectly

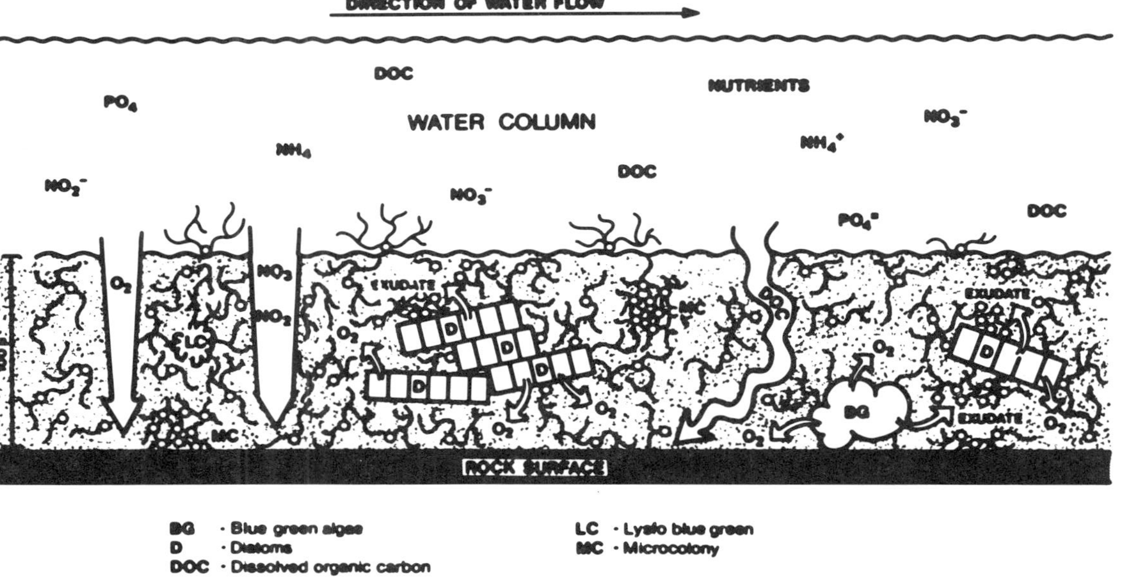

FIGURE 3. Diagrammatic representation of a mixed natural biofilm in which bacteria and algae are present as microcolonies within a homogeneous matrix.

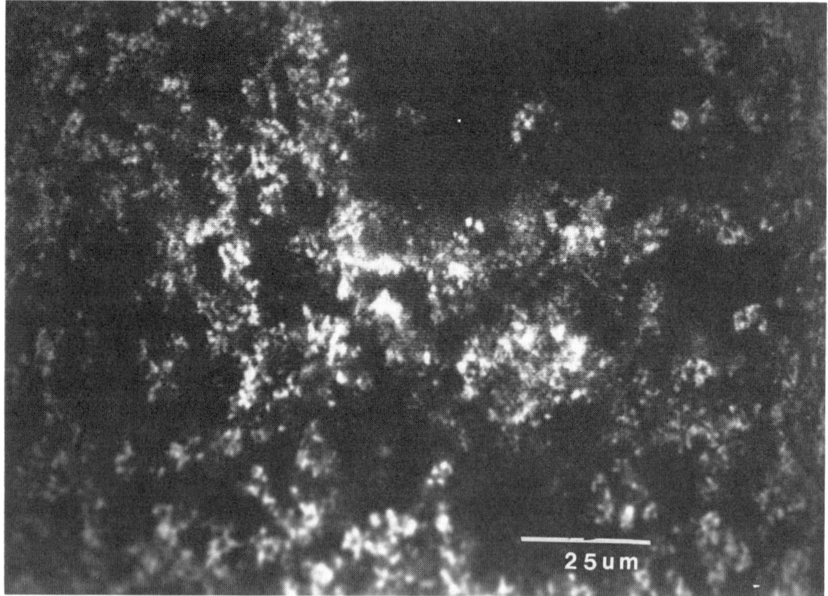

FIGURE 4. Confocal scanning laser micrograph showing the development of microcolonies and water channels within a 24-h biofilm formed by *Escherichia coli*.

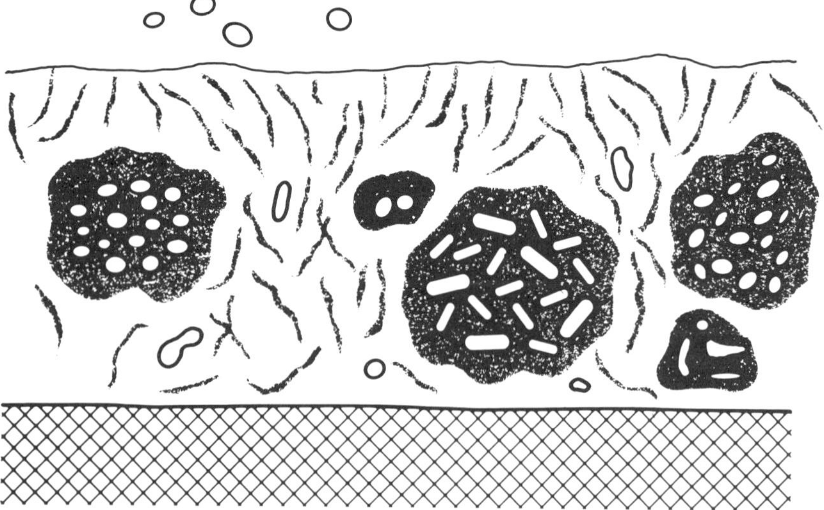

FIGURE 5. Diagrammatic representation of the confocal scanning laser microscope data showing that most biofilm bacteria grow in microcolonies surrounded by dense matrix material interspersed between water channels within which the matrix material is less dense.

for life in the complex, highly structured, physiologically cooperative world of the biofilm — binary fission and phenotypic plasticity.

III. BINARY FISSION AND PHENOTYPIC PLASTICITY

Organisms that reproduce exclusively by sexual process are essentially restricted by their environments, and a shallow pond that freezes to near the bottom may offer poor fishing even though the ecosystem teems with trout food on a brilliant summer day. A finite number of propagules survive the least permissive state of the environment and these survivors can fatten in response to positive changes in the environment, but they cannot increase their number until the next breeding cycle. In contrast, bacteria can respond to positive changes in their environment in a matter of minutes, and reproduction by asexual binary fission persists until the environment again becomes limiting. Oligotrophic aquatic systems are heavily populated with ultramicrobacteria[9] that are dormant because of nutrient limitation. When nutrients suddenly enter the ecosystem, the dormant bacteria resuscitate[10] and adhere avidly to this source of organic nutrients, eventually forming biofilms of macroscopic proportions. At the cellular level within a biofilm, a bacterial cell that finds itself in a nonpermissive niche simply persists as a single cell, while another identical cell that encounters permissive conditions proliferates to produce a microcolony. Figure 6 illustrates the response of two identical cells of a hypothetical bacterium that find themselves in different microenvironments. In part A, the cell persists in spite of nutrient limitation, while in part B, the cell is provided with an essential organic substrate by its proximity to a neighboring microcolony of cooperative organisms and proliferates, favoring the nutrient gradient, until it forms a functional consortium within the biofilm. An extensive morphological survey of the highly structured biofilms found in natural and industrial aquatic systems clearly indicates that the vast majority of sessile bacteria grow within microcolonies that are juxtaposed to microcolonies of cooperative organisms or to the surface of a nutritive substrate.[8] It is the property of rapid reproduction by binary fission, in response to favorable changes in the microenvironment, that builds the remarkably efficient consortia seen in biofilms.

Because the small size and mode of growth of bacterial cells generally deny them homeostasis, the cells must respond to rapidly changing environments in order to survive. The extreme phenotypic plasticity exhibited by bacteria remained unknown by microbiologists for centuries because of the custom of removing organisms from complex natural biofilms and growing them as Pasteur did, as planktonic cells in monospecies cultures in rich media. Brown and Williams[11] showed that they

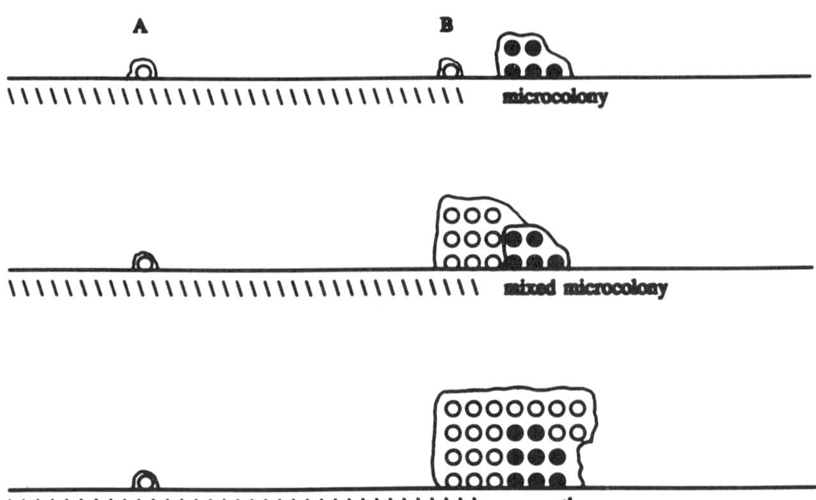

FIGURE 6. Diagrammatic representation of the different responses of two identical, newly adherent cells that find themselves in (A) a nutrient-poor surface area and (B) a surface area supplied with nutrient by a microcolony of a physiologically cooperative organism. The progeny of the cell at B eventually form a mixed microcolony, and then a functional consortium, with the cooperative organism.

could take virtually any microbial genotype, including the redoubtable *Escherichia coli* K12, and alter most of its physiological processes and even the structure of its cell wall by altering its environment within a chemostat. This remarkable plasticity in response to changing environmental conditions is achieved by rapid, directed, phenotypic adaptation, in contrast to slow random changes in the genotype, and this property of bacteria serves them very well in growth within biofilms. Marshall's group[12] has now shown that adherence to a surface itself constitutes an environmental change that depresses certain genes and triggers the production of specific gene products, and Brown et al.[13] have noted profound phenotypic differences between planktonic and biofilm cells of the same genotype.

These two bacterial properties — reproduction by binary fission and phenotypic plasticity — introduce a phenomenal, inherent heterogeneity into bacterial biofilms that stands in stark contrast to the relative homogeneity of planktonic cells floating or swimming in a bulk fluid. Biofilm bacteria are so responsive to changes in their microenvironments that some cells of a particular genotype (species) may be effectively dormant, while other cells of the same genotype living only a few microns away may be proliferating wildly (see Figure 6) and assuming phenotypic forms that differ profoundly from their dormant counterparts and even more profoundly from planktonic cells of the same species. These statements are not extrapolative, but are based on direct structural and physiological observations of living natural systems. We can now begin to question the usefulness of continuing conventional microbiological studies of planktonic bacteria in monospecies cultures in rich media when the intent of the researchers is to extrapolate their data to explain microbial processes in natural, industrial, and medical ecosystems.

IV. DIFFERENCES BETWEEN BIOFILM CELLS AND PLANKTONIC CELLS

The direct demonstration that most bacteria in aquatic systems live in highly structured biofilms and the perception that bacteria are capable of reacting very precisely to microenvironments within these complex sessile communities suggest that biofilm bacteria may differ very substantially from planktonic organisms of the same genotype. As in all sweeping generalizations, it may be useful to examine individual factors that contribute to these potential differences and to attempt a correlation of their cumulative effects in a summary.

A. Adhesion Events

Many microbiologists have attempted to model the process of bacterial adhesion using physical principles that apply to the adhesion of inert colloidal particles to surfaces, but these data have not been useful in understanding the adhesion of wild strains of bacteria in natural ecosystems. This is because wild bacteria are not regular spheres and rods, but are "hairy" cells whose actual surfaces are composed of carbohydrate (glycocalyx) and protein (pili and flagella) appendages, and because bacteria are capable of "sensing" the surface and responding by phenotypic change and by specific behaviors.[4] The responses of bacteria during the process of adhesion range from the activation of the algC gene by *Pseudomonas aeruginosa*[14] to the complete loss of flagella and the initiation of stalk formation by *Caulobacter crescentus*. Furthermore, in the short period immediately following adhesion, bacterial cells carry out lateral movements[7] involving specific patterns of exopolysaccharide secretion and which are species specific and clearly related to their new microenvironment.

B. Matrix Effects

The chemical nature of the biofilm matrices produced by bacteria varies widely between species,[2] but most are carbohydrate in nature and many are highly anionic, uronic, acid-containing polymers. The production and extrusion of these polymers is a very exergonic process that involves many genes whose clustering and conservative patterns of regulation suggest that their cumulative product — the matrix — is extremely important to the survival and success of the species. Planktonic cells may be surrounded by relatively small exopolysaccharide glycocalyces, but cells of most species have been shown to elaborate very large amounts of this highly hydrated exopolymer during and immediately following adhesion.[7] Within minutes of the initiation of colonization, adherent bacteria can be seen to be surrounded by matrix material and, within hours, matrix-enclosed microcolonies are formed that eventually coalesce to produce confluent biofilms.

The loss of Brownian movement is one of the first visible signs that a bacterial cell has adhered to a surface or has been entrapped in a biofilm. The immobilization within an ionic polymer radically changes the state of the surface of a cell that has heretofore been free-floating in bulk fluid. Instead of the relatively free diffusion of molecules and ions to and from the cell surface of a mobile cell, the biofilm cell experiences the effects of a matrix that closely resembles an ion-exchange resin. The biofilm cell is stationary within a fibrous matrix that extends in all directions within which large amounts of ions (including divalent cations) and structured water are associated. Molecules and ions that would readily approach planktonic cells must satisfy all of the spatial and chemical requirements of the matrix in order to even approach the biofilm cell. Similarly, molecules and ions produced by the biofilm cell must satisfy the same criteria to diffuse away from the cell, and some of these molecules and ions may be retained as a "halo" around the biofilm cell by virtue of their size or charge density. The direct observation of a ± 0.5-μm zone of reduced pH around living cells in a biofilm[6] may indicate that protons extruded by these bacterial cells form a "halo" in the vicinity of the biofilm cells, whereas they would be lost to the menstruum by planktonic bacteria. In this sense, a biofilm cell can be considered to live in a microniche that is defined by the biofilm matrix and partially conditioned by its own metabolic activity. Further heterogeneity is introduced into the matrix component of the microniche in which biofilm bacteria grow by the fact that the physical nature of the matrix may change from a sol to a gel when the concentration of Ca^{2+} in the matrix-associated water is raised.[15] This Ca^{2+}-mediated physical change in the alginate matrix of biofilms of *P. aeruginosa* radically alters the penetration of tobramycin into these biofilms, indicating that the diffusion limitation imposed by the matrix varies with changes in the concentration of this divalent cation. These data suggest that growth within a structured exopolysaccharide matrix makes the microenvironment of biofilm cells radically different from the bulk-fluid environment of their planktonic counterparts, and that these differences produce many important physiological differences between cells in these very different modes of growth.

Because the extent to which the biofilm matrix constitutes a barrier to diffusion has been called into question,[16] it may be useful to carefully separate speculation from data and to designate two unequivocal examples of diffusion limitation. The diffusion of oxygen into biofilms is limited to the extent that fastidious anaerobes grow very vigorously in the depths of well-developed biofilms in aerobic systems.[8] Bacterial cells within biofilms are notably resistant to very high concentrations of biocides and antibiotics,[17] but these cells return to exquisite antibiotic sensitivity just 2 to 3 min after these biofilms have been scraped from the colonized surface and dispersed by vortex mixing and ultrasonication. Brown and colleagues[18] have suggested that growth rate has a profound effect on the sensitivity of bacterial cells to various antibiotics, and they have also suggested that bacteria within a biofilm exhibit a very wide range of growth rates, depending on their local microenvironments. It is clear that the matrix of the biofilm excludes both antibodies[19] and phagocytic cells,[20] and biofilm bacteria are protected from these host defense mechanisms, even when a developing population of adherent bacteria consists only of a few scattered glycocalyx-enclosed microcolonies.[20]

C. Microcolony Effects

The direct examination of monospecies bacterial biofilms by CSL microscopy has clearly shown that the cells of most species tend to form microcolonies.[4] Dextran penetration studies have shown that these microcolonies are surrounded by dense matrix material and are separated by "water channels" where the matrix material is less dense (see Figure 5). The existence of these water channels was confirmed by Fourier transform infrared spectroscopy studies of antibiotic penetration,[21] which showed very rapid penetration of these agents to the bottom of the biofilm without significant killing of the bacteria in well-protected microcolonies. This predominance of microcolonies within monospecies biofilms has been demonstrated by direct morphological and physiological examinations, and it provides a very useful model for the understanding of the essential differences between planktonic and biofilm bacteria. Microcolonies are formed when individual cells within a biofilm find themselves in a permissive microenvironment, and these monospecies aggregates are really the basic units of sessile life. They have access to growth substrates from the water channels and from neighboring microcolonies (see Figure 5), and they have the corporate ability to generate large amounts of enzymes, which remain in the microenvironment[8] and digest available nutrient substrata (e.g., cellulose). The growth of a successful microcolony is limited only by the tendency to reduce the access of individual cells to essential nutrients if large size creates diffusion limitations. A very wide range of physiological cooperations has clearly been recorded within established biofilms in natural and industrial aquatic ecosystems, and each of these cooperations involves the interaction of bacteria in microcolonies that may even reach macroscopic size. A case in point is the methanogenic granules that form spontaneously in wastewater treatment systems[22] and consist of macroscopic microcolonies of methanogens surrounded by concentric layers of acetogenic and heterotrophic organisms. These natural microbial consortia carry out organic degradation and methanogenesis at truly phenomenal rates. Similarly, the microbial digestion of cellulose to produce volatile fatty acids and biomass in rumen systems is dependent on bacterial microcolonies within highly structured biofilms.[23] In the biofilms that form on the cellulose, the rate of cellulose digestion is dependent on the development of consortia in which molecular products (butyrate) are removed from the primary cellulolytic organisms to prevent feedback inhibition and promote maximum rates of digestion.[24] A survey of very large numbers of mixed biofilms formed in natural systems leads us to conclude that most of their component bacteria live in microcolonies within which they are juxtaposed to specific substrates (e.g., cellulose) and/or to microcolonies of physiologically cooperative bacteria. All of the studies of monospecies biofilms confirm that bacteria have evolved the adhesion mechanisms and the rapid plastic responses to favorable environments that enable them to fit themselves into the highly structured physiological mosaic of functioning developed biofilms. Biofilms in the form of algal mats, soil crusts, and aquatic biofilms cover much of the surface of the earth, often in very challenging ecosystems. If these adherent populations were multicellular organisms, which they often resemble in their structural and physiological complexity, they would be considered to be phenomenally successful.

V. CONCLUSION

A biofilm cell is different from its planktonic counterpart because it lives in a microniche that limits its contact with the bulk fluid. It is protected from phagocytic cells, bacteriophage, and antibodies, and nutrients and antagonists must reach it via a *de facto* ion-exchange resin that limits their diffusion. Molecules and ions produced by the cell are similarly diffusion limited and the microniche may be heavily conditioned by protons, organic molecules, and bacterial enzymes. Because each cell lives in a distinct microenvironment, even within a microcolony, adjacent cells of the same species may have very different growth rates and very different phenotypic characteristics in response to environmental conditions in its own microniche. Biofilms are seen to be very highly structured, and we can begin to understand how these very plastic and responsive cells can develop together to form multispecies consortia within which each individual cell is provided with nutrients and stimulated by end-product removal. None of the thousands of microenvironments within a mature biofilm would be expected to approximate that of a planktonic cell, and we must therefore consider biofilm cells and planktonic cells to be profoundly different.

REFERENCES

1. **Costerton, J. W., Geesey, G. G., and Cheng, K.-J.,** How bacteria stick, *Sci. Am.*, 238, 86, 1978.
2. **Sutherland, I. W., Ed.,** *Surface Carbohydrates of the Prokaryotic Cell*, Academic Press, London, 1977.
3. **Costerton, J. W.,** The role of electron microscopy in the elucidation of bacterial structure and function, *Annu. Rev. Microbiol.*, 33, 459, 1979.
4. **Lawrence, J. R., Korber, D. R., Hoyle, B. D., Costerton, J. W., and Caldwell, D. E.,** Optical sectioning of microbial biofilms, *J. Bacteriol.*, 173, 6558, 1991.
5. **Caldwell, D. E., Korber, D. R., and Lawrence, J. R.,** Imaging of bacterial cells by fluorescence exclusion using scanning confocal laser microscopy, *J. Microbiol. Methods*, 15, 249, 1992.
6. **Caldwell, D. E., Korber, D. R., and Lawrence, J. R.,** Confocal laser microscopy and computer image analysis in microbial ecology, in *Advances in Microbial Ecology*, Vol. 12, Marshall, K. C., Ed., Plenum Press, New York, 1992, 1.
7. **Lawrence, J. R., Korber, D. R., and Caldwell, D. E.,** Behavioral analysis of *Vibrio parahaemolyticus* variants in high and low viscosity microenvironments using digital image processing, *J. Bacteriol.*, 174, 5732, 1992.
8. **Costerton, J. W., Cheng, K.-J., Geesey, G. G., Ladd, T. I., Nickel, J. C., Dasgupta, M., and Marrie, T.J.,** Bacterial biofilms in nature and disease, *Annu. Rev. Microbiol.*, 41, 435, 1987.
9. **Kjelleberg, S. and Hermansson, M.,** Starvation-induced effects on bacterial surface characteristics, *Appl. Environ. Microbiol.*, 48, 497, 1984.
10. **Lappin-Scott, H. M., Cusack, F., MacLeod, A., and Costerton, J. W.,** Starvation and nutrient resuscitation of *Klebsiella pneumoniae* isolated from oilwell waters, *J. Appl. Bacteriol.*, 64, 541, 1988.
11. **Brown, M. R. W. and Williams, P.,** The influence of environment on envelope properties affecting survival of bacteria in infections, *Annu. Rev. Microbiol.*, 38, 527, 1985.
12. **Marshall, K.,** proceedings of the American Chemical Society Symposium on Biofouling/Bioerrosion in Water Systems.
13. **Brown, M. R. W., Costerton, J. W., and Gilbert, P.,** Leading articles — extrapolating to bacterial life outside the test tube, *J. Antimicrob. Chemother.*, 27, 565, 1991.

14. **Davies, D. G., Chakrabarty, A. M., and Geesey, G. G.,** Exopolysaccharide production in biofilms: substratum activation of alginate gene expression, *Appl. Environ. Microbiol.*, in press.
15. **Hoyle, B. D. and Costerton, J. W.,** Transient exposure to a physiologically-relevant concentration of calcium confers tobramycin resistance upon sessile cells of *Pseudomonas aeruginosa, FEMS Microb. Lett.*, 60, 339, 1989.
16. **Nichols, W. W., Evans, M. J., Slack, M. P. E., and Walmsley, H. L.,** The penetration of antibiotic into aggregates of mucoid and non-mucoid *Pseudomonas aeruginosa, J. Gen. Microbiol.*, 135, 1291, 1989.
17. **Nickel, J. C., Ruseska, I., Wright, J. B., and Costerton, J. W.,** Tobramycin resistance of *Pseudomonas aeruginosa* cells growing as a biofilm on urinary catheter material, *Antimicrob. Agents Chemother.*, 27, 619, 1985.
18. **Brown, M. R. W., Allison, D. G., and Gilbert, P.,** Resistance of bacterial biofilms to antibiotics: a growth-rate related effect?, *J. Antimicrob. Chemother.*, 22, 777, 1988.
19. **Jensen, E. T., Kharazmi, A., Lam, K., Costerton, J. W., and Hoiby, N.,** Human polymorphonuclear leukocyte response to *Pseudomonas aeruginosa* grown in biofilms, *Infect. Immun.*, 58, 2383, 1990.
20. **Ward, K. H., Olson, M. E., Lam, K., and Costerton, J. W.,** Mechanism of persistent infection associated with peritoneal implants, *J. Med. Microbiol.*, 36, 406, 1992.
21. **Jass, J., Hoyle, B. D., Mantsch, H. H., and Costerton, J. W.,** Examination of *Pseudomonas aeruginosa* biofilm by attenuated total reflectance Fourier transform infrared spectroscopy, unpublished data, 1991.
22. **MacLeod, F. A., Guiot, S. R., and Costerton, J. W.,** Layered structure of bacterial aggregates produced in an upflow anaerobic sludge bed and filter reactor, *Appl. Environ. Microbiol.*, 56, 1598, 1990.
23. **Cheng, K.-J., Fay, J. P., Howarth, R. E., and Costerton, J. W.,** Sequence of events in the digestion of fresh legume leaves by rumen bacteria, *Appl. Environ. Microbiol.*, 40, 613, 1980.
24. **Kudo, H., Cheng, K.-J., and Costerton, J. W.,** Interactions between *Treponema bryantii* and cellulolytic bacteria in the *in vitro* degradation of straw cellulose, *Can. J. Microbiol.*, 33, 244, 1987.

Section I
BIOFILM STRUCTURE AND FUNCTION

2

Analysis of Bacterial Behavior During Biofouling of Surfaces

Kevin C. Marshall, Kaye N. Power, Mark L. Angles, René P. Schneider, and Amanda E. Goodman

I. INTRODUCTION

Biofilms usually consist of a heterogeneous array of microorganisms immobilized near a substratum surface by being embedded in an organic polymer matrix of microbial origin.[1] Whereas biofilms are essentially ubiquitous in flowing aqueous environments, they are not necessarily uniform in space and time in any particular situation. The process of biofilm formation begins immediately following exposure of a clean substratum surface to a natural aquatic ecosystem. Organic molecules rapidly adsorb to the surface followed by adhesion of microorganisms, particularly by a variety of bacteria. It is the continued adhesion and growth of these bacteria, followed by excessive excretion of extracellular polymers, that lead to the formation of a mucilagenous biofilm (or slime) at the substratum surface.

It is our intention in this chapter to discuss certain aspects of bacterial behavior in relation to biofilm development, as well as to emphasize the importance of the process of immobilization of bacteria at surfaces on the physiological functions of bacteria exposed to such conditions. It is our contention that reliance on physiological studies of bacteria in aqueous environments does not always provide an accurate assessment of the behavior of the bacteria in a biofilm environment.

0-87371-928-X/94/$0.00+$.50

II. CELL GROWTH AND DIVISION AT SURFACES

A. Conditioning Films

A range of organic molecules rapidly adsorbs to a solid surface following its immersion in an aqueous environment. These molecules include macromolecules, such as proteins, polysaccharides, nucleic acids, and humic acids, and smaller hydrophobic molecules, such as fatty acids, lipids, and pollutants such as DDT, polyaromatic hydrocarbons, and polychlorinated biphenyls. These adsorbed molecules alter the physicochemical properties of the substratum surface and, hence, are termed conditioning films (CFs). In particular, alterations to the surface charge and the surface free energy of the immersed substratum are readily demonstrated.[2]

From the point of view of biofilm formation, some effects of different CFs might be to alter the numbers of adherent bacteria (Figure 1), their relative strength of adhesion, or the nature of the adherent populations, or to provide a concentration of nutrients that will encourage the subsequent growth of adherent bacteria. In this latter sense, biofilm bacteria may be particularly effective in the degradation of pollutant molecules adsorbed at surfaces.

B. Adhesion of Starved Bacteria

Bacteria in many natural habitats are confronted with (1) very low nutrient concentrations and (2) an unbalanced concentration of the few nutrients that are available. Hence, such bacteria may exhibit small size and reduced metabolic activity in order to enhance their chances for survival.[3] In some bacteria, this starvation-survival phase is associated with an increase in adhesiveness, and it has been postulated that adhesion of starved bacteria to surfaces possessing a CF provides an opportunity for growth of some of these starved bacteria under conditions where little or no growth occurs in the aqueous phase.[4] The primary colonizing bacteria of surfaces exposed in marine waters were found to be very small bacteria,[5] and the subsequent appearance of normal size bacteria was later assumed to be the result of growth of starved bacteria to normal size at the nutrient-enriched surface.[6] This possibility has been tested experimentally as detailed below.

C. Cell Growth

Cell growth of starved marine bacteria at surfaces has been observed in a dialysis flow chamber by means of phase-contrast microscopy combined with time-lapse video techniques.[7,8] In one instance, starved cells of *Vibrio* DW1 were allowed to adhere to the dialysis membrane of the

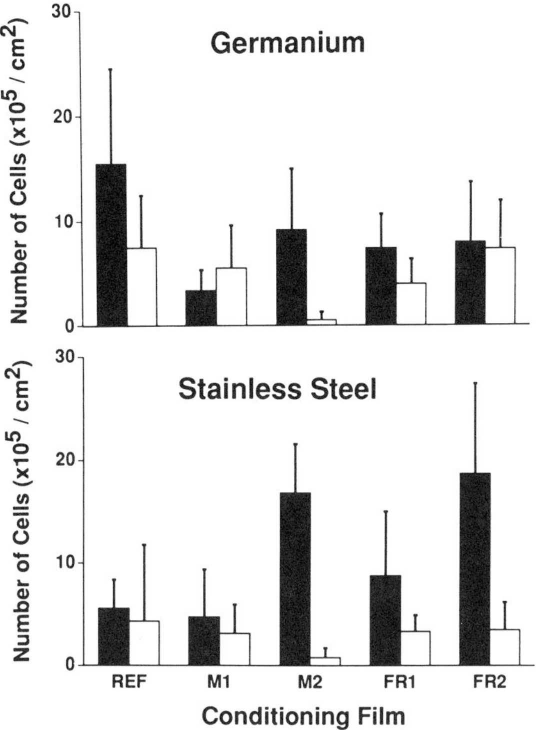

FIGURE 1. Effects of conditioning films (CFs) on adhesion on carbon (solid bars) or nitrogen-limited (open bars) SW8 to different substrata. Bars represent the mean number of cells retained on a particular substratum-CF combination. The standard deviation for each experiment is indicated by the error bars. CF: M1, marine site 1 (Coogee Beach); M2, marine site 2 (Balmoral Beach); FR1, freshwater lake 1 (Centennial Park); FR2, freshwater lake 2 (Centennial Park).

chamber, and a medium containing 1/1000th the normal C and N concentrations was perfused through the apparatus.[7] Although the cells remained in a starvation condition in the aqueous medium, the individual bacteria at the surface were found to grow to normal size as a result of nutrient concentration at the membrane surface. Similarly, cell growth of starved bacteria at the surface was observed when the only source of energy substrate in the chamber was stearic acid adsorbed to the membrane surface.[8] This "surface feeding" response was shown for a highly adhesive organism, *Pseudomonas* JD8, as well as for a completely nonadhesive organism, *Vibrio* MH3.[8,9]

D. Cell Division

Biofilm development is ultimately dependent upon population growth of the initial surface-colonizing bacteria and upon copious production of extracellular polymer by these colonizing organisms. In the examples intensively studied for growth responses at surfaces, all have exhibited population growth (active cell division) following growth of initially starved cells in response to nutrients accumulated at a surface. What is remarkable are the different reproductive strategies employed by the different test organisms when they are confined at a surface. *Vibrio* DW1 adhered to the surface in a perpendicular manner and, while division resembled binary fission, in reality it was asymmetric in the sense that the adhering "mother" cell remained attached and released an unlike, motile "daughter" cell.[7] *Pseudomonas* JD8 attached to the surface in a face-to-face manner and certainly divided by binary fission, but the two daughter cells then proceeded to migrate slowly away from each other across the membrane surface.[8] This behavior was interpreted as a purely physicochemical effect, whereby the cells utilized the fatty-acid substrate bound at the substratum surface, leaving a more hydrophilic surface to which the bacteria did not bind strongly. Limited movement of the cells resulted in a further encounter with the hydrophobic fatty acid-coated surface and a relatively firmer adhesion until that limited amount of substrate was utilized. Repetition of this cycle of events would account for the observed slow migration of cells across the surface.

Vibrio MH3, on the other hand, was nonadhesive; however, cells adjacent, but not attached, to the surface exhibited constriction of the bacteria following cell growth as expected for classical binary fission. At this stage, however, the apparently dividing cells always drifted away from the surface (i.e., out of focus at the plane of the membrane surface).

It was assumed that these cells completed the division cycle in the aqueous phase, because daughter-size cells were regularly observed returning to the surface (all other cells in the aqueous phase were starving and, thus, very much smaller than the daughter cells).

III. DETACHMENT STRATEGIES

In addition to purely physical means of removal of microorganisms from surfaces, bacteria have evolved a variety of strategies to ensure a regular release of cells from the immobilized state to the aqueous phase. These strategies are detailed below.

A. Direct Release of Daughter Cells

Many bacteria in nature attach to surfaces in a perpendicular (or edge-to-face) fashion, such as reported for *Flexibacter* CW7[10] and *Vibrio* DW1.[7] Division in these bacteria is best described as a "mother-daughter" relationship, because the "mother" cell remains at the surface and continues to produce further progeny[7] while the "daughter" cell is motile and swims away from the surface. This strategy ensures the return of numerous bacteria to the aqueous phase, where the motile cells are free to seek other suitable sites for colonization.

B. Alteration in Cell Properties

Detachment of the hydrophobic bacterium *Acinetobacter calcoaceticus* from the surface of oil droplets has been correlated with the production of a hydrophilic extracellular polymer, emulsan, by the bacteria.[11] Fattom and Shilo[12] showed that the attachment of benthic cyanobacteria similarly was associated with hydrophobic surface properties of the organisms and that the same bacteria in the planktonic phase possessed hydrophilic surface properties. Consequently, some bacteria are able to control the relative extent of attachment and rate of detachment by altering their cell surface hydrophobicity.

C. Alteration in Substratum Properties

Substratum properties may be modified as a result of molecular adsorption to the surface (see Section II.A) or by subsequent utilization of adsorbed organic molecules by bacteria present on the surface. As discussed in Section II.D, *Pseudomonas* JD8 utilized the hydrophobic surface-bound substrate, stearic acid, resulting in a limited degree of detachment of the individual cells.[8] After prolonged exposure to the surface, the bacteria utilized most of the limited amount of surface-bound substrate with the result that the majority of the bacteria detached from the surface. Thus, some bacteria are capable of modifying the substratum properties to a point where they engineer their own release from the surface.

D. Reversible Adhesion

As noted in Section II.D, *Vibrio* MH3 is an example of a nonadhesive bacterium that can approach a surface and can scavenge surface-bound nutrients for metabolism[9] and for both cell and population growth.[8] Once cell division is well under way, these bacteria appear to detach from the surface and return to the aqueous phase. It would seem that this particular mechanism of feeding without ever attaching firmly to the surface is a particularly efficient means of taking advantage of surface concentrations of nutrients with maximum opportunities for return to the water phase to seek alternative sources of nutrients.

E. Shear Forces

In flowing aquatic systems, the most obvious mechanism for detachment of bacteria from surfaces is the shear forces generated by the water motion past the surface. Such forces may be moderate under conditions of laminar flow, but increase rapidly as the flow becomes more turbulent.[13]

IV. GENE EXPRESSION AT SURFACES

A. Logic Involved

The boundary properties between two different phases in natural habitats are entirely different from those of the individual phases themselves.[14] Consequently, it might be expected that these different properties exert some control over the expression of certain genes in bacteria confined at an interface. Numerous reports are found in the literature indicating that certain physiological characteristics of bacteria on a surface are different from those found in the same organisms suspended in an aqueous phase. Lorian[15] has documented numerous examples of distinctive changes in bacterial properties when grown on solid media. An example of such differences is the change from swimmer to swarmer cell type that several species of *Vibrio, Proteus,* and *Serratia* undergo. In liquid, the swimmer cells are motile by means of one or several flagella, often polar, whereas on an agar surface the cells differentiate to the swarmer state by the production of numerous lateral flagella. McCarter and Silverman[16] have elucidated the molecular control of this phenomenon in *V. parahaemolyticus*. They have shown that the single polar flagellum acts as a dynamometer which senses the slowing in the rotation of the polar flagellum in a viscous environment and triggers the expression of the *laf* gene, resulting in the production of numerous lateral flagella. Expression of the *laf* gene simultaneously requires conditions of iron limitation.[16]

We considered that agar is not a solid surface in the normal sense and, consequently, set out to show that some genes not expressed in liquid or on solid media can be expressed at an inert solid surface.

B. Technique Employed

We have employed the plasmid vector pJO100, described by Östling et al.,[17] to transfer the mini-Mu transposon containing a promoterless reporter gene (*lacZ*)[18] into the marine *Pseudomonas* S9.

Transposon mutants that failed to express β-galactosidase (the gene product of *lacZ*) on agar or in liquid media containing the substrate X-gal were selected.[19] These mutants were then tested on a solid surface by placing 50 μl of each culture grown in liquid medium containing X-gal into a polystyrene microtiter well in order to provide a very large surface-to-volume ratio.

C. Results Obtained

Typically, about 1 in 200 to 300 mutants expressed β-galactosidase in the presence of the polystyrene surface, but still failed to express the gene on agar medium containing X-gal.[19] The behavior of the mutants was checked by incubating the mutants in liquid medium containing X-gal (15 ml in a polystyrene tube) for 3 h, at which time there was no evidence of the blue reaction product of β-galactosidase and X-gal. A 200-μl aliquot was placed in a polystyrene petri dish, and both the dish and the tube were incubated for a further 3 h, after which an intense blue color was observed in the droplet on the dish, but no color developed in the tube despite a total incubation of 6 h.[19]

Such a result indicates that some genes are turned on at a solid surface, despite not being expressed in liquid or on semisolid media. It is likely that other genes are turned off at surfaces. By the use of transposon-induced mutants, it should be possible to clone the genes involved and to use the mutants to study the mechanism(s) involved in the triggering process.

V. MECHANISMS OF INDUCTION OF GENE EXPRESSION

A. Nutrients

As discussed in Section II.A, organic molecules rapidly adsorb to freshly immersed surfaces, and these molecules are subject to metabolism by bacteria adhering to the surfaces (see Sections II.C and II.D). Consequently, bacteria requiring specific substrates that partition at surfaces may exhibit adaptation to these substrates only when in contact with surfaces where the substrate is available.

B. Surfactants

Surfactants localized at surfaces have been shown to stimulate endogenous respiration, oxygen uptake, and heat output in starved bacteria at surfaces in the absence of exogenous substrate.[20] Since many bacteria excrete surfactants in natural habitats, it is feasible that such surfactants adsorb at surfaces and modify the physiological behavior of bacteria when they adhere to these surfaces.

C. pH Changes

Since most surfaces in nature have or rapidly acquire a net negative surface charge, cations, including protons, are attracted to the surfaces as counter ions.[14] Substantial accumulation of protons at a surface will result in a significant reduction in the pH near the surface as opposed to that in the bulk aqueous phase. This lower pH may be directly involved in influencing gene expression at a surface, or it may have an indirect effect by altering the transmembrane potential of bacteria adhering to the surface.[21] Molecular genetic studies, employing transposon mutagenesis, have identified acid-inducible genes in *Escherichia coli* and *Salmonella typhimurium*, as well as alkali-inducible genes in *E. coli*.[22]

D. Other Physicochemical Effects

As noted in Section IV.A, increased viscosity may play a role in determining gene expression in some bacteria associated with a surface.[16] This effect may also be linked with a reduction in water activity or change in water structure in the vicinity of a surface. Another possibility is a deformation of the bacterial cell wall as a result of forces operative at the solid-liquid interface.[23]

VI. GENE TRANSFER IN BIOFILMS

Gene transfer between bacteria has been demonstrated in soil and freshwater habitats[24,25] and in microcosms in the presence of an appropriate selection pressure.[26] Since biofilms represent a high concentration of essentially immobilized microorganisms, we have attempted to demonstrate plasmid transfer in marine biofilms in the absence of any selection pressure.[27]

A. System Employed

The recipient organism, a hydrophobic marine bacterium termed SW5, was employed to establish biofilms on hydrophilic or hydrophobic beads in a flowing reactor system using artificial seawater plus a carbon source. Following biofilm formation, the hydrophilic donor strain, *Vibrio* S142, harboring the conjugative broad host range plasmid RP1, was introduced into the reactor. Transconjugants were selected on a medium containing streptomycin to counterselect the donors and kanamycin to counterselect the recipients.

B. Gene Transfer without Selection Pressure

In the absence of any antibiotic selection pressure, transconjugants were detected in the aqueous phase, among the loosely attached cells, and among the firmly attached biofilms in reactors containing either hydrophilic or hydrophobic beads (Table 1).[27] Despite the fact that the numbers of transconjugants recovered from each phase were similar, transfer frequencies/(recipient × donor), as described by Jones et al.,[28] were highest among cells firmly attached to both hydrophobic (20-fold) and hydrophilic (100-fold) beads. The unstable nature of the biofilm became evident when loosely and firmly attached cells were expressed as a percentage of the total number of cells in the biofilm. It was found that 96 and 97% of donors, 87 and 98% of recipients, and 89 and 94% of transconjugants were loosely attached in the biofilm on hydrophobic and hydrophilic beads, respectively. These results demonstrate that a

TABLE 1. Plasmid Transfer Among Bacteria in a Marine Biofilm Environment

Reactor	Treatment	Source	Cell numbers			Plasmid transfer frequency/ (recipient × donor)
			SW5H	S142(RP1)	Transconjugants	
1	Hydrophilic glass beads; biofilm formed for 170 h	Cells in aqueous phase[a]	1.5×10^8	1.4×10^7	7.2×10^2	3.4×10^{-13}
		Cells loosely attached in biofilm[b]	2.3×10^8	7.3×10^8	5.4×10^3	3.2×10^{-12}
		Cells firmly attached in biofilm[b]	4.6×10^6	2.5×10^5	3.3×10^2	2.9×10^{-10}
2	Hydrophobic glass beads, biofilm formed for 170 h	Cells in aqueous phase[a]	4.6×10^7	1.5×10^8	2.5×10^2	3.6×10^{-14}
		Cells loosely attached in biofilm[b]	1.4×10^8	5.1×10^6	8.5×10^2	1.2×10^{-12}
		Cells firmly attached in biofilm[b]	2.1×10^7	2.1×10^5	1.0×10^2	2.3×10^{-11}

[a] Colony-forming units (cfu) ml^{-1}.
[b] cfu cm^{-2} of bead.

poorly adhesive plasmid donor strain is capable of interacting with potential recipient cells in a biofilm to yield transconjugants, even in the absence of any imposed selection pressure.

VII. CONCLUSIONS

Biofilm formation at surfaces is an inevitable consequence of a series of events beginning with adsorption of organic molecules to the substratum surface, followed in turn by bacterial adhesion, cell growth, and, finally, cell reproduction and exocellular polymer formation to build up the biofilm matrix. Although a variety of strategies exist for cell detachment and, hence, return of some bacteria to the aqueous phase, the rates of adhesion and reproduction of the adhering cells must exceed the detachment rate to ensure the observed rapid biofilm development.

It is becoming increasingly obvious that bacteria immobilized at a surface often exhibit physiological properties that are not found in the same organisms in the aqueous phase. Further investigations on the nature of these changes and the conditions at surfaces inducing them are urgently required.

Biofilms are essentially concentrations of microorganisms in natural environments and, as such, they represent ideal sites for gene transfer in nature, especially where a selection pressure exists, but this does not appear to be an obligatory requirement.

VIII. ACKNOWLEDGMENTS

This work was supported in part by grants to K.C.M., R.P.S., and A.E.G. from the Australian Research Council.

REFERENCES

1. **Characklis, W. G. and Marshall, K. C.,** Biofilms: a basis for an interdisciplinary approach, in *Biofilms,* Characklis, W. G. and Marshall, K. C., Eds., Wiley-Interscience, New York, 1990, chap. 1.
2. **Baier, R. E.,** Substrate influence on adhesion of microorganisms and their resultant new surface properties, in *Adsorption of Microorganisms to Surfaces,* Bitton, G. and Marshall, K. C., Eds., Wiley-Interscience, New York, 1980, chap. 3.
3. **Morita, R. Y.,** Starvation-survival of heterotrophs in the marine environment, *Adv. Microbial Ecol.,* 6, 171, 1982.

4. **Dawson, M. P., Humphrey, B. A., and Marshall, K. C.,** Adhesion: a tactic in the survival strategy of a marine vibrio during starvation, *Curr. Microbiol.*, 6, 195, 1981.
5. **Marshall, K. C., Stout, R., and Mitchell, R.,** Selective sorption of bacteria from seawater, *Can. J. Microbiol.*, 17, 1413, 1971.
6. **Marshall, K. C.,** Growth at interfaces, in *Strategies of Microbial Life in Extreme Environments,* Shilo, M., Ed., Verlag Chemie, Weinheim, 1979, 281.
7. **Kjelleberg, S., Humphrey, B. A., and Marshall, K. C.,** The effect of interfaces on small starved marine bacteria, *Appl. Environ. Microbiol.*, 43, 1166, 1982.
8. **Power, K. and Marshall, K. C.,** Cellular growth and reproduction of marine bacteria on surface-bound substrate, *Biofouling,* 1, 163, 1988.
9. **Hermansson, M. and Marshall, K. C.,** Utilization of surface localized substrate by non-adhesive marine bacteria, *Microbial Ecol.*, 11, 91, 1985.
10. **Marshall, K. C. and Cruickshank, R. H.,** Cell surface hydrophobicity and the orientation of certain bacteria at interfaces, *Arch. Mikrobiol.*, 91, 29, 1973.
11. **Rosenberg, E., Kaplan, N., Pines, O., Rosenberg, M., and Gutnick, D.,** Capsular polysaccharides intefere with adherence of *Acinetobacter calcoaceticus* to hydrocarbons, *FEMS Microbiol. Lett.*, 17, 157, 1983.
12. **Fattom, A. and Shilo, M.,** Hydrophobicity as an adhesion mechanism of benthic cyanobacteria, *Appl. Environ. Microbiol.*, 47, 135, 1984.
13. **Characklis, W. G., Turakhia, M. H., and Zelver, N.,** Transport and interfacial transfer phenomena, in *Biofilms,* Characklis, W. G. and Marshall, K. C., Eds., Wiley-Interscience, New York, 1990, chap. 9.
14. **Marshall, K. C.,** *Interfaces in Microbial Ecology,* Harvard University Press, Cambridge, 1976, chap. 3.
15. **Lorian, V.,** *In vitro* simulation of *in vivo* conditions: physical state of the culture medium, *J. Clin. Microbiol.*, 27, 2403, 1989.
16. **McCarter, L. and Silverman, M.,** Surface-induced swarmer cell differentiation of *Vibrio parahaemolyticus, Mol. Microbiol.*, 4, 1057, 1990.
17. **Östling, J., Goodman, A., and Kjelleberg, S.,** Behaviour of IncP-1 plasmids and a miniMu transposon in a marine *Vibrio* sp.: isolation of starvation inducible *lac* operon fusions, *FEMS Microbiol. Ecol.*, 86, 83, 1991.
18. **Belas, R., Mileham, A., Simon, M., and Silverman, M.,** Transposon mutagenesis of marine *Vibrio* spp., *J. Bacteriol.*, 158, 890, 1984.
19. **Dagostino, L., Goodman, A. E., and Marshall, K. C.,** Physiological responses induced in bacteria adhering to surfaces, *Biofouling,* 4, 113, 1991.
20. **Humphrey, B. A. and Marshall, K. C.,** The triggering effect of surfaces and surfactants on heat output, oxygen consumption and size reduction of a starving marine *Vibrio, Arch. Microbiol.*, 140, 166, 1984.
21. **Ellwood, D. C., Keevil, C. W., Marsh, P. D., Brown, C. M., and Wardell, J. N.,** Surface-associated growth, *Phil. Trans. R. Soc. London,* B, 297, 517, 1982.
22. **Slonczewski, J. L.,** pH-regulated genes in enteric bacteria, *ASM News,* 58, 140, 1992.
23. **Fletcher, M.,** Comparative physiology of attached and free-living bacteria, in *Microbial Adhesion and Aggregation,* Marshall, K. C., Ed., Springer-Verlag, Berlin, 1984, 223.
24. **Stotzky, G.,** Gene transfer among bacteria in soil, in *Gene Transfer in the Environment,* Levy, S. B. and Miller, R. V., Eds., McGraw-Hill, New York, 1989, 165.
25. **Fry, J. C. and Day, M. J.,** Plasmid transfer in the epilithon, in *Bacterial Genetics in Natural Environments,* Fry, J. C. and Day, M. J., Eds., Chapman and Hall, London, 1990, 55.
26. **Fulthorpe, R. R. and Wyndham, R. C.,** Transfer and expression of the catabolic plasmid pBRC60 in wild bacterial recipients in a freshwater ecosystem, *Appl. Environ. Microbiol.*, 57, 1546, 1991.

27. **Angles, M. L., Marshall, K. C., and Goodman, A. E.,** Plasmid transfer between marine bacteria in biofilms and the aqueous phase and biofilms in reactor microcosms, *Appl. Environ. Microbiol.*, 59, 843, 1993.
28. **Jones, G. W., Baines, L., and Genthner, F. J.,** Heterotrophic bacteria of the freshwater neuston and their ability to act as plasmid recipients under nutrient deprived conditions, *Microbial Ecol.*, 22, 15, 1991.

Section I
BIOFILM STRUCTURE AND FUNCTION

3

Metabolic Interactions and Environmental Microniches: Implications for the Modeling of Biofilm Processes

W. Allan Hamilton

I. INTRODUCTION

On the occasion of the Dahlem Workshop on Biofilms, I had the privilege of writing an article with Bill Characklis on the subject of the relative activities of cells in suspension and in biofilms.[1] In that article we explored particular aspects of surface attachment and biofilm heterogeneities that might be expected to influence the activities of the constituent cellular populations. What I should now like to do is to develop some of these ideas further, and to bring them to a wider audience.

Broadly speaking, there are two very different ways of studying biofilms. The first approach (**the inductive approach**) takes as its starting point a relatively simple system whose critical parameters can be both clearly defined and readily subjected to quantitative measurement and mathematical analysis. From this firm base, increasing complexity can then be built into both experimental and mathematical models so that scientific rigor is maintained as a closer approximation to natural biofilms is reached. The primary focus is on the analyses of the fundamental cellular processes of growth, maintenance, and death, while stress is

0-87371-928-X/94/$0.00+$.50

put on determining stoichiometries and kinetics and on achieving mass balances. The interplay between experimental observation and modeling both advances basic knowledge of the system under study and serves the highly practical role of being predictive, both with regard to the characteristics of the biofilm itself and in suggesting potential control measures. As is manifest from this present symposium, the Center for Interfacial Microbial Process Engineering has consistently advocated this approach to biofilm studies and in so doing has been directly responsible for much of our present day understanding of these intriguing systems.

As with any system of experimentation and analysis, however, modeling of biofilms in this manner also has its limitations, and the recognition of these is equally as important as is the power to use any model to the maximum of its capability.

1. Only where quantitative experimental data are available can a particular process be included in a mathematical model. This might therefore lead to the exclusion of such key parameters as cell age or the influence of neighboring cells.[2] Events which occur in a random or stochastic manner can also not be readily incorporated into a model. This applies, for example, in the case of biofilm sloughing.

2. All values obtained are averages from the chosen biofilm volume element. Coincident with this is the concept of scale within biofilm systems.[3] At the microscale is the biofilm volume element itself comprised of microbial cells, water, extracellular polymeric substances (EPS), and any inorganic inclusions such as corrosion products. The mesoscale represents the biofilm, per se, along with its substratum and surrounding bulk liquid, while the macroscale constitutes the complete reactor, heat exchanger, or soured hydrocarbon reservoir.

3. Consequent on the above, mechanistic modeling does not attempt to describe the structure and dynamics of biofilm populations at the level of the individual cell.

This last point contrasts directly with the rationale and objectives of the other principal approach to the study of biofilms (**the deductive approach**), which takes as its starting point complex biofilms, either as they occur in nature or as they are generated in a laboratory simulation. Here, experimentation is directed toward attempting to describe the biofilm both in terms of the organisms present and their structural and temporal organization within the biofilm, and with regard to the physicochemical characteristics of the biofilm. These last will include organic (e.g., EPS) and inorganic (e.g., corrosion products) inclusions, and the presence of microenvironments created by nutrient and product gradients arising from the interaction between diffusion and microbial utilization and/or production. One is therefore moving from a real but complex system and trying to develop an understanding of its operation by a process of dissection and simplification.

Although the two approaches to biofilm research as outlined here are quite different in both philosophy and the practical techniques applied, they are in no sense contradictory, nor can one be said to be "right" and the other "wrong". In fact, they are truly complementary in the respect that clarification of components of the complexity of natural biofilms serves to direct the development of experimental and mathematical models as they seek to build from their defined, but relatively simple, base.

It is against this background, therefore, that in the remainder of this article I shall describe certain features of biofilm structure and function that appear to be of particular significance. In so doing I shall stress the importance of mixed microbial communities and microenvironments (consortia), the qualitative changes this can induce in the activity of individual species, the nonequivalence of biochemical activity and growth, and the dynamic nature of the quasi-steady state of the so-called mature biofilm.

II. BIOFILMS AND MICROBIALLY INFLUENCED CORROSION

Microbially influenced corrosion (MIC) has been one of the major foci of interest and stimuli to biofilm research in recent years.[4–7] Although it is clear that there are a number of quite separate mechanisms determining a wide range of corrosion processes[8] and that most of these mechanisms are individually complex, involving both biotic and abiotic factors, there are certain characteristics that appear to be of general, and perhaps of universal, significance.

1. Corrosion is often associated with **discontinuities in biofilm structure**. Where these occur in the horizontal dimension relative to the metal substratum (patchiness), they can give rise to local differences in metabolic products, pH, or dissolved oxygen (differential aeration cell), all of which can generate active electrochemical corrosion cells. This particular aspect of corrosion processes, and their prevention, is further explored by Costerton and by Lee and colleagues in the present volume. Where the discontinuity is in the vertical dimension, it generally takes the form of gradients of nutrients and ions, created by the combined action of mass transfer into the biofilm, and of microbial utilization within the biofilm. By far the most significant component in

this regard is dissolved oxygen, with the creation of an aerobic/anaerobic interface within a biofilm being the single most important factor determining and controlling the activities associated with that biofilm. It is this feature, for example, which makes possible the growth of sulfate-reducing bacteria in corroding biofilms within an aerobic bulk phase.

2. In the great majority of cases, biofilms constitute **communities of interacting microorganisms**, where the nature of these interactions and their attendant microenvironments (e.g., the aerobic/anaerobic interface; O_2/AnO_2) are at least as important as the structure of the biofilm, *per se*. It is by this means that the strictly anaerobic sulfate-reducing bacteria with their restricted nutritional spectra can nonetheless develop high levels of activity in biofilms, or sediments, associated with, for example, marine environments where the primary nutrients may be materials such as cellulose or hydrocarbons that cannot be degraded directly by the sulfate-reducing bacteria themselves.[4] In such instances the growth and corrosive activity of the sulfate-reducing bacteria are critically dependent on the presence of other heterotrophic, fermentative, and acetogenic species, which both create the required nutrients (primarily hydrogen and acetate) and generate the necessary reducing environment at the base of the biofilm.[1]

These questions of structural heterogeneities within the biofilm and the obligate interdependencies characteristic of consortia are therefore of fundamental importance to MIC in general, and to the biological component of sulfate-reducing bacterial corrosion in particular. Evidence is now accumulating, however, that the central event in sulfide-induced pitting corrosion is abiotic and results from the action of an electrochemical cell established between unreacted metal and iron sulfide corrosion products.[9–11] The physical and chemical nature of these corrosion products is critical to this mechanism and is directly influenced by the presence of oxygen (see Lee et al., Chapter 13). Both with regard to the access of oxygen and the localized formation of the primary sulfide corrosion products, horizontal discontinuities in the structure of the biofilm are again going to be absolutely central to any model of the overall corrosion process.

III. MONOSPECIES BIOFILMS WITH *PSEUDOMONAS AERUGINOSA*

In earlier studies of *P. aeruginosa* biofilms,[1,12] Characklis and colleagues clearly demonstrated that in biofilms up to 50 μm thick where mass transfer and diffusional resistances were shown to be negligible, the intrinsic properties $\mu_{max} = 0.37\ h^{-1}$ and $K_s = 2$ mg glucose·l^{-1}, as determined in the chemostat, applied equally to the biofilm. It was further noted that under both conditions there existed the same linear relationship between growth rate, μ, and the specific cellular substrate removal rate,q_s. The constancy of this relationship did not hold, however, when it was applied to total biomass in the biofilm; i.e., cells plus EPS. Although this last point suggests that cells within a biofilm may not be identical with freely suspended cells with respect to their overall metabolic activity, these experiments do indicate that such a thin biofilm is homogeneous, with all cells throughout its depth showing the same pattern of metabolic activity and growth.

These data immediately beg the questions: What happens when diffusional resistance is significant and cells in the deeper layers of a biofilm become limited with respect to carbon and energy nutrients and/or terminal electron acceptors? How comparable are cells in the surface layers of a biofilm with those below the critical or active thickness?[13] Are any differences merely quantitative or are there qualitative differences in the patterns of metabolic activity and/or growth?

At least partial answers are now available to these questions from the studies of thick (up to 500-μm) biofilms by Wimpenny and colleagues[14–16] (and personal communication). Monospecies *P. aeruginosa* biofilms were grown to quasi-steady state (approximately 100 h) and then freeze-sectioned in the presence of dextran as a cryoprotectant. The cryostat sections were 12 μm thick and allowed the measurement of two indicators of the physiological state of the cells throughout the depth of the biofilm. The assays carried out were cell viability and adenylate energy charge. Although controversy surrounds the quantitative validity of the absolute values of both parameters when applied to microbial ecosystems, they do allow meaningful comparisons to be made between like systems. Whereas viable counts give a measure of *in situ* viability, they do not indicate whether or not the natural population is actively growing within the ecosystem under study. Energy charge, on the other hand, does reflect the actual physiological state of the cells within the biofilm.

What Kinniment and Wimpenny[16] observed was that maximum values of both cell viability and energy charge occurred either just below or at the surface of the biofilm. Viability showed a sharp peak in the upper third of the biofilm with counts rising from 1–10 $\times$ 10^4 to 4–8 $\times$ 10^6. Adenylate energy charge increased more evenly across the thickness of the biofilm with values from three replicates ranging from 0.17–0.35 to 0.36–0.6. Although these values are comparatively low for an active population of cells, they do accord closely with values already obtained by the same group from their study of a multispecies dental biofilm, where only in young (24-h) biofilms at high glucose concentrations were values of energy charge approaching the maximum of 1 obtained.[17]

These data can be interpreted at two levels:

1. With regard to physiological state and growth potential, all cells are not equal even within a monospecies biofilm. It remains to be established whether this effect is gradual or is coupled to a transition zone within the biofilm, whether it results from carbon and/or oxygen limitation, and whether it coincides with the O_2/AnO_2 interface.

2. What is the nature of the differences between cells in the "active" surface layers and those in the "inactive" deeper layers of the biofilm? Do the different populations merely show variations in rates of, for example, substrate utilization and growth, or are the changes more qualitative in character; growth/no growth or viable/nonviable? If cells are active/growing in the surface layers, but inactive/nongrowing closer to the substratum, does this require two separate models, each with different assumptions, to describe a single biofilm? Does the coupling between substrate utilization and growth remain constant?

A. Cellular Processes: Growth and Maintenance

This last question raises a rather wider issue with regard to the assumptions underlying most biofilm models; namely, that respiratory and metabolic activities can be equated with growth and that maintenance energy is a constant requirement at all growth rates. Careful physiological experimentation and analysis suggest, in fact, that neither situation holds.[18–20] While the studies leading to this conclusion have been carried out over many years using various species of the facultative enteric bacteria, there is no reason to think it applies with any less conviction to other groups of microorganisms.[21]

The energy yield from catabolism in the form of ATP is invariably in excess of the requirements for biosynthesis and cell growth. Where there may be a close approximation between the two under conditions of optimal growth, μ_{max}, at slow growth rates this no longer holds. The most striking evidence of this phenomenon is the high respiratory activity that can be shown by so-called resting, i.e., nongrowing, cell suspensions. A number of mechanisms have been recognized whereby cells bring into balance their net requirements for ATP. ATP yield can be reduced by the use of branched terminal respiratory pathways where one or more sites of energy coupling may be lost. Branched fermentation pathways can also lead to variable ATP yields. Incomplete metabolism with the excretion of metabolic intermediates also reduces the amount of ATP resulting from catabolism. Various so-called futile cycles have also been proposed as a means of increasing ATP turnover; for example, an induced K^+ leakage current would have this effect. Non-growth-associated biosynthesis, for example, of EPS in carbon-sufficient cultures, is also a common mechanism linked to the obligatory cellular requirement for a balanced turnover of ATP. What is particularly important to appreciate is that there is no single or common mechanism whereby this objective is attained. The specific mechanism(s) employed vary both with the organism in question and, more importantly in the present context, with the nature of the factor limiting growth: carbon, energy, nitrogen or phosphorus, terminal electron acceptor, etc.

The concept of maintenance energy was introduced to account for the nonequivalence of substrate removal (or respiratory) rate and growth rate in microbial populations. Although the processes involved have never been clearly defined, they presumably are a composite of the various activities described in the previous paragraph. The value of maintenance is obtained by extrapolating the plot of substrate removal rate against growth rate to zero growth, and it is assumed that this level of maintenance will be present at all growth rates. As compared to growth under glucose limitation, growth of *Escherichia coli* shows maintenance requirements 25- or 30-fold higher under phosphate or acetate limitation. In careful analyses, Tempest[18] and Marr,[20] respectively, have demonstrated that these figures are at least partially artefactual. Since the same values of μ_{max} are obtained whatever the nature of the growth limitation at slower growth rates, the high levels of maintenance noted with noncarbon limitation cannot be a constant and must vary with growth rate. In the case of acetate limitation, our knowledge of the molecular mechanism for the control of flux through isocitrate dehydrogenase to the tricarboxylic acid cycle for energy generation, and through isocitrate lyase to the glyoxylate pathway for carbon synthesis, makes it clear that the graphical extrapolation of substrate removal rate to zero growth rate is negated by a qualitative change in the physiological function of the cells at low growth rates.

This discussion of the analyses of thick monospecies *P. aeruginosa* biofilms tells us that whether our approach to biofilm study is inductive through modeling or deductive through dissection and characterization, our existing level of understanding remains as yet a relatively unconvincing approximation to the truth.

IV. MATURE BIOFILMS: THE DYNAMIC NATURE OF THE STEADY STATE

It is instructive to note that the likely nature of the limitation of the primary nutrient and the development of anaerobic conditions, which are so deleterious in the case of monospecies *P. aeruginosa* biofilms, are exactly the conditions that lead to the generation of active sulfate reduction in many naturally occurring biofilms. In fact, the monospecies biofilm, especially one formed by an obligate aerobe, is essentially a laboratory artefact. In most natural or industrial environments, one would expect the development of anoxic and reducing environments within the biofilm to facilitate the growth of facultative organisms and of a terminal oxidizing species such as a sulfate reducer or methanogen. Although this statement may constitute a generalization of wide applicability and of practical utility in biofilm research, it must also be stressed that each biofilm system is to a greater or lesser extent unique and that extrapolation from one to another can only be done on the basis of sufficient initial characterization.

These facets have considerable impact on temporal heterogeneities within biofilms and on the nature of the so-called steady state. In an aerobic environment, are there a few anaerobic organisms, possibly in some resting state, which are initially trapped within the biofilm but only able to grow at a later stage after the development of anaerobiosis? Or is there migration of anaerobes into and through the "mature" aerobic biofilm?

With respect to the steady state of a mature biofilm, it is generally assumed that there is an equilibrium between the rates of growth and of detachment. What is not firmly established for any one biofilm, however, is whether these rates reflect an actively growing microbial population, or one more characteristic of resting cells. Again, this is likely to be a property uniquely characteristic of each individual biofilm.

Where a biofilm has clearly defined regions on either side of an O_2/AnO_2 interface, it is important to consider the relative growth rates of the two or more microbial populations. In the case of *P. aeruginosa* biofilms at least, these rates are clearly not the same. How can this be accommodated in modeling the process of biofilm detachment? Does it form the basis for an explanation of the phenomenon of sloughing, commonly experienced both in monospecies and in more complex biofilms?

Implicit in the argument being developed throughout this article is the quite separate identity and treatment of biofilm heterogeneities in the horizontal and vertical dimensions relative to the underlying substratum. The data and the discussions of their relevance being presented by Costerton (Chapter 1) and Lee et al. (Chapter 13) suggest that this may not be the case, with anaerobic microenvironments owing as much to colonial growth across the surface of the biofilm as to differentiation throughout its depth.

V. CONCLUSION

The scientific study of biofilms and the identification of their major impact on environmental and engineering systems have necessitated the development of new experimental techniques and of novel approaches to the definition and solution of a scientific problem. In considering some of these, I have sought to identify questions that remain unresolved and assumptions that may require modification. If I have found it easier to ask questions than to supply answers, perhaps this truly reflects the early, and exciting, stage of biofilm research at the present time.

VI. ACKNOWLEDGEMENTS

I wish to express my gratitude to Julian Wimpenny for permission to present and discuss his data, published and unpublished. The period that I spent at the Center for Interfacial Microbial Process Engineering during the summer of 1991, and the various discussions I had at that time with many colleagues and friends, formed the basis for the ideas I have tried to formulate here. As a friend and mentor I owe much to Bill Characklis; he is sorely missed.

REFERENCES

1. **Hamilton, W. A. and Characklis, W. G.,** Relative activities of cells in suspension and in biofilms, in *Structure and Function of Biofilms,* Characklis, W. G. and Wilderer, P. A., Eds., John Wiley & Sons, Chichester, 1989, 199.
2. **Wanner, O.,** Modeling population dynamics, in *Structure and Function of Biofilms,* Characklis, W. G. and Wilderer, P. A., Eds., John Wiley & Sons, Chichester, 1989, 91.
3. **Characklis, W. G. and Wilderer, P. A., Eds.,** *Structure and Function of Biofilms,* John Wiley & Sons, Chichester, 1989.
4. **Hamilton, W. A.,** Sulphate-reducing bacteria and anaerobic corrosion, *Annu. Rev. Microbiol.,* 39, 195, 1985.
5. **Dowling, N. J., Mittleman, M. W., and Danko, J. C., Eds.,** *Microbially Influenced Corrosion and Biodeterioration,* NACE, Knoxville, TN, 1991.
6. **Flemming, H.-C. and Geesey, G. G., Eds.,** *Biofouling and Biocorrosion in Industrial Water Systems,* Springer-Verlag, Berlin, 1991.
7. **Videla, H. A. and Gaylarde, C. C., Eds.,** *Microbially Influenced Corrosion,* Special Issue, *Int. Biodeterior. Biodegrad.,* 29, 1992.
8. **Tatnall, R. E.,** Case histories: biocorrosion, in *Biofouling and Biocorrosion in Industrial Water Systems,* Flemming, H.-C. and Geesey, G. G., Eds., Springer-Verlag, Berlin, 1991, 165.
9. **Hardy, J. A. and Bown, J.,** The corrosion of mild steel by biogenic sulfide films exposed to air, *Corrosion,* 40, 650, 1984.
10. **Hamilton, W. A.,** Sulphate-reducing bacteria and their role in biocorrosion, in *Biofouling and Biocorrosion in Industrial Water Systems,* Flemming, H.-C. and Geesey, G. G., Eds., Springer-Verlag, Berlin, 1991, 187.

11. **McKenzie, J. and Hamilton, W. A.,** The assay of *in-situ* activities of sulphate-reducing bacteria in a laboratory marine corrosion model, *Int. Biodeterior. Biodegrad.*, 29, 285, 1992.
12. **Bakke, R., Trulear, M. F., Robinson, J. A., and Characklis, W. G.,** Activity of *Pseudomonas aeruginosa* in biofilms, *Biotech. Bioeng.*, 26, 1418, 1984.
13. **Characklis, W. G.,** Biofilm development: a process analysis, in *Microbial Adhesion and Aggregation*, Marshall, K. C., Ed., Springer-Verlag, Berlin, 1984, 137.
14. **Peters, A. C. and Wimpenny, J. W. T.,** A constant depth laboratory model film fermenter, *Biotech. Bioeng.*, 32, 263, 1988.
15. **Kinniment, S. L. and Wimpenny, J. W. T.,** Biofilms and biocides, *Int. Biodeterior.*, 26, 181, 1990.
16. **Kinniment, S. L. and Wimpenny, J. W. T.,** Measurements of the distribution of adenylate concentrations and adenylate energy charge across *Pseudomonas aeruginosa* biofilms, *Appl. Environ. Microbiol.*, 58, 1629, 1992.
17. **Wimpenny, J. W. T., Peters, A. C., and Scourfield, M.,** Modelling spatial gradients, in *Structure and Function of Biofilms*, Characklis, W. G. and Wilderer, P. A., Eds., John Wiley & Sons, Chichester, 1989, 111.
18. **Tempest, D. W.,** The biochemical significance of molar growth yields: a reassessment, *Trends Biochem. Sci.*, 3, 180, 1978.
19. **Tempest, D. W. and Neijssel, O. M.,** Growth yield and energy distribution, in *Escherichia coli and Salmonella typhimurium Cellular and Molecular Biology*, Neidhardt, F. C., Ed., American Society of Microbiology, Washington, D.C., 1987, 797.
20. **Marr, A. G.,** Growth rate of *Escherichia coli*, *Microbiol. Rev.*, 55, 316, 1991.
21. **Thauer, R. K. and Morris, J. G.,** Metabolism of chemotrophic anaerobes: old views and new aspects, *Symp. Soc. Gen. Microbiol.*, 36, 123, 1984.

Section I
BIOFILM STRUCTURE AND FUNCTION

4

Modeling of Mixed-Population Biofilm Accumulation

Oskar Wanner

I. INTRODUCTION

Mathematical models are important tools for the investigation of complex systems that are difficult to examine experimentally. Biofilms, in which a considerable number of dissolved and particulate components interact with each other and where strong spatial gradients and fast changes in time are observed, are such systems. General objectives of modeling are to gain insight into the structure and behavior of a complex system and to evaluate experimental observations. Since our interest focuses on corrosion, our specific objective is to make predictions about the physicochemical conditions at the interface between the biofilm and the solid surface (substratum) on which the biofilm grows. In order to make such predictions it is necessary to establish a quantitative relationship between the conditions in the bulk fluid and those in the biofilm depth. This can be achieved by means of mechanistic mathematical models that include all relevant processes and components of the biofilm system.

The development of mathematical biofilm models started more than 10 years ago. The first models were designed to calculate spatial concentration profiles for a single dissolved substrate (e.g., chemical oxygen demand) as a function of diffusive transport and substrate transformation in the biofilm. In these models the microorganisms were merely

0-87371-928-X/94/$0.00+$.50

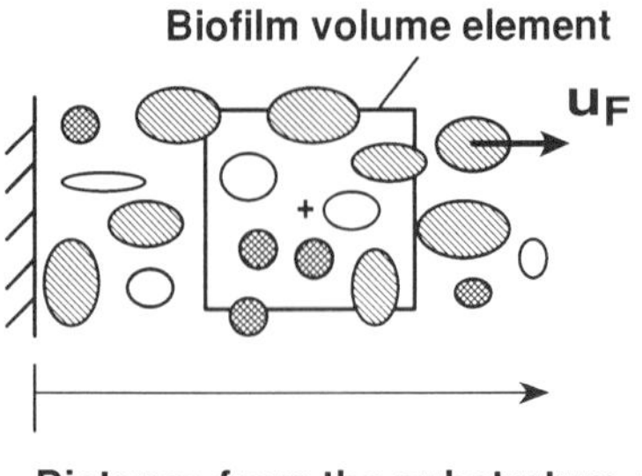

FIGURE 1. Schematic representation of a biofilm cross section.

treated as "biomass" and assumed to be homogeneously distributed without any time-dependent properties. Multisubstrate models were then developed in a subsequent step, thereby enabling concentration profiles of several dissolved components to be calculated simultaneously. However, even though microbial species were differentiated in these models, they were treated in very much the same way as in the previous models. About 10 years ago, the first attempts were made to mechanistically model the spatial distribution and time development of microbial species in biofilms.[1–3] Mechanistic mixed-population biofilm models have lately been applied successfully to various systems.[4–6]

This chapter summarizes the mathematical framework available for mechanistic modeling of mixed-population biofilms and applies it to an artificial example of a biofilm model with two microbial species. Mathematical problems and numerical solution techniques will only briefly be discussed here, but are thoroughly treated in a separate paper. The objective of this chapter is to illustrate the practical application of mixed-population biofilm models, to provide a complete list of the experimental data required for biofilm modeling, and to present and discuss the results that can be obtained.

II. MIXED-POPULATION BIOFILM MODELING

A. Modeling Concept

Electron micrographs provide us with an idea of the structure of the biofilm solid matrix. A schematic representation of such a solid matrix is given in Figure 1. As can easily be seen, a biofilm is a discrete system, i.e., it is made up of a distinct number of objects. However, let us consider the small biofilm volume element marked in Figure 1. If the cells of a given microbial species are counted for this element, and if their mass is calculated and related to the element volume, a concentration is obtained which is a representative measure for the mass of the microbial species in this volume element. If this procedure is repeated for a volume element that has been shifted slightly in space, it will not yield a drastically different concentration. This fact is the basis for the continuum concept to be applied to biofilm modeling.[7]

If microbial cells in the biofilm depth grow and divide, they will require more space. If, for the time being, we assume that the fraction of the liquid phase between the cells remains constant, the space requirement will result in a displacement of the cells located closer to the biofilm surface. This displacement is represented in Figure 1 by the arrow u_F, indicating the shift in space per unit time of the cell to which it is affixed. The arrow u_F can thus be interpreted as advective velocity. Since the various cells of the biofilm are connected to each other and form together a spongelike structure, it can be assumed that the directly adjacent cells are displaced with about the same velocity. Therefore, the velocity u_F, varying with time and space, is called "velocity of the biofilm solid matrix".

It is assumed that everywhere in the biofilm u_F points in the direction perpendicular to the substratum. This assumption is based on microelecrode data, which reveal that over a distance of less than a hundred micrometers in the biofilm the concentrations of dissolved substances can change several orders of magnitude, whereas gradients in the bulk fluid are usually orders of magnitude smaller. For a biofilm surface area with uniform bulk fluid conditions, it can be assumed that the development of the biofilm will also be uniform and that no significant gradients will be formed parallel to the substratum. A mathematical description by one-dimensional models is therefore possible for most biofilm systems.[7]

B. Particulate Components in the Biofilm

1. Mass Balance Equation

On the basis of the assumptions made in the last paragraph, one-dimensional mass balances can be established for the particulate components that form together the biofilm solid matrix. The particulate components may be made up of microbial species, extracellular polymeric substances, and organic and inorganic particles. Each particulate component that is to be included in the biofilm model requires a mass balance equation,

$$\frac{\partial X_{F,i}}{\partial t} = -\frac{\partial j_{F,Xi}}{\partial z} + r_{Xi} \tag{1}$$

where t is the time (T), z is the distance from the substratum (L),* $X_{F,i}$ is the concentration ($M\ L^{-3}$), j_{F,X_i} is the mass flux ($M\ L^{-2}\ T^{-1}$), and r_{X_i}

* See the list of symbols following the text.

is the transformation rate ($M\ L^{-3}\ T^{-1}$) of the particulate component i. The first term on the right side of the Equation 1 describes the effect of transport on $X_{F,i}$. An important feature of the presented biofilm models is the universal description of this term, whereas the second term, describing the net effect of all transformation processes on $X_{F,i}$, is specific for each model application. There are numerous transformation processes and rate laws to describe their kinetics. Examples of transformation processes of particulate components are microbial growth, product formation, maintenance, endogenous metabolism, death, lysis, and hydrolysis.

2. *Transport of Particulate Components in the Biofilm*

According to the above assumptions, the mass flux of each particulate component is directly proportional to the concentration of the component and to the velocity of the solid matrix, u_F ($L\ T^{-1}$). It is described as an advective transport process by

$$j_{F,X_i} = u_F X_{F,i} \tag{2}$$

The velocity u_F is obtained as the net result of all transformation processes of the n_X particulate components considered in the model. It is calculated by superposition of the n_X mass balance equations of the particulate components, solution for $\partial u_F/\partial z$ and integration over z as

$$u_F = \frac{1}{1-\epsilon_l} \int_0^z \sum_{i=1}^{nx} \frac{r_{X_i}}{\rho_{S_i}}\, dz' \tag{3}$$

where ϵ_l is the liquid-phase volume fraction in the biofilm (−) and ρ_{S_i} is the density of the particulate component i ($M\ L^{-3}$).[8] The density ρ_{S_i} is defined as mass of i per unit volume of the solid phase as formed by i. For microbial cells it is usually experimentally determined as dry or wet cell weight per unit cell volume and then converted to the units used in the model for the particulate components.[9] The density is related to the concentration of i by

$$X_{F,i} = \epsilon_{S_i}\, \rho_{S_i} \tag{4}$$

where ϵ_{S_i} is the volume fraction of the solid phase i (−). The concentration and the volume fraction change with time and space, whereas the density is considered a constant property of the component. In other models, density is often defined as mass per unit biofilm volume and therefore is not constant.

3. Progression of the Biofilm Thickness

Biofilm accumulation is defined as change of the biofilm thickness over a period of time. This change is the net result of the total net production of particulate components in the biofilm, detachment of particulate components from the biofilm surface to the bulk fluid, and attachment of suspended cells and particles to the biofilm surface. Mathematically this change is defined as

$$\frac{dL_F}{dt} = u_L \tag{5}$$

where L_F is the biofilm thickness (L) and u_L is the velocity (L T^{-1}) at which the interface between the biofilm and the bulk fluid moves in space. This velocity can be expressed as

$$u_L = u_F(z = L_F) - u_{de} + u_{at} \tag{6}$$

where u_{de} and u_{at} are the detachment and attachment velocities (L T^{-1}), respectively. The attachment velocity is described as

$$u_{at} = \frac{1}{1-\epsilon_l} \sum_{i=1}^{n_X} \frac{k_{at,X_i} X_{L,i}}{\rho_{s_i}} \tag{7}$$

where $X_{L,i}$ is the concentration of the suspended cells or particles of component i at the bulk-fluid side of the biofilm surface (M L^{-3}). Attachment rate coefficients k_{at,X_i}(L T^{-1}) are usually assumed to be constant. Detachment velocities have been observed to vary with biofilm thickness, density, growth rates, time, and hydraulic conditions in the bulk fluid.[10–13] They are modeled by empirically established relationships such as

$$u_{de} = c_{de} L_F^2 \tag{8}$$

where c_{de} (T^{-1} L^{-1}) is an empirical detachment rate coefficient.

4. Boundary Conditions

If the substratum is assumed to be inert, the boundary condition at the interface between the biofilm and the substratum (z = 0) is

$$j_{F,X_i} = 0 \tag{9}$$

This no-flux condition assumes that no particulate material is exchanged between the biofilm and the substratum. However, this does not mean that processes such as adsorption and desorption of dissolved components at the substratum are excluded.

The interface between the biofilm and the bulk fluid moves in space with the growth of the biofilm. At this interface the concentrations of the particulate components are discontinuous. The attached cells and particles, which together form the biofilm solid matrix, exist only at the biofilm side of the interface, and the suspended cells and particles are assumed to exist only at the bulk-fluid side. To model this situation, three processes have to be considered. The first process describes the mass flux across the interface at the biofilm side, i_{F,X_i} ($M\,L^{-2}\,T^{-1}$), and is given by

$$i_{F,X_i} = (u_F - u_L)\,X_{F,i} \tag{10}$$

The second process models the transfer of particulate components between the solid phases in the biofilm and the liquid phase of the bulk fluid. It is expressed as an interfacial transfer rate, r''_{X_i} ($M\,L^{-2}\,T^{-1}$), and is described by the following empirical rate law:

$$r''_{X_i} = u_{de} X_{F,i} - k_{at,X_i} X_{L,i} \tag{11}$$

The third process describes the mass flux across the interface into a completely mixed bulk-fluid compartment, i_{L,X_i} ($M\,L^{-2}\,T^{-1}$). This flux is modeled by the following empirically determined expression:

$$i_{L,X_i} \kappa_{X_i} = X_{L,i} - X_{B,i} \tag{12}$$

where $X_{B,i}$ is the concentration of the cells or particles suspended in the bulk fluid ($M\,L^{-3}$) and κ_{X_i} is the mass-transfer resistance coefficient ($T\,L^{-1}$) of component i. The values of the function u_{de} and of the parameters k_{at,X_i} and κ_{X_i} have to be determined experimentally.

The boundary condition at the interface between the biofilm and the bulk fluid ($z = L_F$) consists of a continuity condition for the biofilm side of the interface,

$$i_{F,X_i} = r''_{X_i} \tag{13}$$

and of a continuity condition for the bulk-fluid side of the interface,

$$r''_{X_i} = i_{L,X_i} \tag{14}$$

C. Dissolved Components in the Biofilm

1. *Mass Balance Equations*

The spatial profiles and time development of the dissolved components in the biofilm are described by the well-established one-dimensional mass balance equation,

$$\frac{\partial(\epsilon_l C_{F,i})}{\partial t} = -\frac{\partial j_{F,C_i}}{\partial z} + r_{C_i} \tag{15}$$

where $C_{F,i}$ is the concentration ($M\,L^{-3}$), j_{F,C_i} is the mass flux ($M\,L^{-2}\,T^{-1}$), and r_{C_i} is the transformation rate ($M\,L^{-3}\,T^{-1}$) of the dissolved component i. Note that $C_{F,i}$ is the concentration of component i in the biofilm liquid phase and is related to the concentration $S_{F,i}$ ($M\,L^{-3}$), defined as mass of i per unit biofilm volume, by

$$C_{F,i} = \frac{S_{F,i}}{\epsilon_l} \tag{16}$$

It is assumed that dissolved components are transported in the liquid phase of the biofilm by molecular diffusion. Their mass flux is described as

$$j_{F,C_i} = -u_F(1 - \epsilon_l)C_{F,i} - fD_{C_i}\frac{\partial C_{F,i}}{\partial z} \tag{17}$$

where D_{C_i} is the diffusivity of the dissolved component i in pure water ($L^2\,T^{-1}$) and f is the ratio of the diffusivities in pure water and in the biofilm (−).[9,14] The first term on the right side of Equation 17 originates from the advective flow induced by the conversion of dissolved components to biomass in the biofilm.

Examples of transformation reactions of dissolved components are oxygen consumption, substrate utilization and conversion by microorganisms, and biotic or abiotic transformations such as redox or acid-base reactions.

2. *Boundary Conditions*

At the interface between the biofilm and the substratum ($z = 0$), various situations can be distinguished, of which two are described here. If the substratum is inert and impermeable, the boundary condition is

$$j_{F,C_i} = 0 \tag{18}$$

i.e., the mass flux is zero at this interface.

If there is adsorption and desorption of dissolved components at the substratum, the boundary condition consists of a continuity condition for the biofilm side of the interface, where

$$j_{F,C_i} = r''_{C_i} \tag{19}$$

j_{F,C_i} is the mass flux as given by Equation 17 for $z = 0$ and r''_{C_i} is the interfacial transfer rate ($M\,L^{-2}\,T^{-1}$) describing the adsorption-desorption reaction. This reaction is modeled here by first-order kinetics as

$$r''_{C_i} = k_{ds,C_i}S_{S,i} - k_{as,C_i}C_{F,i} \tag{20}$$

where k_{as,C_i} and k_{ds,C_i} are the adsorption and desorption rate constants ($L\,T^{-1}$), respectively, and where $S_{S,i}$ is the concentration of component i in the substratum ($M\,L^{-3}$). The concentration $S_{S,i}$ must either be known and assumed not to change significantly in time, or its time development must be modeled by an additional differential equation.

At the interface between the biofilm and the bulk fluid ($z = L_F$), there are two continuity conditions which must be fulfilled. The first condition requires that the concentration in the liquid phase is continuous. This condition is expressed by

$$C_{F,i} = C_{L,i} \tag{21}$$

where $C_{L,i}$ is the concentration of the dissolved component i ($M\,L^{-3}$) at the bulk-fluid side of the biofilm surface. The second condition requires that the mass flux in the liquid phase is continuous. This condition is expressed by

$$i_{F,C_i} = i_{L,C_i} \tag{22}$$

where i_{F,C_i} and i_{L,C_i} are the mass fluxes at the biofilm side and bulk-fluid side of the interface, respectively. These mass fluxes are given by

$$i_{F,C_i} = [-u_F + (u_F - u_L)\epsilon_l]C_{F,i} - fD_{C_i}\frac{\partial C_{F,i}}{\partial z} \tag{23}$$

and by the empirically found expression

$$i_{L,C_i}\kappa_{C_i} = C_{L,i} - C_{B,i} \tag{24}$$

where $C_{B,i}$ is the concentration of the dissolved component i in the bulk fluid ($M\,L^{-3}$) and κ_{C_i} is the mass-transfer resistance coefficient ($T\,L^{-1}$), which must be determined experimentally.

D. Dissolved and Suspended Particulate Components in the Bulk Fluid

In the simplest applications of biofilm models it can be assumed that the biofilm grows in a completely mixed aquatic environment. This environment is usually referred to as bulk fluid and can be described by the mass balance equations,

$$\frac{d}{dt}(V_B X_{B,i}) = Q(X_{in,i} - X_{B,i}) + A i_{L,X_i} + V_B r_{X_i} \tag{25}$$

and

$$\frac{d}{dt}(V_B C_{B,i}) = Q(C_{in,i} - C_{B,i}) + A i_{L,C_i} + V_B r_{C_i} \tag{26}$$

where A is the area covered with biofilm (L^2), Q is the volumetric flow rate ($L^3\ T^{-1}$), and $C_{in,i}$ and $X_{in,i}$ are the influent concentrations of the dissolved and particulate components ($M\ L^{-3}$), respectively. Note that in Equation 26 the liquid phase volume fraction ϵ_l has been assumed to equal one, i.e., the volume of the suspended cells and particles has been neglected. Thus, the volume of the bulk fluid V_B (L^3) is calculated as

$$V_B = V_R - AL_F \tag{27}$$

where V_R is the volume of the reactor or water body (L^3).

E. Mathematical Treatment

In order to apply the mixed-population biofilm model, several substantial mathematical problems have to be solved, such as the severe stiffness of the differential equation system, the moving interface between biofilm and bulk fluid, and the extreme spatial gradients encountered in the biofilm.[8] These problems cannot be discussed here; however, they will be treated in a future paper on numerical solution techniques for biofilm models. Computer programs such as BIOSIM,[15,16] developed specifically for the simulation of mixed-population biofilms on personal computers, or the more sophisticated program AQUASIM, which will soon be available, help to solve the model equations.

III. EXAMPLE: BIOFILM REACTOR WITH TWO MICROBIAL SPECIES

An example is used to illustrate the application of the mixed-population biofilm model. This example does not describe an existing biofilm

TABLE 1. Parameters of the Biofilm Reactor and Physical Properties

Parameter	Symbol	Value[a]	Unit
Biofilm area	A	0.1	m^2
Volumetric flow rate	Q	1.0	$m^3\ d^{-1}$
Reactor volume	V_R	0.001	m^3
Liquid-phase volume fraction	ϵ_l	0.9	—
Ratio of the diffusivity in the biofilm and in pure water	f	0.8	—
Detachment velocity	u_{de}	$c_{de}L_F^2$	$m\ d^{-1}$
Detachment coefficient	c_{de}	500	$m^{-1}\ d^{-1}$

[a] Typical values as found in the literature are used.[4,14]

system, but is based on experimental data and literature originating from various sources. It was created to exemplify the important aspects of biofilm modeling and to yield a comprehensive list of all the experimental data required for model calibration.

A. Biofilm Reactor and Physical Properties of the Biofilm

The walls of a completely mixed stirred-tank reactor (CSTR) are assumed to be evenly colonized by a biofilm. Table 1 contains a list of the data necessary to describe the geometry of the CSTR and the parameters required to characterize the physical properties of the biofilm.

B. Transformation Processes

1. *Microbial Kinetics*

Two microbial species are assumed to be the only particulate components contributing significantly to the mass of the biofilm solid matrix and interacting with the dissolved components present in the reactor. Species A is an *Acetobacter sp.* that grows aerobically on acetate; species B is a glucose fermenter. The growth kinetics of these two species is described by the following rate laws:

$$\mu_A = \mu_{m_A} \frac{C_{Ace}}{K_{Ace} + C_{Ace}} \frac{C_{O_2}}{K_{O_2} + C_{O_2}} F_{pH_A} \tag{28}$$

and

$$\mu_B = \mu_{m_B} \frac{C_{Glu}}{K_{Glu} + C_{Glu}} \frac{K_{O_2}}{K_{O_2} + C_{O_2}} F_{pH_B} \tag{29}$$

TABLE 2. Parameters of the Microbial Species

Parameter	Symbol	Value[a] Species A	Species B	Unit
Maximum specific growth rate	μ_m	8	2	d^{-1}
Decay rate constant	b	0.4	0.05	d^{-1}
Half-saturation constant for acetate	K_{Ace}	2	—	$g_{COD}\ m^{-3}$
Half-saturation constant for glucose	K_{Glu}	—	1.5	$g_{COD}\ m^{-3}$
Half-saturation constant for dissolved oxygen	K_{O_2}	0.2	0.2	$g_{O_2}\ m^{-3}$
True yield coefficient	Y	0.667	0.167	$g_{COD_x}\ g_{COD_c}^{-1}$
Density (cell mass per unit cell volume)	ρ_s	2.9×10^5	2.9×10^5	$g_{COD}\ m^{-3}$
Attachment rate constant	k_{at}	0	0	$m\ d^{-1}$
Mass-transfer resistance coefficient	κ	0	0	$d\ m^{-1}$
Influent concentration	X_{in}	0	0	$g_{COD}\ m^{-3}$

[a] Typical values as found in the literature are used.[17–22]

where μ_A and μ_B are the specific growth rates of A and B (T^{-1}), respectively. The other parameters used in Equations 28 and 29 are explained in Table 2. The pH effect on microbial growth is described by the mathematical functions

$$F_{pH_A} = \frac{1}{(1 + 10^{5.5-pH})\,(1 + 10^{pH-9.5})} \tag{30}$$

and

$$F_{pH_B} = \frac{1}{(1 + 10^{5-pH})\,(1 + 10^{pH-9})} \tag{31}$$

The functions F_{pH_A} and F_{pH_B} have a maximum of about one at pH 7.5 and 7, respectively, and both decrease at higher and lower pH values. The parameters of F_{pH_A} and F_{pH_B} have been obtained by a fit to experimental data.[6,14]

Maintenance and endogenous metabolism are included in the model as "decay". It is assumed that this decay can be described as a first-order process with a rate constant b (T^{-1}). Superposition of microbial growth and decay yields the net specific growth rate μ_{net_i} (T^{-1}), given as

$$\mu_{net_i} = \mu_i - b_i \quad (32)$$

which plays a key role if the spatial distribution and time development of the microbial population of a mixed-population biofilm are to be examined.

2. *Modeling of pH*

The pH value is determined in this example by dissolved carbon dioxide, bicarbonate, and a phosphate buffer added to the reactor influent. It is calculated as net result of the chemical reactions

$$H^+ + HCO_3^- \underset{k_{b_1}}{\overset{k_{f_1}}{\rightleftarrows}} CO_2 + H_2O \quad (33)$$

and

$$H^+ + HPO_4^{-2} \underset{k_{b_2}}{\overset{k_{f_2}}{\rightleftarrows}} H_2PO_4^- \quad (34)$$

At a pH greater than 7, these are the only reactions and reactants which contribute significantly to the pH value. The reaction rate constants of Equations 33 and 34 are given in Table 3. Values can be found in the literature for most abiotic process parameters. If the kinetic data needed are unavailable, modeling can be based on equilibrium constants, as the abiotic chemical reactions are usually orders of magnitude faster than the microbial processes considered in the model. The unknown dihydrogen phosphate loss rate k_{b_2} (T^{-1}) was assumed to have a high value of 10^4 d^{-1} and was then used along with the known equilibrium constant K_2 ($M\ L^{-3}$) to calculate a value for the dihydrogen phosphate formation rate k_{f_2} ($L^3\ M^{-1}\ T^{-1}$) as indicated in Table 3.

3. *Stoichiometric Matrix*

As shown in Table 4, the transformation processes used in the model are represented by a stoichiometric matrix comprising the biological and chemical processes, dissolved and particulate components considered in the model, and the process rates p_j ($ML^{-3}\ T^{-1}$), as well as the stoi-

TABLE 3. Parameters of the Dissolved Components

Parameter	Symbol	Value	Unit	Ref.
Carbon dioxide equilibrium constant	K_1	4.47×10^{-4}	mol m^{-3}	23
Carbon dioxide loss rate	k_{b1}	2.6×10^3	d^{-1}	23
Carbon dioxide formation rate	k_{f1}	k_{b1}/K_1	m^3 mol^{-1} d^{-1}	
Dihydrogen phosphate equilibrium constant	K_2	6.17×10^{-5}	mol m^{-3}	23
Dihydrogen phosphate loss rate	k_{b2}	10^4[a]	d^{-1}	
Dihydrogen phosphate formation rate	k_{f2}	k_{b2}/K_2	m^3 mol^{-1} d^{-1}	
Acetate influent concentration	$C_{in_{Ace}}$	5	g_{COD} m^{-3}	
Glucose influent concentration	$C_{in_{Glu}}$	10	g_{COD} m^{-3}	
Dissolved oxygen influent concentration	$C_{in_{O_2}}$	4	g_{O2} m^{-3}	
Carbon dioxide influent concentration	$C_{in_{CO_2}}$	0.084	mol m^{-3}	
Bicarbonate influent concentration	$C_{in_{HCO_3^-}}$	3.75	mol m^{-3}	
Hydrogen ion influent concentration	$C_{in_{H^+}}$	6.65×10^{-5}	mol m^{-3}	
Hydrogen phosphate influent concentration	$C_{in_{HPO_4^{-2}}}$	0.3	mol m^{-3}	
Dihydrogen phosphate influent concentration	$C_{in_{H_2PO_4^-}}$	0.5	mol m^{-3}	
Diffusivity of acetate in pure water	D_{Ace}	0.94×10^{-4}	m^2 d^{-1}	24
Diffusivity of glucose in pure water	D_{Glu}	0.60×10^{-4}	m^2 d^{-1}	24
Diffusivity of dissolved oxygen in pure water	D_{O_2}	2.1×10^{-4}	m^2 d^{-1}	24
Diffusivity of carbon dioxide in pure water	D_{CO_2}	1.7×10^{-4}	m^2 d^{-1}	24
Diffusivity of bicarbonate in pure water	$D_{HCO_3^-}$	1.02×10^{-4}	m^2 d^{-1}	24
Diffusivity of hydrogen ions in pure water	D_{H^+}	5.83×10^{-4}	m^2 d^{-1}	24
Diffusivity of hydrogen phosphate in pure water	$D_{HPO_4^{-2}}$	0.83×10^{-4}	m^2 d^{-1}	24
Diffusivity of dihydrogen phosphate in pure water	$D_{H_2PO_4^-}$	0.68×10^{-4}	m^2 d^{-1}	24
Mass-transfer resistance coefficient	κ_{C_i}	0	d m^{-1}	

[a] Value assumed

chiometric coefficients ν_{ij} (M M^{-1}), which are used to calculate the transformation rates r_i (M L^{-3} T^{-1}). The matrix format readily allows checking the stoichiometry of the transformation processes with regard to mass, electrical charges, etc.[19] The stoichiometric matrix also reveals that both biotic and abiotic processes can be included in the model in the same way. Note that the process rates p_j are defined as mass per unit biofilm volume. This necessitates that the rate laws of the abiotic processes are multiplied by the biofilm liquid-phase volume fraction ϵ_l.

TABLE 4. Transformation Processes

Process		particulate		dissolved								Process rate law
		Stoichiometric matrix with coefficients ν_{ij} — Component[a]										
j	i =	1 A	2 B	3 Ace	4 Glu	5 O_2	6 CO_2	7 HCO_3^-	8 H^+	9 HPO_4^{-2}	10 $H_2PO_4^-$	P_j $(M\ L^{-3}T^{-1})$
1	Growth of A	1		$\frac{-1}{Y_A}$		$1-\frac{1}{Y_A}$	$\frac{2}{64}\left(\frac{1}{Y_A}-1\right)$		$\frac{-1}{64Y_A}$			$\mu_A X_A^b$
2	"Decay" of A	−1		1					$\frac{1}{64}$			$b_A X_A$
3	Growth of B		1	$\frac{1}{Y_B}-1$	$\frac{-1}{Y_B}$				$\frac{1}{64}\left(\frac{1}{Y_B}-1\right)$			$\mu_B X_B^b$
4	"Decay" of B		−1		1							$b_B X_B$
5	Formation of CO_2						1	−1	−1			$k_{f1}\epsilon_l C_{H^+} C_{HCO_3^-}$
6	Loss of CO_2						−1	1	1			$k_{b1}\epsilon_l C_{CO_2}$
7	Formation of $H_2PO_4^-$								−1	−1	1	$k_{f2}\epsilon_l C_{H^+} C_{HPO_4^{-2}}$
8	Loss of $H_2PO_4^-$								1	1	−1	$k_{b2}\epsilon_l C_{H_2PO_4^-}$

Transformation rate $(M\ L^{-3}T^{-1})$: $r_i = \sum_{j=1}^{np} \nu_{ij} p_j$.

[a] Units for microbial species A and B, acetate, glucose, and O_2 are gCOD m^{-3}, units for CO_2, HCO_3^-, H^+, HPO_4^{-2}, and $H_2PO_4^-$ are mol m^{-3}.

[b] μ_A and μ_B according to Equations 28 and 29.

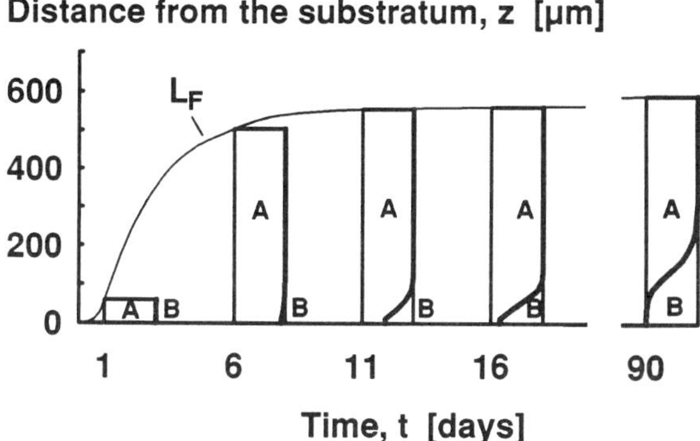

FIGURE 2. Progression of the biofilm thickness and the relative abundance of the microbial species A and B between the substratum-biofilm interface (bottom) and the biofilm-bulk fluid interface (top). The width of the bars represents the biofilm volume fraction ($1 - \epsilon_l$) occupied by the two microbial species.

C. Simulation

A computer simulation was performed based on the above example and equations of the mixed-population biofilm model. The simulation was started with an initial biofilm thickness of $L_{F_0} = 1\ \mu m$, uniform concentrations of both microbial species of $X_{0_i} = 14{,}500$ gCOD m^{-3}, and concentrations of the dissolved components in the biofilm and bulk fluid equal to the reactor influent concentrations as given in Table 3. For the sake of simplicity the mass transfer resistance at the biofilm-bulk fluid interface was neglected ($\kappa_{C_i} = 0$ and $\kappa_{X_i} = 0$). The substratum was assumed to be inert ($k_{ds,C_i} = 0$ and $k_{as,C_i} = 0$). Detachment was modeled according to Equation 8 with $c_{de} = 500\ m^{-1}\ d^{-1}$ and attachment was neglected ($k_{at,X_i} = 0$).

The simulation was performed on a SUN-SPARCstation IPX with the Beta version of the AQUASIM program.[25] The computer time required for the calculation of the spatial profiles in the biofilm with 27 grid points and for 90 d was about 40 min.

IV. RESULTS AND DISCUSSION

A. Microbial Species in the Biofilm

1. Biofilm Accumulation

Figure 2 shows the progression of the biofilm thickness L_F. For about 1 d the biofilm grows exponentially, then, due to oncoming diffusion

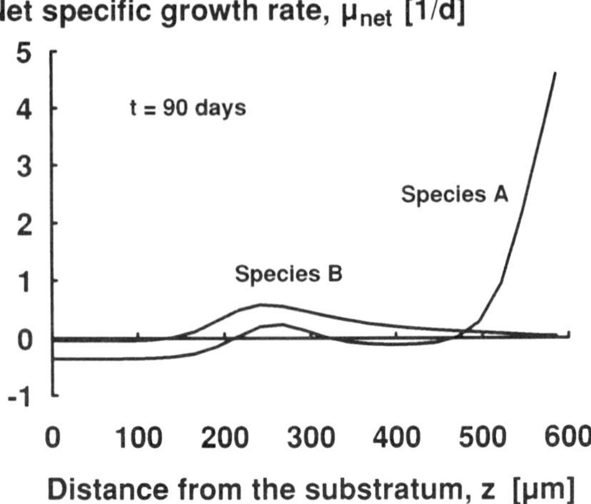

FIGURE 3. Spatial profile of the net specific growth rate of the microbial species in the biofilm.

limitation in the biofilm, the increase of L_F slows down. After about 11 d the biofilm thickness approaches a steady state caused by an equilibrium between biomass production in the biofilm and erosion of microbial cells from the biofilm surface.

2. *Spatial Profiles and Development in Time*

As can be seen in Figure 2, the aerobic microbial species A completely outgrows the anaerobic species B during the first 6 d. At the 6th day the mass of species A is large enough to utilize all of the dissolved oxygen in the biofilm and create an anaerobic environment in the biofilm depth, where species B then starts to develop. On the 90th day when a steady state is eventually reached, the two microbial species each form a homogeneous layer in the biofilm.

3. *Net Specific Growth Rates*

The microbial distribution results obtained are quite obvious. In less transparent biofilm systems, the net specific growth rate μ_{net} can provide insight into the spatial distribution and time development of the microbial species in the biofilm. As illustrated in Figure 3, the spatial profiles of the net specific growth rate show a maximum μ_{net_A} near the biofilm surface and a maximum μ_{net_B} at a distance of about 250 μm from the substratum. At this distance, μ_{net_B} is greater than μ_{net_A}. As shown by a mathematical analysis, coexistence of species B with species A in the

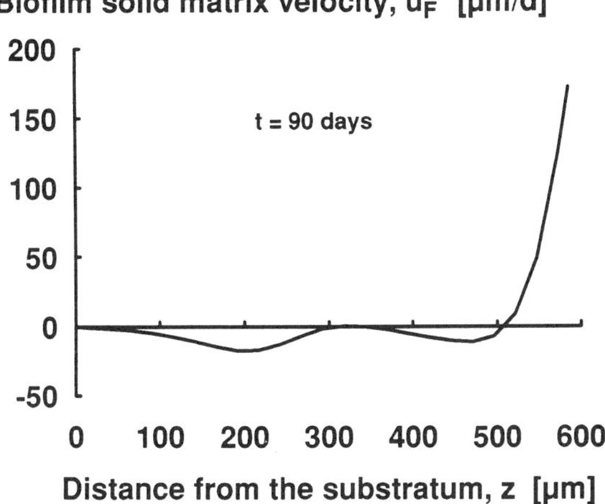

FIGURE 4. Spatial profile of the biofilm solid matrix velocity.

biofilm is not possible if μ_{net_B} is smaller than μ_{net_A} throughout the biofilm.[8] The biofilm solid matrix shrinks in biofilm zones where all net specific growth rates are negative.

4. Velocity of the Biofilm Solid Matrix

Progression of the biofilm solid matrix is reflected by the solid matrix velocity u_F, as introduced in Equation 2 and displayed in Figure 4. In zones of the biofilm where u_F is positive, the particulate components are displaced towards the biofilm surface. In zones where u_F is negative, the particulate components are transported to the opposite direction. At locations where u_F equals zero, there is no displacement, i.e., these locations separate biofilm layers in which the particulate components develop independently. The exact mathematical condition for coexistence of particulate components requires the net specific growth rates or production rates μ_{net_i} to be equal in a biofilm or biofilm layer at locations where $u_F = 0$ and $\mu_{net_i} > 0$.[8] If this condition is applied to the data given in Figures 3 and 4, the microbial population in this biofilm has not yet reached a steady state. However, considering the variety and complexity of biofilms, conclusions based on a purely mathematical analysis should be drawn with caution.

B. Dissolved Components in the Biofilm

1. Dissolved Oxygen, Acetate, and Glucose

Figure 5 illustrates the spatial concentration profiles of dissolved oxygen in the biofilm over a period of several days. These profiles confirm the

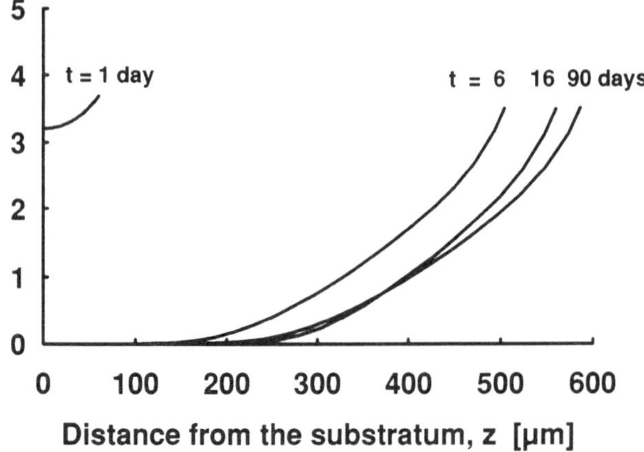

FIGURE 5. Spatial concentration profiles of dissolved oxygen in the biofilm.

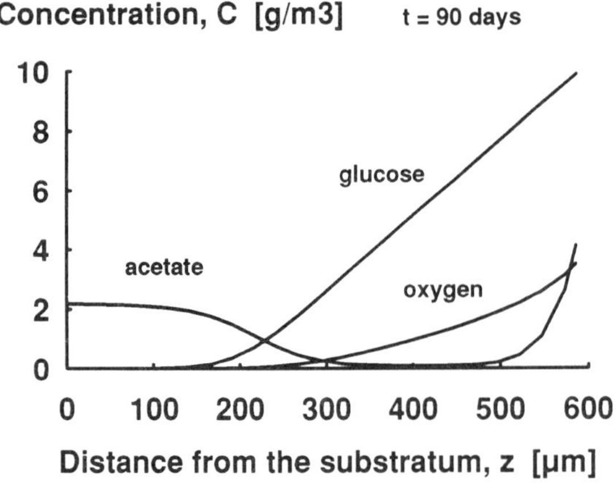

FIGURE 6. Spatial concentration profiles of dissolved oxygen, acetate, and glucose in the biofilm.

development of an anaerobic zone in the biofilm depth between the 1st and the 6th day. After the 6th day, the profiles are merely shifted due to the movement of the biofilm-bulk fluid interface in space, and their shape hardly undergoes further changes. The profiles indicate that the depth of the active zone in the biofilm is usually very thin. Figure 6 displays spatial concentration profiles of dissolved oxygen, acetate, and glucose. They reveal that the local maximum of μ_{net_A} at $z = 250$ μm (see Figure 3) is due to the availability of acetate produced in the biofilm depth and dissolved oxygen originating from the bulk fluid. Figure 6

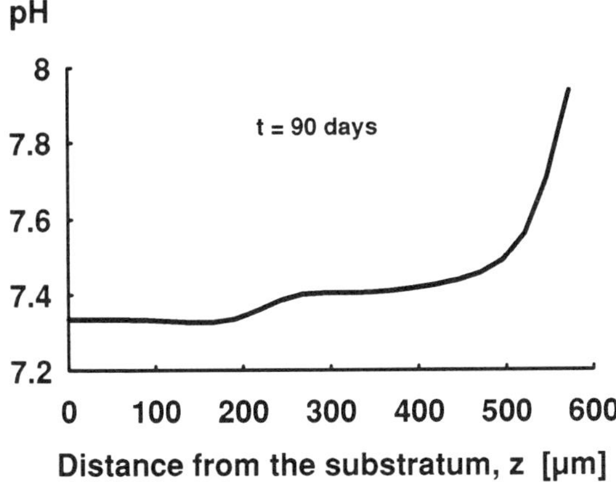

FIGURE 7. Spatial pH profile in the biofilm.

also shows that the growth of species B must always remain limited, since species A acts as a diffusion barrier for glucose.

2. *Role of pH*

As can be seen in Figure 7, the value of the pH changes from 7.3 in the biofilm depth to 8 near the biofilm surface. The high pH in this zone of the biofilm is almost completely due to the activity of the microbial species A. Figure 7 shows clearly how much the physicochemical conditions between the biofilm surface and the biofilm-substratum interface can change, and how important it is to dispose of accurate experimental data or model predictions of the conditions at this interface.[26]

C. Effect of Detachment

Detachment of cells and particles from the biofilm surface to the bulk fluid is a very important process which is not yet fully understood. Figure 8A, containing the results of three simulations performed with alternative models of the detachment velocity, shows some consequences of this process. Case a shows the result of the simulation discussed so far, where u_{de} was assumed to be proportional to L_F^2. Case b is based on an initial biofilm thickness of 600 μm and was calculated with a detachment velocity $u_{de} = u_F(z = L_F)$. In this case, the concentration of the anaerobic microbial species B is larger than that of the other two cases (Figure 8B), since the anaerobic zone in the biofilm depth already existed here at day 0. Case c was calculated with $u_{de} = 0$. In this case, a steady state of the biofilm thickness is brought about exclusively by an equilibrium of growth

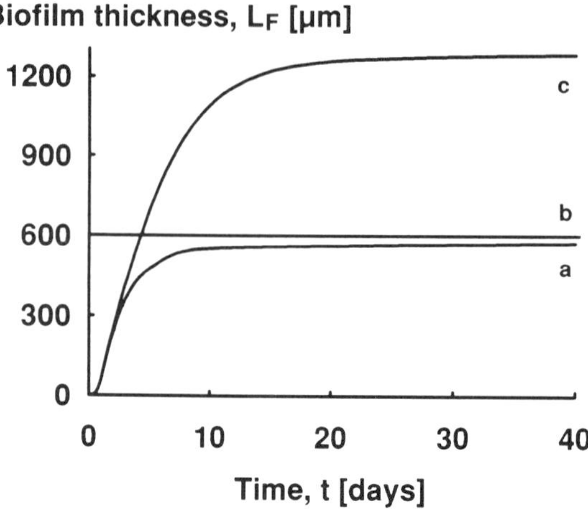

FIGURE 8A. Progression of the biofilm thickness for the detachment velocity u_{de} assumed to be proportional to L_F^2 (a), equal to the biomass production rate (b), and equal to zero (c).

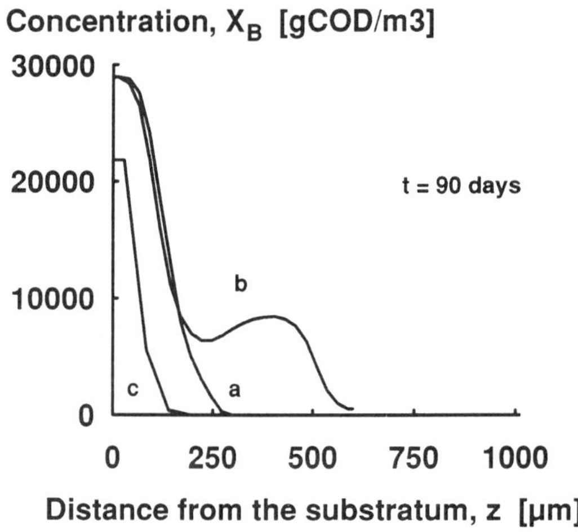

FIGURE 8B. Spatial concentration profile of the microbial species B in the biofilm for all three cases of Figure 8A.

and decay of the microbial species in the biofilm. In this case, the growth of species B is hindered by the stronger diffusion limitation for glucose.

D. Volume Fraction of the Liquid Phase in the Biofilm

In the mixed-population biofilm model it is assumed that the biofilm liquid-phase volume fraction ϵ_l, as defined in Equation 16, does not

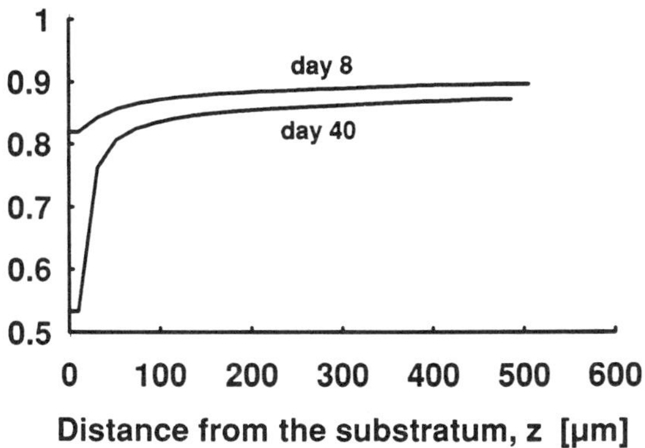

FIGURE 9A. Spatial profile of the biofilm liquid-phase volume fraction at the 8th and 40th days.

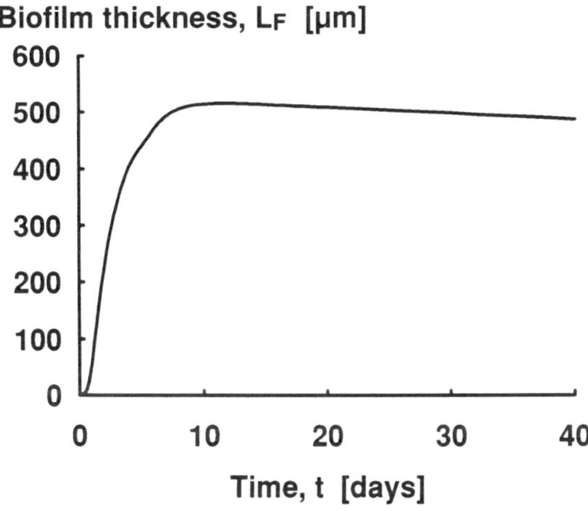

FIGURE 9B. Progression of the biofilm thickness for the biofilm liquid-phase volume fraction that varies as shown in Figure 9A.

change with time and space. This assumption was always controversial; in fact, some experimental data indicate that ϵ_l is not a constant.[9] The equations of the mixed-population biofilm models were therefore recently extended to allow for changes of ϵ_l. Figures 9A and 9B reveal the results of a simulation where ϵ_l is slowly decreasing with time, particularly in the biofilm depth. The resulting spatial profiles of ϵ_l for the 8th and 40th days are given in Figure 9A. In this simulation, the production of particulate mass in the biofilm leads only partly to a displacement of the biofilm solid matrix as described by the solid matrix velocity u_F, and partly to a decrease of the void space between the microbial cells. Figure 9B illustrates one of the consequences resulting from this change in the model assumptions: after the 10th day, the biofilm thickness slowly starts to decrease due to a reduction of available liquid-phase volume for the diffusive transport of the dissolved components in the biofilm and a consequent decrease in available substrate for microbial growth in the biofilm depth.

V. SUMMARY AND CONCLUSIONS

This chapter presents a concept and mathematical framework for modeling of mixed-population biofilms. An example of a biofilm with two microbial species was used to illustrate the application of the model, to show which experimental data are required for mixed-population biofilm modeling, and to discuss the results obtained.

With the presented mixed-population biofilm model it is possible to reproduce most of the phenomena observed in experiments and representing universal characteristics of biofilm structure and behavior. At present, attempts to refine the model are being made in order to incorporate experimental data which have so far remained unexplained. However, the primary need today is not the development of more refined mathematical models but rather the performance of carefully designed experiments enabling verification (confirmation or rejection) of the hypotheses on which the models are based and which thereby provide reliable numerical data for the parameters and processes used in the models. These data, which are partly missing today, include values for rate constants of attachment and detachment processes, information on changes in time and space of the composition of the solid matrix, liquid-phase volume fraction and diffusivities in the biofilm, as well as kinetic data about specific microbial processes in the biofilm.

VI. ACKNOWLEDGMENT

I wish to thank my colleague Peter Reichert, who is a dedicated scientist and an inspiring partner, for his contribution to this work.

LIST OF SYMBOLS

Dimensions are expressed by L (length), M (mass), and T (time).

A	Biofilm area (L^2)
b	Decay rate constant (T^{-1})
C	Dissolved component concentration, $C = S/\epsilon_l$ (mass per unit liquid-phase volume) (ML^{-3})
c_{de}	Detachment coefficient ($L^{-1}\ T^{-1}$)
D	Diffusivity ($L^2\ T^{-1}$)
F_{pH}	Function describing the effect of pH (−)
f	Ratio of the diffusivity in the biofilm and in pure water (−)
i	Interfacial mass flux (mass per unit total area and unit time) ($ML^{-2}\ T^{-1}$)
j	Mass flux (mass per unit total area and unit time) ($ML^{-2}\ T^{-1}$)
K	Half-saturation or equilibrium constant (ML^{-3})
k_{as}	Adsorption rate constant (LT^{-1})
k_{at}	Attachment rate constant (LT^{-1})
k_b	Backward-reaction rate constant (T^{-1})
k_{ds}	Desorption rate constant (LT^{-1})
k_f	Forward-reaction rate constant ($L^3\ M^{-1}\ T^{-1}$)
L_F	Biofilm thickness (L)
n_P	Total number of processes (−)
n_X	Total number of particulate components (−)
p	Process rate (mass produced per unit total volume and unit time) ($ML^{-3}\ T^{-1}$)
Q	Volumetric flow rate ($L^3\ T^{-1}$)
r	Production rate (mass produced per unit total volume and unit time) ($ML^{-3}\ T^{-1}$)
r''	Interfacial transfer rate (mass transferred per unit total area and unit time) ($ML^{-2}\ T^{-1}$)
S	Dissolved component concentration, $S = \epsilon_l C$ (mass per unit total volume) (ML^{-3})
t	Time (T)
u_{at}	Attachment velocity (LT^{-1})
u_{de}	Detachment velocity (LT^{-1})
u_F	Advective velocity of biofilm solid matrix (LT^{-1})
u_L	Velocity of the biofilm-bulk fluid interface (LT^{-1})
V_B	Bulk fluid volume (L^3)
V_R	Reactor volume (L^3)

X Particulate component concentration, $X = \rho_s \epsilon_s$ (mass per unit total volume) (ML^{-3})
Y True yield coefficient ($M_X M_C^{-1}$)
z Space coordinate perpendicular to biofilm surface with origin at the substratum-biofilm interface (L)
z′ Integration variable (L)

Greek Letters

ϵ_l Liquid-phase volume fraction (ratio of liquid-phase volume to total volume) (−)
ϵ_s Solid-phase volume fraction (ratio of solid-phase volume to total volume) (−)
κ Mass-transfer resistance coefficient (TL^{-1})
μ Specific growth rate (T^{-1})
μ_m Maximum specific growth rate (T^{-1})
μ_{net} Net specific growth rate, $\mu_{net} = \mu - b$ (T^{-1})
ν Stoichiometric coefficient (MM^{-1})
ρ_s Particulate component density (mass of particulate component per unit solid-phase volume) (ML^{-3})

Subscripts

B Bulk-fluid compartment
C Dissolved component
F Biofilm compartment
i Component
in Influent
j Process
L Liquid boundary layer
l Liquid phase
S Substratum
s Solid phase
X Particulate component
0 Initial value

REFERENCES

1. **Wanner, O. and Gujer, W.,** Competition in biofilms, *Water Sci. Technol.*, 17, 27, 1984.
2. **Kissel, J. C., McCarty, P. L., and Street, R. L.,** Numerical simulation of a mixed-culture biofilm, *J. Environ. Eng.*, 110, 393, 1984.
3. **Wanner, O. and Gujer, W.,** A multispecies biofilm model, *Biotechnol. Bioeng.*, 28, 314, 1986.

4. **Fruhen, M., Christan, E., Gujer, W., and Wanner, O.,** Significance of spatial distribution of microbial species in mixed culture biofilms, *Water Sci. Technol.*, 23, 1365, 1991.
5. **Arcangeli, J. P. and Arvin, E.,** Modelling of toluene biodegradation and biofilm growth in a fixed biofilm reactor, *Water Sci. Technol.*, 26, 617, 1992.
6. **Debus, O. and Wanner, O.,** Degradation of xylene by a biofilm growing on a gas-permeable membrane, *Water Sci. Technol.*, 26, 607, 1992.
7. **Wanner, O.,** Modeling population dynamics, in *Structure and Function of Biofilms*, Characklis, W. G. and Wilderer, P. A., Eds., John Wiley & Sons, New York, 1989, 91.
8. **Gujer, W. and Wanner, O.,** Modeling mixed population biofilms, in *Biofilms*, Characklis, W. G. and Marshall, K. C., Eds., John Wiley & Sons, New York, 1990, chap. 11.
9. **Christensen, B. E. and Characklis, W. G.,** Physical and chemical properties of biofilms, in *Biofilms*, Characklis, W. G. and Marshall, K. C., Eds., John Wiley & Sons, New York, 1990, chap. 4.
10. **Rittmann, B. E.,** The effect of shear stress on biofilm loss rate, *Biotechnol. Bioeng.*, 24, 501, 1982.
11. **Bryers, J. D.,** Biofilm formation and chemostat dynamics: pure and mixed culture considerations, *Biotechnol. Bioeng.*, 26, 948, 1984.
12. **Rittmann, B. E.,** Detachment from biofilms, in *Structure and Function of Biofilms*, Characklis, W. G. and Wilderer, P. A., Eds., John Wiley & Sons, New York, 1989, 49.
13. **Bakke, R., Characklis, W. G., Turakhia, M. H., and Yeh, An-I,** Modeling a monopopulation biofilm system: *Pseudomonas aeruginosa*, in *Biofilms*, Characklis, W. G. and Marshall, K. C., Eds., John Wiley & Sons, New York, 1990, chap. 13.
14. **Siegrist, H.,** Stofftransportprozesse in festsitzender Biomasse, Ph. D. thesis Nr. 7726, Swiss Federal Institute of Technology, Zurich, Switzerland, 1985.
15. **Ruchti, J., Wanner, O., and Reichert, P.,** BIOSIM: simulation of the dynamics of mixed culture biofilm systems, *Environ. Software*, 6, 29, 1991.
16. **Reichert, P., Ruchti, J., and Wanner, O.,** BIOSIM: interactive program for the simulation of the dynamics of mixed culture biofilm systems, User Manual, EAWAG, CH-8600 Duebendorf, Switzerland, 1989.
17. **Gottschal, J. C. and Thingstad, T. F.,** Mathematical description of competition between two and three bacterial species under dual substrate limitation in the chemostat: a comparison with experimental data, *Biotechnol. Bioeng.*, 24, 1403, 1982.
18. **Robinson, J. A., Trulear, M. G., and Characklis, W. G.,** Cellular reproduction and extracellular polymer formation by *Pseudomonas aeruginosa* in continuous culture, *Biotechnol. Bioeng.*, 26, 1409, 1984.
19. **Henze, M., Grady, C. P. L., Jr., Gujer, W., Marais, G. v. R., and Matsuo, T.,** Activated Sludge Model Nr. 1, Scientific and Technical Reports Nr. 1, IAWPRC, London, 1987.
20. **Dolfing, J.,** Acetogenesis, in *Biology of Anaerobic Microorganisms*, Zehnder, A. J. B., Ed., John Wiley & Sons, New York, 1988, chap. 9.
21. **Hooijmans, C. M., Geraats, S. G. M., van Neil, E. W. J., Robertson, L. A., Heijnen, J. J., and Luyben, K. Ch. A. M.,** Determination of growth and coupled nitrification/denitrification by immobilized *Thiosphaera pantotropha* using measurement and modeling of oxygen profiles, *Biotechol. Bioeng.*, 36, 931, 1990.
22. **Grotenhuis, J. T. C., Smit, M., Plugge, C. M., Yuansheng, X., van Lammeren, A. A. M., Stams, A. J. M., and Zehnder, A. J. B.,** Bacteriological composition and structure of granular sludge adapted to different substrates, *Appl. Environ. Microbiol.*, 57, 1942, 1991.

23. **Stumm, W. and Morgan, J. J.,** *Aquatic Chemistry,* 2nd ed., John Wiley & Sons, New York, 1981.
24. **Reid, R. C., Prausnitz, J. M., and Poling, B. E.,** *The Properties of Gases and Liquids,* 4th ed., McGraw-Hill, New York, 1987.
25. **Reichert, P. and Ruchti, J.,** AQUASIM: computer program for the analysis of aquatic ecosystems, User Manual, EAWAG, CH-8600 Duebendorf, Switzerland, 1992.
26. **Szwerinski, H., Arvin, E., and Harremoës, P.,** pH-Decrease in nitrifying biofilms, *Water Res.,* 20, 971, 1986.

5

Effects and Extent of Biofilm Accumulation in Membrane Systems

Hans-Curt Flemming, Gabriela Schaule, Richard McDonogh, and Harry F. Ridgway

I. INTRODUCTION

Microorganisms are present in nearly all water systems. They tend to adhere to surfaces and grow, mainly at the expense of nutrients accumulated from the water phase. The attached microorganisms excrete extracellular polymeric substances (EPS), in which they are embedded, and form biofilms (see Chapter 1). Thus, the originally dissolved organic and inorganic nutrients are now locally immobilized and converted from solution to a semisolid state. This is the principle of biofilm reactors utilized in wastewater and drinking water purification, and in these cases biofilms are highly desirable. However, biofilms form without regard to human needs or convenience; thus, they also occur in the wrong place. This phenomenon is quite well known in water treatment and is called biofouling. Clogging of filters and aquifers, contamination of treated water and other interferences with technical processes have been reported.[1] In this chapter we will focus on the effects and extent of biofilm formation on the surfaces of reverse osmosis (RO) and ultrafiltration membranes, although we acknowledge that biofilms in other compart-

0-87371-928-X/94/$0.00+$.50

ments of a membrane system can further interfere with overall performance of a separation plant.

Membrane systems are widely used in water purification. Dissolved and particulate matter from the feed stream is deposited on the membrane surface during the separation process and contributes to the overall resistance of the separation system. This phenomenon is known as "fouling", an expression adopted from heat-exchanger technology.[2,3] In membrane technology, the most important types of fouling include:

1. Crystalline fouling ("mineral scaling", deposition of minerals due to the excess of solution product)
2. Organic fouling (deposition of dissolved humic acids, oil, grease, lipids, etc.)
3. Particle and colloid fouling (deposition of clay, silt, particulate humic substances, debris, silica)
4. Microbiological fouling ("biofouling", adhesion and accumulation of microorganisms, forming biofilms)

The first three types of fouling generally can be controlled by appropriate pretreatment of the water phase. If the foulant concentration in the water is decreased, fouling on a surface is reduced. Biofouling of a surface, however, is hard to control by reducing the number of microorganisms in the water phase. Nevertheless, this is the common approach at present for biofouling control.[4] Microorganisms can multiply and will do so if they have nutrients. As microorganisms are ubiquitous in technical systems, and as disinfection in a technical system neither leads to sterility nor is maintained over a long period of time, microorganisms will always invade and colonize the system. In low-nutrient systems, adhesion seems to be the strategy of bacteria to survive,[5] forming biofilms capable of scavenging nutrients even when they are in very low concentration in the feed water flowing past. Biofilms are reported even in high-purity water systems with a conductivity resistance of 18 MΩ.[6] Effective prevention of microbial growth in membrane systems is only achieved when a continuous, sufficiently high chlorine concentration is maintained. This, however, cannot be viewed as an ultimate solution due to growing environmental concerns and stricter legislative regulations regarding the discharge of chlorinated brines. Membrane materials with reduced bacterial affinity are currently under research, but such membranes are not yet commercially available.

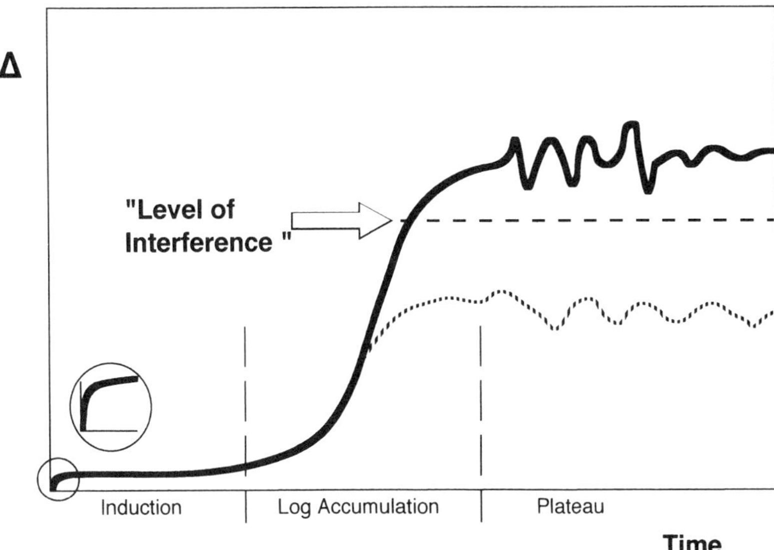

FIGURE 1. Time-dependent development of biofilm accumulation. Δ = biofilm growth parameter (thickness, weight, etc.). Inset: primary colonization. "Level of interference": arbitrary extent of biofilm development above which the biofilm interferes with the performance of a membrane system. The dotted line symbolizes biofilm accumulation below the "level of interference"; the bold line symbolizes biofilm accumulation above "level of interference". (Modified from Characklis, W. G., *Biofilms,* Characklis, W. G. and Marshall, K. C., Eds., John Wiley & Sons, New York, 1990, 195. With permission.)

II. EFFECTS OF BIOFILM FORMATION IN MEMBRANE SYSTEMS

As indicated above, biofouling may be considered as "a biofilm reactor in the wrong place". Biofouling usually does not occur in a spectacular way, but rather develops in a more subtle manner. This is due to the gradual accumulation of the biological deposit: the biofilm. The general development of a biofilm can be schematically depicted by a three-phase development (Figure 1).[7]

The three general phases of biofilm development may be described as follows:[7]

Phase 1 The induction phase, with primary colonization (important for the prevention of biofilm development; the induction phase represents the time between two cleaning measures). The inset represents the usually rapid primary colonization, followed by a primary plateau. Adhesion is essentially proportional to the cell density in the water phase and occurs due to weak physicochemical interactions such as hydrogen bonds and hydrophobic and electrostatic interactions.[8]

Phase 2 The exponential growth phase, when cell growth on the surface contributes more to biofilm accumulation than the adhesion of raw water cells.

Phase 3 The plateau phase, when biofilm growth (adhesion and cell multiplication), cell detachment, and cellular senescence are in balance. This phase is mainly controlled by nutrient concentration and the resultant growth rate, the mechanical stability of the biofilm, and the effective shear forces. It is independent of the concentration of cells in the raw water. Original surface properties of the membrane are masked by the biofilm.

Technical systems have usually reached the plateau phase when operating under stable conditions. The term "biofouling" is operational, referring to that extent of biofilm accumulation which exceeds a "level of interference" as hypothetically depicted in Figure 1. Thus, the level of the plateau in relation to the level of interference is of crucial importance in terms of process stability, energy consumption, and economics. Presently, we have no information about the actual growth rates of biofilms in membrane systems. The concentration of assimilable organic carbon is crucial for the level of the plateau. The implications of this fact will be discussed later.

Biofilms usually exhibit a more or less slimy consistency which reflects a highly hydrated gel polymer matrix, typical of a hydrogel, consisting of the following components:

- Water (75 to 95 % of wet weight)
- EPS ("slime", "organic glue"), excreted by the microorganisms (75 to 95% of dry weight)
- Microorganisms
- Entrapped particles
- Sorbed dissolved substances (the EPS acts as a molecular sieve or ion exchanger and can sorb organic and inorganic molecules from the water phase)

Biofilm composition may vary greatly, depending on the kind of microorganisms involved, the composition of the raw water, the temperature, and the hydrodynamic conditions. However, the high water content and the slimy consistency are characteristic of most biofilms.

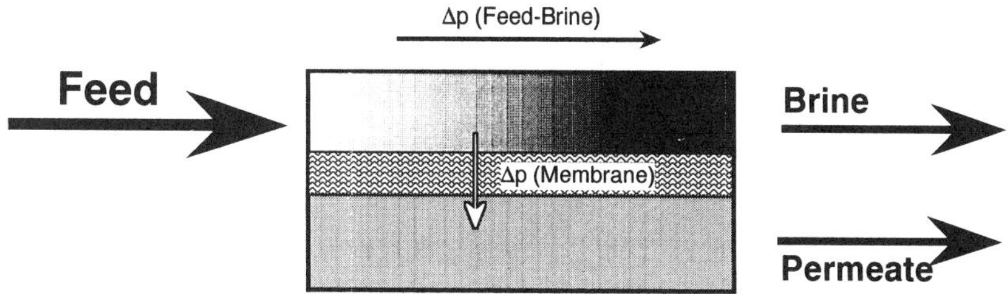

FIGURE 2. Schematical depiction of the pressure situation in a membrane system.

In membrane systems, biofilm thickness ranges between 10 and 100 μm.[9–11]

The greatest impact of biofilms on membrane systems may be attributed to the physical properties of the EPS matrix generated by the embedded microorganisms. Relevant characteristics of the EPS matrix are

- The gel-like structure, which prevents convectional water transport and allows diffusional transport only
- The porosity of this gel, which controls the permeation rate of water through a biofilm
- The fluid frictional resistance, caused by the rough, viscoelastic properties of biofilms

An additional physiological impact of the biofilm is the possible biodeterioration of membrane modules by microbial attack on membrane material and glue lines, which will be discussed below.

To understand better the effects of biofilms, a schematic depiction of the pressure and flux situation in a membrane system is helpful (Figure 2). The biofilm gel will cause a transmembrane pressure drop ($\Delta p_{membrane}$) and the rough, viscoelastic surface of the biofilm will cause a feed-brine pressure drop ($\Delta p_{feed/brine}$):

A. Increased Separation System Resistance by the Biofilm

The pressure drop imposed on the membrane system by the biofilm as a secondary membrane will lead to a flux decline. This effect can substantially affect the performance,[12] as illustrated in Figure 3.

Note that membrane flux is temporarily restored following each cleaning event (indicated by arrows). Also note the biphasic flux decline kinetics (a = rapid decline phase, b = slow decline phase).[12]

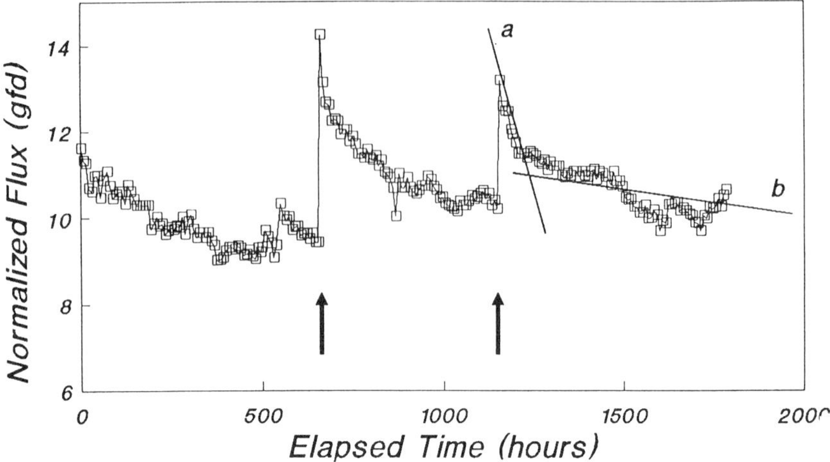

FIGURE 3. Example of flux decline data for subunit 2A at Water Factory 21. Note that membrane flux is temporarily restored following each cleaning event (indicated by arrows). Also note biphasic flux decline kinetics. *a* = rapid decline phase, *b* = slow decline phase. (From Ridgway, H. F. and Safarik, J., in *Biofouling and Biocorrosion in Industrial Water Systems,* Flemming, H.-C. and Geesey, G. G., Eds., Springer-Verlag, Heidelberg, 1991, 81. With permission.)

In most cases, the system pressure will be increased by an increase of the pump performance in order to compensate the flux decline due to the biofilm. This will result in an increase in energy consumption. However, the membrane modules have pressure limits, above which failure occurs. Above the pressure limit, spiral modules will telescope and be damaged irreversibly. Thus, the permeation properties of the biofilm as a secondary membrane adjacent to the original separation membrane are of major importance. Further investigations have revealed that a biofilm forms very rapidly after putting a membrane in operation. It was shown that after only 3 d of operation a biofilm had developed uniformly over the membrane surface.[13] Figure 4 is a scanning electron micrograph (SEM) of the original membrane; note the roughness of the surface. Although the dimension of the roughness is below that of bacteria, it will provide a better contact surface for the adhesive material by which bacteria attach themselves. Figures 5 and 6 show the biofilm on this membrane respectively, after 1 and 3 d. Figure 7 shows the time-dependent colonization of the surface and the corresponding flux decline.

The results presented above indicate that the separation system generally includes the biofilm, which starts forming from the very beginning

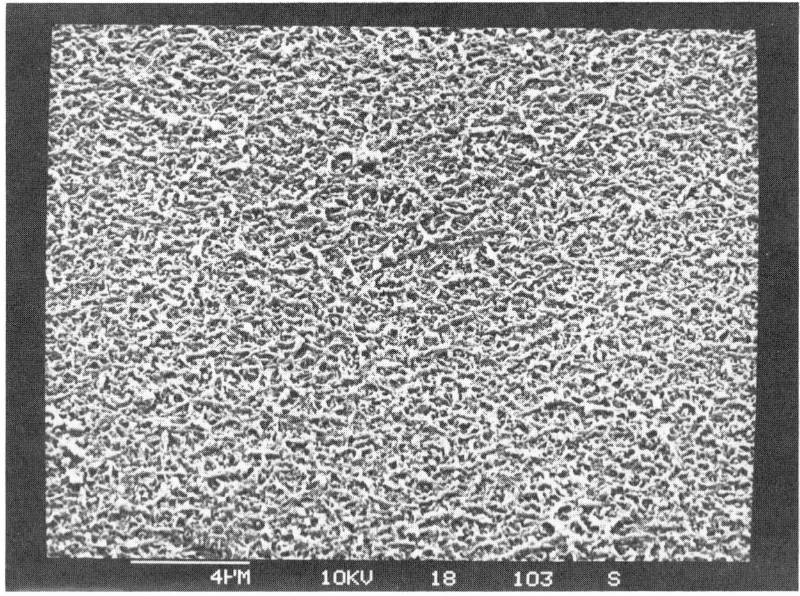

FIGURE 4. Original FT 30 polyamide membrane (FT 30).

of operation, acting as a secondary membrane. It is very likely that all nonsterile membrane systems will carry a microbial layer which participates in the separation process, influencing the performance parameters in an as yet undetermined way. This secondary layer is unnoticed, unless it interferes with membrane performance. By the time membrane performance drops below some threshold value, fouling is well established. It is important to be aware that a change in membrane performance is not a sudden event, rather a gradual response to changing conditions on the membrane surface, such as an increase of biofilm thickness and/or a decrease of biofilm porosity. Further research is needed to identify the factors that lead to biofilm evolution.

B. The Biofilm as a Gel Phase between Water and Membrane Surface

Separation processes based on the pressure-driven transport of water through a membrane inherently lead to concentration polarization of dissolved substances retained by the membrane at the raw water side. This effect will increase the pressure required for further separation of water from the concentrate. In addition, the solubility product of certain salts can be exceeded, leading to precipitation of crystals and to mineral scaling. The biofilm may provide a supersaturated microcosm with

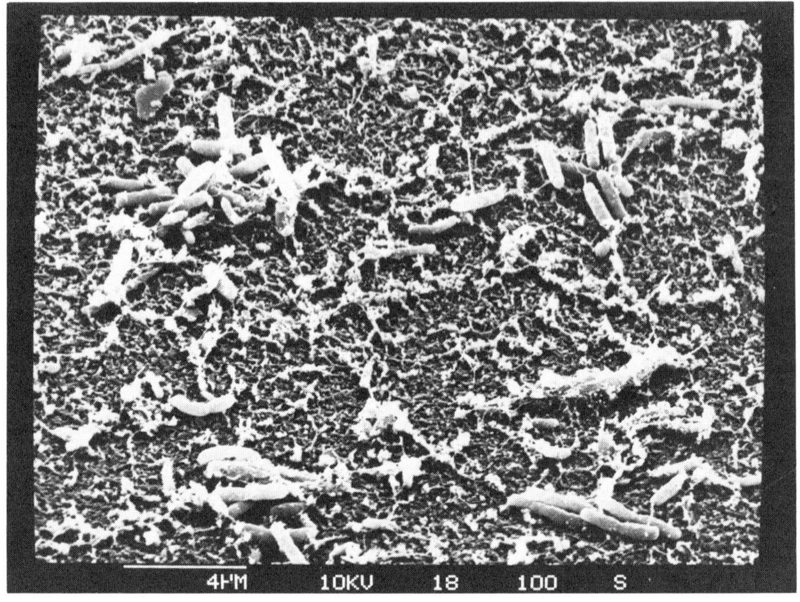

FIGURE 5. Polyamide membrane after 1 d of operation with clean tap water.

FIGURE 6. Same membrane as in Figure 5 after 3 d of operation with clean tap water.

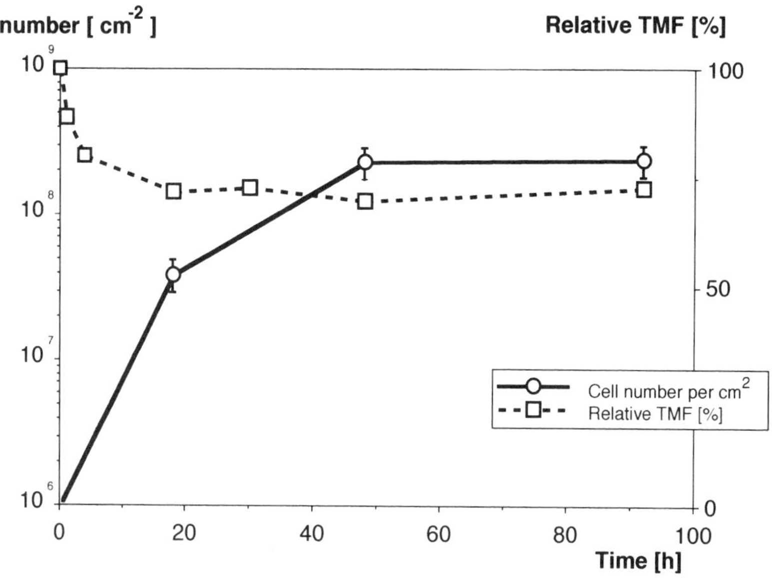

FIGURE 7. Flux decline and surface colonization of a membrane operated with clean tap water. TMF = relative transmembrane flux in %.

crystallization nuclei, favoring the accumulation of precipitated salts in its matrix. This combined fouling effect has not been investigated in membrane technology. From dental caries research, however, it is known that the biofilm on teeth acts as a diffusional barrier which favors the precipitation of insoluble phosphates.[14]

Membrane technology has addressed the problem of concentration polarization by application of a cross flow tangential to the membrane. Thus, the solutes concentrated by the separation process are diluted again by cross-flowing water from the bulk liquid. In spiral-wound modules, this mixing effect is enhanced by the turbulence created by the "spacer". The spacer is a network of plastic fibers which separates the spiral-wound membrane sheets from each other. There is a need for improved spacer design to achieve better turbulent mixing in membrane modules.

If a biofilm forms on the membrane surface, the gel matrix prevents convectional transport. Thus, turbulence generated by the spacer will not occur at the actual site where concentration polarization occurs. The biofilm not only shields the membrane surface from convectional flow, it also exerts a fluid frictional resistance due to its viscoelastic properties[15] and increases the pressure drop as symbolized in Figure 2 as $\Delta p_{feed\text{-}brine}$. In some instances, the biofilm will overgrow the spacer completely, negating its hydraulic function. Figure 8 shows an SEM of an irreversibly

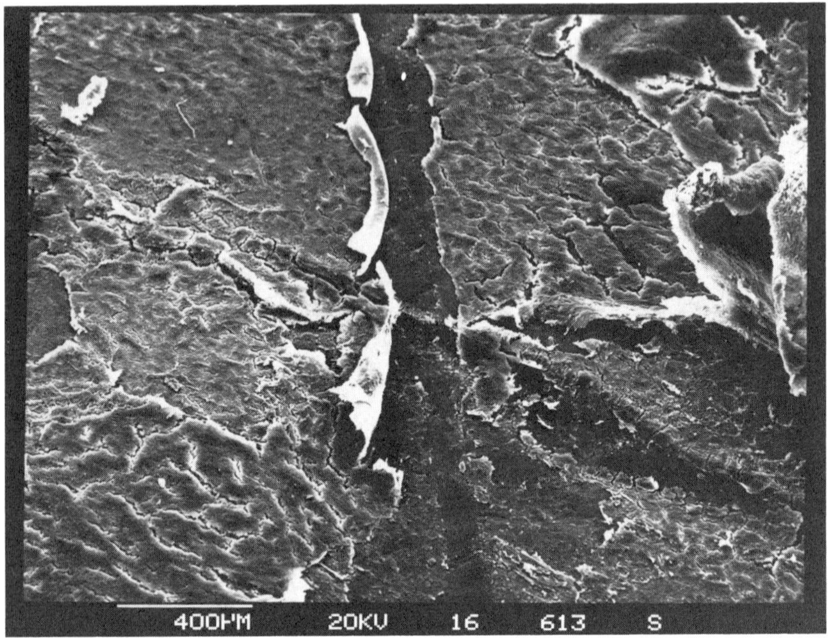

FIGURE 8. SEM of irreversibly biofouled membrane at a location where spacer was removed. Note that this biofilm has been treated many times with biocides and cleaners.

biofouled membrane. The spacer has been removed and its original position can be seen as the break in the biofilm. Figure 9 shows the spacer overgrown by biofilm.

The salt rejection is a function of the difference between the salt concentration at the raw and the permeate water side. Concentration polarization will increase the salt content in the permeate. Thus, a biofilm indirectly leads to a decrease of the quality of the produced water. In some instances, microorganisms of the biofilm will penetrate the membrane and settle on the permeate water side. Figure 10 shows the SEM of the permeate water side of a RO membrane. Note the presence of microorganisms. RO membranes are theoretically impermeable to particles, such as microorganisms. How these microorganisms got to this location is not clear.

C. Microbial Attack

Biofilms can attack membranes by excretion of acids or exoenzymes which attack the support material. This effect is called "biodeterioration".[16] There are numerous reports suggesting that bacteria attached to

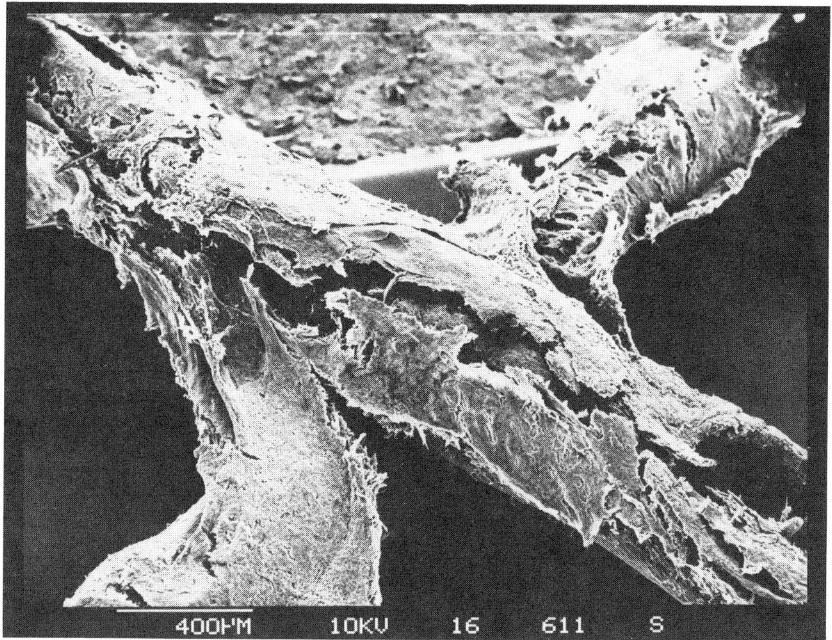

FIGURE 9. Spacer that has been removed from fouled membrane surface shown in Figure 8.

cellulose acetate RO membranes may produce enzymes or other substances which can attack and hydrolyze the membrane directly.[17–21] However, enzyme-catalyzed hydrolytic cleavage of the fully substituted cellulose acetate polymer has not yet been unequivocally demonstrated under controlled laboratory conditions. Therefore, the true significance of this potential mechanism of RO membrane degradation is not clear at this time. Ho et al.[22] reported that bacteria and fungi isolated from fouled RO membranes were unable to degrade cellulose acetate membranes and/or free polymer *in vitro*, though some of the fungal isolates were capable of enzymatically hydrolyzing unsubstituted cellulose. The inability of the fouling organisms to degrade cellulose acetate was attributed to the high degree of acetate substitution of the cellulose polymer. Reese[23] first provided experimental evidence that various cellulose derivatives (like cellulose acetate) became more resistant to microbial attack with increasing degrees of substitution. A number of cellulolytic fungi were found to completely degrade a water-soluble cellulose acetate with a degree of substitution (DS value) of 0.76 through the action of a cellobiose octaacetase. However, organisms possessing cellobiose octaacetase were unable to degrade cellulose triacetate having a DS value of 2.86. Presumably, the cellobiose octaacetase enzyme is sterically

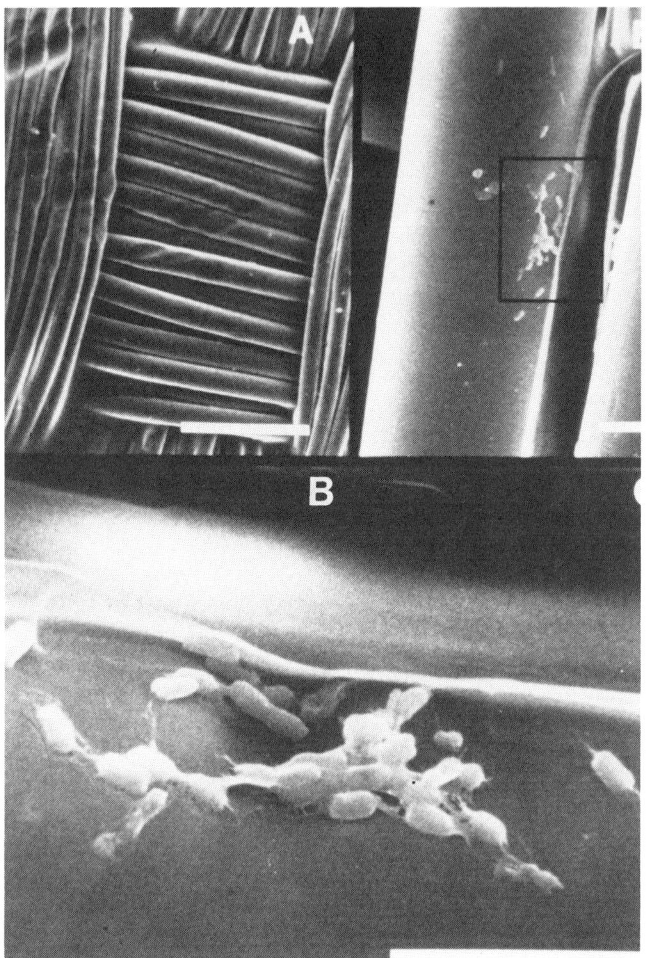

FIGURE 10. (A) Permeate side of the membrane shown in Figure 8; (B) enlargement of area in (A) containing colony of bacteria. The microorganisms adhering to the surface may detach and contaminate the product water.

hindered by excessive acetyl substitution, although such a mechanism has never been experimentally demonstrated.

Kutz et al.[24] reported the isolation of 129 bacterial isolates from cellulose acetate enrichment cultures using water from the Wellton-Mohawk canal (located near Yuma, AZ) as an inoculum. Water from this canal serves as feed water for the Yuma Desalting Facility. Of 16 isolates which were found to colonize cellulose acetate membrane material, a total of 7 were able to grow on a solidified basal salts medium supplemented with either cellulose acetate (DS value unspecified), carboxymethyl cellulose, cellobiose octaacetate, or cellobiose. Of the 16 organisms, 6 were identified as Gram-negative, cellulolytic, fluorescent pseudomonads which produced water-soluble green pigments. Although it was not directly demonstrated that the presence of any of the pseudomonads was responsible for a membrane failure (i.e., rapid loss of mineral rejection), their existence in an RO feed water stream suggests the possibility that such a failure might occur given the proper physicochemical and nutritional conditions on the RO membrane surface.

Kutz et al.[24–26] reported that a *Seliberia* sp. was associated with visible micropores or tunnels in cellulose acetate RO membrane surfaces which had apparently been severely biodegraded. The same organism, which displays diverse morphologies when grown on different carbon sources (including cellobiose), was also isolated from a groundwater source that supplied the RO system. When cellulose acetate RO membrane strips were incubated in the presence of a *Seliberia* sp. in a mineral salts basal medium, significant hydrolysis of the membrane polymer was reportedly evident after 4 weeks of incubation at 25°C. However, no direct evidence was presented that linked the observed membrane degradation with enzymatic activity of any kind. Thus, it is possible that nonenzymatic processes, such as localized pH changes due to metabolic activity of the adherent bacteria, could have indirectly caused an acceleration of membrane hydrolysis.

Few, if any, reports exist concerning biodeterioration and performance failure of noncellulosic RO membranes, such as polyamide- or polysulfone-type membranes. While such polymers may indeed be more recalcitrant to biodegradation than substituted cellulose derivatives, they probably should not be considered completely resistant to microbial attack and slow biodeterioration. Polyamide-type polymers provide a potentially rich source of reduced carbon and nitrogen for microorganisms to assimilate.

Similarly, the organically based glues that are employed in the construction of RO membrane modules may also be considered potentially biodegradable, and during extended periods of plant shutdown appropriate precautions should be taken to minimize the likelihood of biologically mediated membrane degradation. An example of severe damage to the membrane module by microbial attack at the glue lines is demonstrated in Figure 11. The membrane module was used in surface water purification. The glue was utilized as a nutrient by fungi, which led to a failure of the module stability.

Typical precautions for the prevention of membrane biodeterioration include the use of one or more broad-spectrum microbiocidal agents (e.g., formaldehyde, glutaraldehyde, quarternary amines, sodium bisulfite, etc.) for soaking RO membranes during long-term storage of elements.[27]

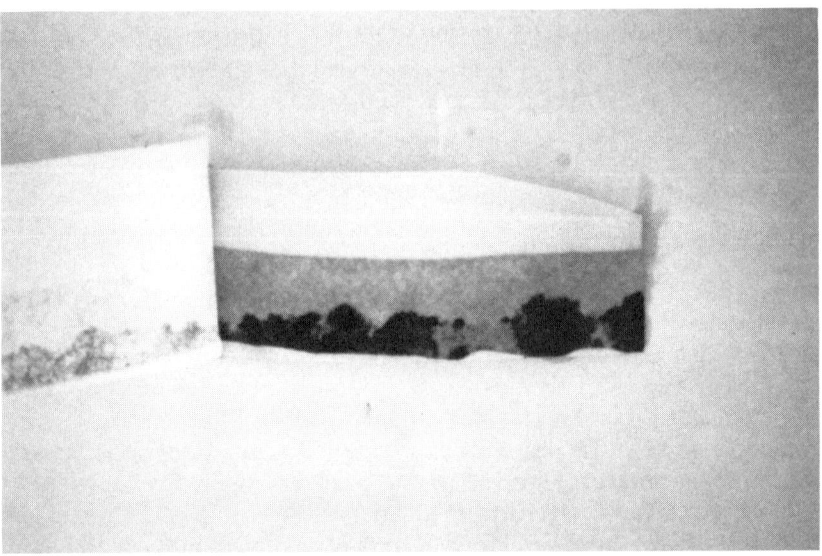

FIGURE 11. Fungal attack at a glue line in a membrane module utilized for surface water purification.

D. Costs

The costs of biofouling are difficult to assess, because they arise both directly and indirectly. They arise from the following effects of biofilms:

- Loss of product quantity
  flux decline due to increased $\Delta p_{membrane}$ (see Figure 2)
- Loss of product quality
  possible microbial contamination of permeate
  decrease of salt rejection because of concentration polarization
- Higher energy demand
  increased $\Delta p_{membrane}$ to overcome
  increased concentration polarization
- Higher pretreatment and cleaning demand
  chemicals (biocides, cleaners)
  labor
  shut down time
  eventual treatment of effluent to meet legal regulations
- Higher replacement costs
  damage by cleaning
  damage by microbial attack

TABLE 1. The Costs Due to Biofouling of the RO Membranes at Water Factory 21, Orange County, CA

Cost area/item	Unit cost/comment	Annualized subtotal ($)
Membrane cleaning		
a) labor	$27/h; 12d/year	7,776
b) chemicals	$696/mo; 1 clean/mo	8,352
Pretreatment		
a) lime[a]	60% recycle; 25% for biofilm control	31,536
b) filters	320 filters; 3x/year; $7/filter	6,720
c) polymers	3 mg/l; 150 lb/d; $1.30/lb	7,414
d) biocides	Chlorine 8 mg/l; $325/ton; 77 ton/year	25,084
e) energy[a]	1660 kwh/AF; 25% for biofilm control	223,000
Membrane flux decline	6 mgd; 1694 kwh/AF; $911,000/year; 150% OP for 80% of 4-year life	242,934
Shortened membrane lifetime	$1 million membrane inventory; 4-year actual life; 8-year theoretical life	125,000
Testing/evaluations	Testing new membranes and pretreatment processes	50,000
Total costs		727,816

Note: Data for Water Factory 21 in southern California, a 15-mgd wastewater reuse facility with 6 mgd RO.

[a] Assumed only 25% of process is used for biofouling control purposes.

A detailed assessment of the costs of biofouling has been carried out for the RO plant at Water Factory 21 in Orange County, CA (Table 1). In this plant, wastewater is reclaimed for reuse.

Membrane cleaning was quite inexpensive, but that is only because the plant formulates its own cleaning solutions. If they were purchased commercially, it would cost three to four times those shown in Table 1. About 8 mg/l combined chlorine is added continuously to the feed water. That amounts to about 77 tons per year at $25,000, all of which is for biofouling control. It was assumed that the membranes at Water Factory 21 operate at about 150% of their initial operating pressure (about 200 psi) over 80% of their 4-year life to compute the added energy cost due to flux decline. This is probably a fairly realistic assumption. Also, it should be remembered that cleaning is damaging to the membranes and probably contributes to their premature aging. It was further assumed that the $1 million membrane inventory should theoretically last 8 years instead of the observed 4 years. This amounts to an added cost of $125,000 per year due to biofouling. As there are substantial research efforts on biofouling in this plant (monitoring and optimization of biofouling control), it is expected that the costs as listed in Table 1 represent the lower limit.[12,28] The "bottom line" is $727,816 spent each year to control membrane biofouling. That represents about 30% *of the total operating costs* for Water Factory 21.

The costs of biofouling are increased by some common mistakes, which originate from the fact that biofouling is not understood as a biofilm process. Inadequate methods of detection and monitoring of countermeasures lead to delayed recognition and to suboptimal or ineffective biofouling control.

Usually, when a biofouling problem occurs, water samples are taken. However, the cell density in the water phase reflects neither the location nor the extent of biofouling in the system. Although biofilms seed the water phase with microorganisms, they do not do this in a predictable way. Portions of the biofilm will slough and lead to high transient cell numbers in the bulk water phase. On the other hand, there may be fairly low cell numbers in a water stream above a biofilm, containing a cell density as high as 10^8 to 10^{10} cells per gram biomass. Thus, sampling in the water phase is inadequate to monitor biofouling, and there are large quantities of water phase data which can never be interpreted. In addition, water phase cell numbers do not reflect the effectiveness of disinfection. Suspended cells are 1 to 3 orders of magnitude more susceptible to biocides than biofilm cells.[29,30] Thus, the absence of microorganisms in the water phase may lead to an overestimation of the effectiveness of disinfection. Biofouling is a biofilm problem, and as such arises from microbial colonization of surfaces. Thus, surfaces are the important sampling sites, although it is acknowledged that in a membrane system they are much more difficult to sample than the water phase. Testing coupons and sacrificial elements must be included into a representative monitoring program.

III. EXTENT OF BIOFILM ACCUMULATION IN MEMBRANE SYSTEMS

While the primary plateau (see inset in Figure 1) is mainly, but not completely, controlled by abiotic factors, the level of the final plateau is strongly dependent on the physiological activity of the microorganisms in the biofilm and the resulting EPS gel matrix properties. Thus, these two plateaus can be clearly differentiated from each other.

During the induction phase (see Figure 1), biofilm accumulation seems to be mostly due to physicochemical effects of accumulation. In this situation, microorganisms can be considered as living colloid particles[31] capable of active and/or passive adhesion.[32,33] Figure 7 shows that a primary plateau of biofilm thickness is reached in a period while no significant accumulation could be observed. Figure 12 shows how the system responds when the cell density in the water phase is increased by external addition of microorganisms. Thus, after being challenged the system carries a stable, but thicker biofilm which participates in the separation process. The thickness in this situation is probably limited by the shear forces.

A. Biofilm Thickness in Operating Systems

1. Calculation of Primary Biofilm Thickness from Biofilm Permeability

From given performance values, the thickness of a fouling layer can be calculated, provided the porosity is known. The following data and calculations are based on data which have been collected from a flat-channel RO test cell. The channel of the cell is 40 mm wide and 200 mm long; the height is 1 mm. The operation pressure was 40 bar in all experiments. The shear forces at the membrane have been calculated from the cross flow (50 and 150 l/h) and resulted in values of 10 and 13 N/m^2, respectively.[34]

A biofilm is heterogeneous, containing cells, scale, and debris of different types which are impermeable, whereas the space in between can be considered permeable. So, approximately, a biofilm is composed of two types of species: water permeable and not water permeable. It can be assumed that the net permeabilities of biofilms will differ, depending on relative proportions of these different species.

$$R = m \cdot \frac{1}{A} \cdot d = m \cdot \frac{h}{V} \cdot \alpha \tag{1}$$

Where R is hydraulic resistance, d is specific resistance of the material, and m is the mass of the layer covering an area A. Alternatively, this can be written in terms of the thickness of the layer h. V is the volume of the layer and is evaluated by experiment.[35] For a layer of two components, one permeable (i.e., finite hydraulic resistance, R) and one not (i.e., infinite hydraulic resistance), where p_f is the fraction of permeable material, the net layer resistance is $p_f R$. From Equation 1, then, the layer thickness is

$$h = p_f R \cdot V \cdot \frac{1}{m} \cdot \frac{1}{\alpha} \tag{2}$$

The specific resistance of a permeable EPS matrix has to be determined. The algal polysaccharide agar was used as a model EPS matrix. Agar is a polyhydrated hydrogel. In addition, the specific resistance of an activated sludge filter cake and a bacterial filter cake composed of

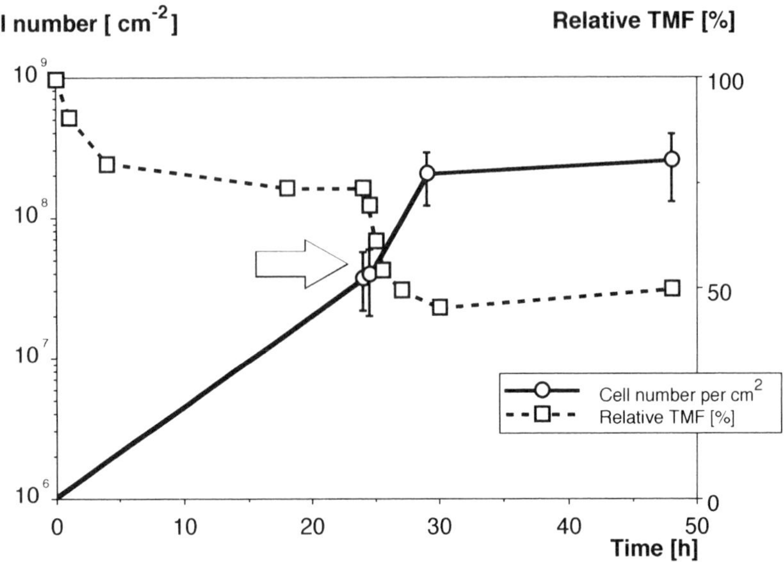

FIGURE 12. Cell density of the membrane after a challenge with high cell numbers in the feed water and responding flux decline.

TABLE 2. Specific Hydraulic Resistance Values for Various Model Biofilms[34]

Model biofilm material	Specific hydraulic resistance
Agar	$(1.3 \pm 0.4) \times 10^{13}$ m/kg
Activated-sludge filter cake	$(12 \pm 2) \times 10^{13}$ m/kg
Bacterial filter cake	$(42 \pm 5) \times 10^{13}$ m/kg

Pseudomonas diminuta was measured. The data are given in Table 2. Translating these values into permeable fractions, based on agar, the activated sludge has 10% permeable material and the bacterial layer is 3.3% permeable.

2. Calculation of Biofilm Thickness from Yield Value

If it is assumed that the biofouling layer thickness is shear controlled, then when the wall shear exceeds the yield strength of the layer, Y, the layer will no longer thicken. By knowing the flow regime and the yield of the accumulating layer, the shear-limited height of this layer can be estimated. For laminar flow in a rectangular channel where Q is the flow

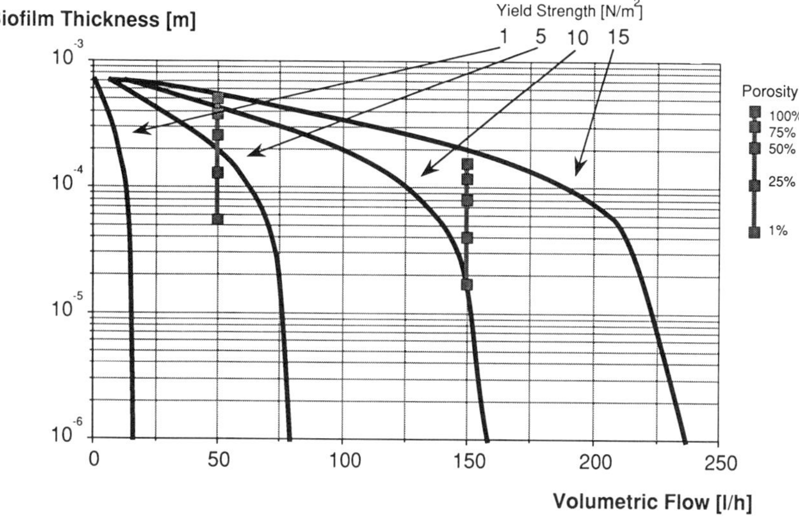

FIGURE 13. Biofilm thickness as a function of volumetric cross-flow rate. The effective module height is assumed $h = h_o - h_{biofilm}$. Thickness estimated from yield values (curves) and permeabilities (bars) can be compared. (From Flemming, H.-C., Schaule, G., Gaveras, E., and Beck, M., in *Herstellung von Reinwasser,* Marquardt, K., Ed., Expert Verlag, Ostfildern, in press.)

rate, w the channel width, and η the liquid viscosity, the height of the biofilm is

$$h_{biofilm} = h_O \cdot \sqrt{9\,\eta\,\frac{Q}{w_{ch}} \cdot \frac{1}{Y}} \tag{3}$$

Indirect evidence indicates that a value of 13 Nm^{-2} for the yield of the biofilm may be reasonable.[34] Figure 13 shows estimates of biofilm thickness in a rectangular RO cell channel. The bars represent estimates of thickness using Equation 2 for the layer resistance at two flow values. The layer resistance can be estimated from the transmembrane flow rates and the transmembrane pressure drop, which can be experimentally evaluated.[36] For example, for a flow of 150 l/h, if the layer were 100% permeable, then the thickness would be around 150 μm; if it were only 10% permeable, then the layer need only be 15 μm thick to allow the observed permeate flux. Using Equation 3, the thickness based on shear yield is calculated. If these layers have shear-dependent thickness, then a difference in yield values from 10 to 15 Nm^{-2} would produce the differences seen with the different volumetric flows.

B. The Role of Biofilm Permeability

As indicated above, thickness is not the only parameter influencing the hydraulic resistance of the separation system. The lower the level of interference the more sensitive is the operation to cross flow, and the more permeable a biofilm is, the greater is the thickness which can be tolerated. Recent investigations using confocal laser microscopy revealed that the structure of biofilms is heterogeneous, with tightly packed microcolonies, areas of dense EPS, and free channels and pores.[36] This indicates that biofilms display a spatially heterogeneous permeability. Changes in the freedom of flow are then described by a tortuosity term. In the following a "flow porosity" is used instead of tortuosity because the permeability of the matrix is considered as a whole and not as flow in channels. "Flow porosity" refers to the area available for flow in any infinitesimal cross section.

1. *Improvement of Fouling Layer Porosity*

In a given system, the biofilm will reach a plateau according to the limiting forces, such as growth and shear forces. In cases when the consequences of increasing biofilm thickness mean the level of interference is exceeded, disinfection and cleaning measures are carried out. Frequently an improvement of performance is achieved, although there is no evidence of biomass removal. In some instances, $\Delta p_{membrane}$ will decrease, whereas $\Delta p_{feed\text{-}brine}$ will stay the same. The latter value, resulting from the fluid frictional resistance between the fouling layer and the flowing medium, indicates that the biofilm is still present. Investigations of the efficiency of cleaning methods have confirmed this in some cases.[10,34,37] The decrease of $\Delta p_{membrane}$ indicates that the permeability has increased. This approach can be optimized in order to better tolerate biofilms which cannot be removed and interfere to an unacceptable extent with the process demands. In other words: if a biofilm cannot be removed, it can be better tolerated when it is possible to improve its permeability.

Preliminary data clearly demonstrate that it is possible to improve biofilm porosity on a membrane, improving flux rate without changing the thickness of the layer.[35,38] Some chemicals, however, can cause a decrease in the permeability, making a tolerable biofilm thickness intolerable. This may explain the deleterious effects of some cleaning measures. Experiments with agar filter cakes have shown that a commercial cleaning mixture consisting of a nonionic and an anionic surfactant at high pH improves the permeability of the hydrogel significantly (Figure 14). Again, agar was used as a model biofilm. The agar was placed on a membrane in a simple pressure cell. The system was pressurized with

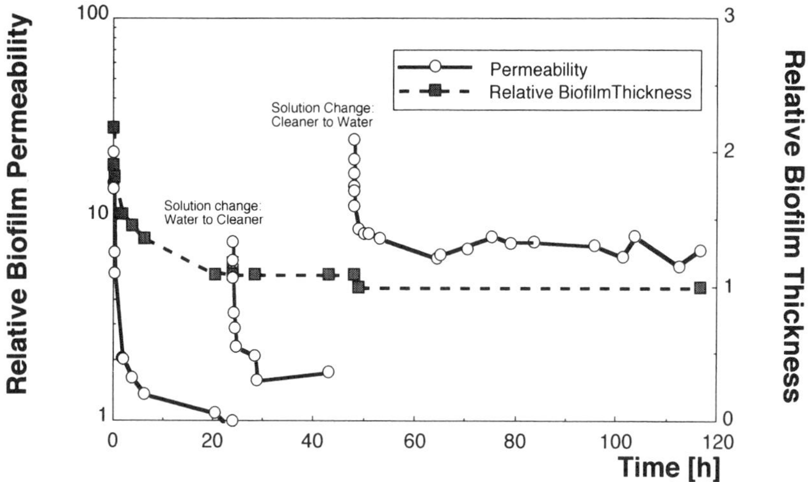

FIGURE 14. Thickness and permeability of a hydrogel (agar) before and after application of a commercial cleaner; note that the gel thickness remained constant.

1 bar from a water-filled reservoir. The permeability was measured until a steady state was reached (10% or less deviation over 10 h of operation). This was taken as the initial permeability of the layer. The application of the cleaner initially increased the permeability only slightly, but after the cleaner had been replaced by the original raw water, permeability remained about five times higher than initially. This indicates that the cleaner molecules have changed the porosity of the layer permanently. The experiment revealed that this effect was not due to a correspondent decrease of the filter layer thickness. The mechanism of this process has to be further investigated, but the implications are clear: it will be of interest to have chemicals which improve the permeability of a given biofilm permanently. The experiments have been extended to other model biofilms, such as a bacterial filter cake as proposed by Fane et al.[39] and Hodgson et al.,[40] and sewage sludge as harvested from a rotating disk reactor. The resulting contact times were between 1 and 3 h. The data are summarized in Table 3. Tannin was selected as a common membrane conditioning agent, formaldehyde represents a common disinfectant. Tetramethyl urea was selected because of its "chaotrophic" properties, i.e., the interference with hydrogen bonds and the water structure. As the interactions of the molecules of a hydrated matrix are expected to be mostly due to hydrogen bonds and primary adhesion can be prevented effectively by 0.1 *M* tetramethyl urea,[8] the chance existed that this agent could improve the porosity of the film.

TABLE 3. Relative Water Permeability of Agar Hydrogels, Bacteria Filter Layers, and Activated Sludge after Flushing with Conditioning Agents

Layer	Agent	Relative permeability J/J_o
Agar	Commercial cleaner (anionic-nonionic surfactants)	5.1
Agar	1% tannin	0.38
Agar	0.1 *M* tetramethylurea	1.0
Bacteria filter cake	Commercial cleaner (anionic-nonionic surfactants)	2.5
Bacteria filter cake	1% tannin	0.8
Activated-sludge filter cake	1% formaldehyde	0.6

C. The Role of Nutrients

In industrial systems, most biofilms will reach the plateau phase (c.f. Figure 1). The extent of biofilm accumulation in this phase will depend on different factors:

- Nutrient concentration, type and availability
- Shear forces
- Mechanical stability of the biofilm matrix, as influenced by
  - oxidizing agents (biocides, etc.)
  - biodispersants
  - mechanical stress
  - temperature
  - type of microorganisms
  - physiological activity (gas production)
  - structure and physical strength of the EPS network
  - grazing organisms

The most important factors are nutrient concentrations and shear forces. Once the plateau phase is reached, cell density in the water seems to be a factor of minor importance, compared to biofilm accumulation by cellular growth.[7,41] Thus, biodegradable material represents potential biomass. However, a biocide treatment mainly addresses suspended cells, while the nutrient concentration is not decreased. In some instances, biocides such as ozone or chlorine may even increase nutrient concentration (i.e., when humic acids are partially oxidized, they become more susceptible to microbial degradation).[42]

As indicated in the introduction, biofouling can be regarded as a biofilm reactor in the wrong place. Thus, if a high surface area is offered *ahead* of the system to be protected against biofouling, biofilms will form there and consume degradable matter from the water stream. This will decrease the extent of biofilm development in compartments downstream. The occurrence and effect of possible sloughing events requires further research. Bott[41] investigated biofouling in heat exchangers and has shown clearly that the extent of biofilm accumulation could be decreased by decreasing the nutrient concentration. Biological filter systems, such as used in drinking water treatment, might provide useful tools in reducing the nutrient concentration and thus in prevention of biofouling. To date, biocides are the primary means of biofouling control on membrane systems. Facing increasing difficulties in the application of biocides in both effectiveness and environmental regulations, manufacturers of membrane technology might be well advised to investigate biocide-free, antifouling strategies. The optimization of nutrient limitation techniques could provide new possibilities. The role of biodegradable dissolved organic carbon[43] will be very important to this strategy. This approach would utilize biofilms in the right place, where they can be handled easily, in order to minimize the extent of biofilm formation at remote sites where biofilm accumulation must be controlled. However, if it is applied where appropriate, it might reduce the need for large amounts of biocides. This strategy needs reliable biofilm monitoring devices for biofilm accumulation on membrane surfaces. Sacrificial module elements, as already proposed,[44,45] will be highly useful for destructive analysis. Considerable research efforts will be required before it is possible to put this strategy into practice.

IV. CONCLUSIONS

The data as presented reveal that every operating RO plant under nonsterile conditions will develop biofilm. This biofilm takes part in the separation process as a secondary membrane. Its separation characteristics will influence the overall system performance. As long as the biofilm does not affect the performance of the system more than can be tolerated, it will not be noticed. As soon as some "level of interference" is reached, the search for the cause and cure will begin. If biofilms are the reason for the trouble, they will be detected only by adequate sampling, i.e., sampling of biofilms. Sampling of the water phase will not indicate the site or extent of biofilm accumulation.

The effects of biofilms on plant performance can be substantial, but mostly hidden. Lime clarification, chlorination (and its side effects), higher energy demand, and shorter lifetime of modules are some of the effects.

Under most system operating conditions, the formation of biofilms is unavoidable, if run unsterile. In some cases, it is possible to control the adhesion of microorganisms by relatively high dosages of chlorine, but this cannot be the solution for all systems. The use of broad-spectrum biocides is gradually falling out of favor due to the direct and indirect damage of biocides to plant components and the hazards of handling the biocides. In addition, stricter environmental regulations are limiting the use of many biocides. More stringent effluent treatment will be required in the future, which increases the costs of water treatment and may negate the economic advantage of membrane filtration.

The control of biofilms can be achieved by nutrient limitation. This will not eliminate them, but keep their accumulation and effects below the "level of interference". The introduction of sacrificial surfaces, on which biofilm growth occurs deliberately ("biofiltration"), depleting the water stream of nutrients, offers an approach to biocide-free, antifouling strategy. Biological filters, as developed already in drinking water treatment, seem to offer clear advantages. Compartments behind these filters should be protected from excessive biofilm accumulation. This approach would balance wanted and unwanted biofilms without the need of biocides and, thus, offer an ecologically sound alternative.

Another approach could be to improve the permeation properties of the biofilm so that a system will tolerate thicker biofilms. This system will exert a higher "level of interference", which is helpful in cases when the biofilm cannot be removed. This approach may give an additional lifespan to "irreversibly fouled" modules in some cases.

The strategies presented here are far from a "cure all". They are potential alternatives to approach and solve biofouling problems.

LIST OF SYMBOLS

A	Area of layer	m^2
A_{Ch}	Cross section area of flow channel	m^2
h	Channel height	m
h_o	Channel height before layer accumulation	m
h_{actual}	Channel height due to layer accumulation	m
h_{bio}	Thickness of biofilm	m
J_o	Pure water flux of membrane	ms^{-1}
J	Permeate flux of membrane	ms^{-1}
L_p	Layer permeability	$ms^{-2}dPa^{-1}$
m	Layer mass	kg
n	Layer compressibility	
P	Pressure	Pa
P_o	Reference pressure	Pa
p_f	Flow porosity	m^2
Q	Volumetric flow rate	m^2s^{-1}

R	Hydraulic resistance	m^{-1}
R_{mem}	Hydraulic resistance of membrane	m^{-1}
R_{bio}	Hydraulic resistance of biofilm	m^{-1}
Re	Reynolds number	
TMF	Relative transmembrane flux	%
V	Volume of biofilm	m^3
w	Flow velocity in the middle of the channel	ms^{-1}
w_{ch}	Width of channel	m
Y	Layer yield strength	Nm^{-2}

GREEK LETTERS

α	Specific resistance	mkg^{-1}
α_o	Specific resistance at reference pressure P_o	mkg^{-1}
λ	Friction factor	
ρ	Density	kgm^{-3}
τ	Shear stress	Nm^{-2}
η	Viscosity	kg

REFERENCES

1. **Flemming, H.-C.,** Biofouling in water treatment, in *Biofouling and Biocorrosion in Industrial Water Systems,* Flemming, H.-C. and Geesey, G. G., Eds., Springer-Verlag, Heidelberg, 1991, 47.
2. **Epstein, N.,** Fouling: technical aspects, in *Fouling of Heat Transfer Equipment,* Somerscales, E. F. C. and Knudsen, J. G., Eds., Hemisphere, Washington, D.C., 1981, 31.
3. **Characklis, W. G.,** Microbial fouling, in *Biofilms,* Characklis, W. G. and Marshall, K. C., Eds., John Wiley & Sons, New York, 1990, 523.
4. **Characklis, W. G.,** Microbial fouling control, in *Biofilms,* Characklis, W. G. and Marshall, K. C., Eds., John Wiley & Sons, New York, 1990, 585.
5. **Marshall, K. C.,** Bacterial adhesion in oligotrophic habitats, *Microb. Sci.,* 2, 321, 1985.
6. **Mittelman, M. C.,** Bacterial growth and biofouling control in purified water systems, in *Biofouling and Biocorrosion in Industrial Water Systems,* Flemming, H.-C. and Geesey, G. G., Eds., Springer-Verlag, Heidelberg, Berlin, 1991, 113.
7. **Characklis, W. G.,** Biofilm processes, in *Biofilms,* Characklis, W. G. and Marshall, K. C., Eds., John Wiley & Sons, New York, 1990, 195.
8. **Schaule, G., Flemming, H.-C., and Poralla, K.,** Forces involved in primary adhesion of *Pseudomonas diminuta* to filtration membranes, 5th Int. Conf. on Microb. Ecol., Barcelona, September 8 to 12, 1992.
9. **Ridgway, H. F., Rigby, M. G., and Argo, D. G.,** Bacterial adhesion and fouling of reverse osmosis membranes, *J. Am. Water Works Assoc.,* 77, 97, 1985.

10. **Flemming, H.-C. and Schaule, G.,** Biofouling auf Umkehrosmose- und Ultrafiltrationsmembranen. Teil II. Analyse und Entfernung des Belages, *Vom Wasser,* 73, 287, 1989.
11. **Schaule, G., Kern, A., and Flemming, H.-C.,** A case study: biofouling of membranes in trickling water purification, *Int. Desal. Water Reuse Q.,* II, Vol. 3/1, 17–23, 1993.
12. **Ridgway, H. F. and Safarik, J.,** Biofouling on reverse osmosis membranes, in: *Biofouling and Biocorrosion in Industrial Water Systems,* Flemming, H.-C. and Geesey, G. G., Eds., Springer-Verlag, Heidelberg, 1991, 81.
13. **Flemming, H.-C., Schaule, G., and McDonogh, R.,** How do membrane parameters respond to initial biofilm formation?, *Vom Wasser,* 80, 177, 1993.
14. **ten Cate, J. M.,** Ed., *Recent Advances in the Study of the Dental Calculus,* IRL Press, Oxford, 1989.
15. **Christensen, B. E. and Characklis, W. G.,** Physical and chemical properties of biofilms, in *Biofilms,* Characklis, W. G. and Marshall, K. C., Eds., John Wiley & Sons, New York, 1990, 93.
16. **Rose, A. H.,** History and scientific basis of microbial biodeterioration of materials, in *Microbial Biodeterioration,* Rose, A. H., Ed., Academic Press, London, 1981, 1.
17. **Guy, D. G.,** Biodegradation of cellulosic membranes at the Roswell and Yuma desalination test facilities, *Pure Water,* Vol. 2, Int. Desalination and Environmental Association, Teaneck, NJ, 1980.
18. **Leahy, T. M., III.,** Recent experience with CTA hollow fiber membranes at the Roswell Test Facility, Proc. 8th Ann. Conf. Water Supply Improvement Assoc., Ipswich, MA, 1980.
19. **Cantor, P. A. and Mechalas, B. J.,** Biological degradation of cellulose acetate reverse osmosis membranes, *J. Polym. Sci.,* 28, 225, 1969.
20. **Motomura, H. and Taniguchi, Y.,** Durability study of cellulose acetate reverse osmosis membrane and adverse circumstances for desalting, in *Synthetic Membranes, Vol. I: Desalination,* Turbak, A. F., Ed., American Chemical Society, Washington, D.C., 1981, 79.
21. **Sinclair, N. A.,** Microbial degradation of reverse osmosis desalting membranes. Operation and Maintenance of the Yuma Desalting Test Facility, Vol. IV, U.S. Department of the Interior, Bureau of Reclamation, Yuma, AZ, 1982.
22. **Ho, L. D. W., Martin, D. D., and Lindemann, W. C.,** Inability of microorganisms to degrade cellulose acetate reverse-osmosis membranes, *Appl. Environ. Microbiol.,* 45, 418, 1983.
23. **Reese, E. T.,** Biological degradation of cellulose derivatives, *Ind. Eng. Chem.,* 49, 89, 1957.
24. **Kutz, S. M., Bently, D. L., Sinclair, N. A., and Kelley, L. M.,** Morphological diversity of a bacterium resembling *Seliberia*: an electron microscopic evaluation of nutritional effects, *Abstr. Annu. Meet. Am. Soc. Microbiol.,* Washington, D.C., 1986, 189.
25. **Kutz, S. M., Bentley, D. L., and Sinclair, N. A.,** Morphology of a *Seliberia*-like organism isolated from reverse osmosis membranes, in *Perspectives in Microbial Ecology,* Megusar, F. and Gantar, M., Eds., Slov. Soc. Microbiol., Ljubljana, 1986, 584.
26. **Kutz, S. M., Sinclair, N. A., and Bently, D. L.,** Characterization of *Seliberia* and its effects on reverse osmosis membranes, *Abstr. Annu. Meet. Am. Soc. Microbiol.,* Las Vegas, Nevada, 1985, 223.
27. **U.S. Department of the Interior,** Reverse osmosis technical manual, Office of Water Research and Technology, U.S. Department of the Interior, Washington, D.C., 1979.
28. **Ridgway, H. F.,** Microbial adhesion and biofouling of reverse osmosis membranes, in *Reverse Osmosis Technology,* Parekh, B. S., Ed., Marcel Dekker, New York, 1988, 429.

29. **Costerton, J. W. and Lashen, E. S.,** Influence of biofilm on efficiacy of biocides on corrosion-causing bacteria, *Mater. Perform.*, 23, 13, 1984.
30. **LeChevallier, M. W., Cawthon, C. D., and Lee, R. G.,** Inactivation of biofilm bacteria, *Appl. Environ. Microbiol.*, 54, 2492, 1988.
31. **Marshall, K. C.,** Mechanisms of bacterial adhesion at solid-water interfaces, in *Bacterial Adhesion,* Savage, D. C. and Fletcher, M. M., Eds., Plenum Press, New York, 1985, 133.
32. **Fletcher, M. M.,** The question of passive versus active attachment mechanisms in nonspecific bacterial adhesion, in *Microbial Adhesion to Surfaces,* Berkeley, R. C. W., Lynch, J. M., Melling, J., Rutter, P. P., and Vincent, B., Eds., Ellis Horwood, Chichester, 1980, 197.
33. **Flemming, H.-C. and Schaule, G.,** Biofouling on membranes — a microbiological approach, *Desalination,* 70, 95, 1988.
34. **Flemming, H.-C., Schaule, G., Gaveras, E., and Beck, M.,** Biofouling von Umkehrosmose-Membranen bei der Reinwasser-Herstellung, in *Herstellung von Reinwasser,* Marquardt, K., Ed., Expert Verlag, Ostfildern, Germany, 1993.
35. **McDonogh, R., Schaule, G., and Flemming, H.-C.,** Permeation properties of biofilms on separation membranes and the problem of biofouling, *J. Membr. Sci.*, 1994, accepted.
36. **Lawrence, J. R., Korber, D. R., Hoyle, B. D., Costerton, J. W., and Caldwell, D. E.,** Optical sectioning of microbial biofilms, *J. Bacteriol.*, 173, 6558, 1991.
37. **Whittacker, C., Ridgway, H. F., and Olson, B. H.,** Evaluation of cleaning strategies for removal of biofilms from reverse-osmosis membranes, *Appl. Environ. Microbiol.*, 48, 395, 1984.
38. **Flemming, H.-C., Schaule, G., and McDonogh, R.,** Biofouling — a biofilm problem, EUROMEMBRANE Conf. Proc., Paris, 1992.
39. **Fane, A. G., Fell, C. J. D., Hodgson, P. H., Leslie, G. L., and Marshall, K. C.,** Microfiltration of biomass and biofluids: effects of membrane morphology and operation conditions, *Filtr. Sep.*, 9/10, 332, 1991.
40. **Hodgson, P. H., Leslie, G. L., Schneider, R. P., Fane, A. G., Fell, C. J. D., and Marshall, K. C.,** Cake resistance and solute rejection in bacterial microfiltration: the role of the extracellular matrix, *J. Membr. Sci.*, in press.
41. **Bott, T. R.,** Bio-fouling, *Fouling of Heat Exchanger Surfaces,* in Bohnet, M., Ed., Conf. Proc., VDI Ges. P.O. Box 1139, 4000 Düsseldorf 1, 5.1, 1990.
42. **LeChevallier, M. W.,** Biocides and the current status of biofouling control in water systems, in *Biofouling and Biocorrosion in Industrial Water Systems,* Flemming, H.-C. and Geesey, G. G., Eds., Springer-Verlag, Heidelberg, 1991, 113.
43. **Block, J. C.,** Biofilms in drinking water distribution systems, in *Biofilms — Science and Technology,* Melo, L., Fletcher, M. M., and Bott, T. R., Eds., Kluwer Academic Publishers, Dordrecht, 1992, 469.
44. **Flemming, H.-C.,** Mechanistic aspects of reverse osmosis membrane biofouling and prevention, in *Membrane Technology,* Vol. I, Amjad, Z., Ed., Van Nostrand Reinhold, New York, 1993, 163.
45. **Winters, H., Isquith, I. R., Arthur, W. A., and Mindler, A.,** Control of biological fouling in seawater reverse osmosis, *Desalination,* 47, 233, 1983.

Section II
BIOFOULING

6

Coliform Regrowth and Biofilm Accumulation in Drinking Water Systems: A Review

Anne K. Camper

I. INTRODUCTION

Coliforms in finished potable water are of concern to the U.S. drinking water industry for a variety of reasons. These bacteria are used as indicator organisms; their presence in water is interpreted as evidence of contamination by fecal material. When coliforms are present in the absence of cross connections or disinfection barrier breakthrough, the microbiological quality of the water is unnecessarily questioned. Since the presence of coliforms is regulated, the utility must initiate a response action, including public notification and boil-water orders, which can seriously erode public confidence. The utility must also select a control measure based on unproven technology which may in fact have little chance of controlling the regrowth event. Coliforms which have proliferated in the system may also conceal the presence of those which arise from sources of genuine public health concern.

A recently completed survey of 164 water utilities[1] indicates that a relatively high number of water-distribution systems in the U.S. experience recurring coliform episodes. Many occurrences have been shown to be independent of obvious disinfection barrier breakthrough or cross contamination. It has generally been assumed that these coliforms grow

0-87371-928-X/94/$0.00+$.50

in biofilms on pipe surfaces at the expense of nutrients in the bulk water. Biofilm-associated coliform growth in oligotrophic environments such as drinking water distribution systems is plausible, as there are sufficient nutrients present to support the growth of surface-associated bacteria, but not suspended organisms.[2–5]

The European drinking water industry is required to limit the presence of heterotrophic plate count bacteria (HPC) in water. The HPC are a much broader class of organisms than the coliforms, and are therefore more numerous and may arise from diverse sources. However, the same phenomenon of increases in HPC counts in finished water after adequate treatment has been seen, and the implication of growth in biofilms has been made. In addition to regulatory concerns, certain bacteria, such as members of the genus *Actinomycetes*, have been found on distribution surfaces[6] and may contribute to taste and odor problems. It may also be hypothesized that nitrification in monochloraminated systems, which may contribute to taste and odor problems as well as increased HPC growth, is the result of bacterial activity on pipe walls.

A biofilm is an accumulation of cells immobilized at a substratum and frequently embedded in an organic polymer matrix of microbial origin.[7] Biofilms are ubiquitous on surfaces in aqueous environments, and growth of coliforms and other nuisance organisms in biofilms occurs under yet undefined environmental conditions; regrowth events are detected when these organisms leave the biofilm and enter the bulk water. Because of the low-nutrient conditions, drinking water distribution system biofilms may be very thin and occur nonuniformly on the pipe walls.[8,9] Heterogeneities also occur in the distribution of species within the biofilm itself.[10–13] For example, coliforms are not uniformly distributed on pipe surfaces, but have been detected in tubercle materials from water-distribution systems.[14,15]

A wealth of circumstantial information exists, generally anecdotal, which relates operational and environmental variables with the appearance of coliforms in finished drinking water. This information is most often system specific and does not establish a relationship between variables and coliform presence. Factors which may contribute to distribution system microbial growth are (1) seasonal variation in water temperature; (2) availability of growth-promoting nutrients; (3) passage of dormant and/or injured organisms through the treatment process; (4) occurrence of distribution-system corrosion products; (5) distribution-system disinfection practices; and (6) distribution-system hydrodynamics.[1] These factors have been addressed in previous research[2,14,16–25] and two review articles.[26,27]

The broad range of experimental and field conditions reported in the literature, together with the inherent heterogeneities of biofilm systems, make it difficult to derive cause-and-effect relationships between variables and coliform/HPC occurrences in finished drinking water. An integrated systems approach with standardized, relevant parameters and analyses will be required to generate information which can be used to successfully control regrowth events. Hence, multiple factors which may be important in regrowth as investigated in field studies, pilot scale studies, and laboratory investigations are addressed in this review.

II. TERMINOLOGY

The current literature uses numerous terms to describe the presence of bacteria in finished drinking water. "Occurrences", "episodes", and "events" are used synonymously to describe the phenomenon wherein coliforms in excess of standards are detected in finished water. Chronic or periodic appearance of bacteria in water has been termed "regrowth". Regrowth is presumed to be associated with the proliferation of bacteria within distribution systems, either in the bulk water or at the pipe walls. This contrasts with "breakthrough", which describes bacteria passing through the treatment and disinfection barrier. Regrowth and breakthrough are related, since bacteria must inoculate the system (breakthrough) in order to proliferate or grow at the expense of nutrients in the system (regrowth).

Other conceptual terms are "operational" and "environmental". Operational variables are those which can be controlled by the operator or system designer, e.g., disinfection and assimilable organic carbon (AOC). Environmental variables or parameters are outside the realm of control by the operator, and include seasonal effects (temperature, consumer demand, raw water quality), hydraulic regime, and pipe composition.

Two terms have been developed to describe the organic carbon fraction in drinking water which is utilized by bacteria as a carbon source. These are operationally based and depend on the type of analysis performed to obtain the information. AOC represents the small, readily degradable fraction of dissolved organic carbon and is traditionally defined as a calculated quantity based on the number of organisms which grow in the test water. Biodegradable organic carbon, biologically degradable organic carbon, and biodegradable dissolved organic carbon (BDOC) refer to the change in total organic carbon in the water after it has supported the growth of microorganisms. The BDOC content of a given water may be greater than the AOC value due to the more diverse microbial population used in the assay. The various procedures used to obtain AOC and BDOC values are summarized in a review article by Huck.[28]

III. FIELD STUDIES

A. Biofilms on Distribution-System Materials

Large numbers of bacteria were shown to be associated with distribution-system materials as early as 1979 when Allen et al.[8] reported that organisms were found in and on water-distribution main tubercles and encrustations. These findings were supported by Tuovinen et al.,[29,30] and more recently by other investigators.[2,14,15] More direct evidence has come through scanning electron microscopic observations of drinking water distribution-system surfaces.[6,9,14,31]

There is little conclusive evidence for the proliferation of coliforms on pipe surfaces, even though significant numbers of these bacteria may be found in the water. This is most likely due to the difficulty in sampling distribution systems and the patchy or heterogeneous manner in which the coliforms may proliferate on these surfaces. However, Olson[6] has recovered *Escherichia coli* from pipe surfaces. LeChevallier et al.[14] have enumerated coliforms from iron tubercles in one distribution system, and have concluded that the indicator organisms found in the bulk fluid originated from surface-associated growth.

Organisms that colonize the pipe materials in distribution systems originate from a variety of sources, but the majority are those which can be found in the source water of the system. This may also include coliform bacteria, especially total coliforms. These bacteria may pass the treatment and disinfectant barriers (breakthrough), enter the finished water in an injured state, and later recover either in the planktonic state or in biofilms. The injured bacteria may be undetected if bacteriological surveillance media such as m-Endo are used to evaluate the quality of drinking water. For example, Bucklin et al.[32] reported that filter effluent produced immediately after filter backwash could be characterized by high levels of turbidity and injured coliforms. Likewise, McFeters et al.[33] demonstrated that a significant number of injured coliforms (5.7 to 67.5 colony-forming units per 100 ml) were present in various drinking water samples while <1 per 100 ml were detected using m-Endo LES medium. Additional information on the significance of injured coliforms may be found in a book chapter by McFeters.[34]

B. Cause-and-Effect Investigations of Regrowth in Distribution Systems

When evaluating field data, it is important to recognize that separation of variables is often impossible. Distribution systems must provide drinking water to their consumers and, as such, cannot be operated as experimental systems. It is difficult to determine direct cause-and-effect relationships in these systems when several variables change at once. For example, elevated summer temperatures are often accompanied by increased nutrient levels and turbidities; all have been implicated in influencing regrowth events. Comparisons between systems are also complicated by differences in hydraulic regime, distribution-system piping materials, raw and finished water quality, and disinfectant type and concentration. Reported general categories of variables which may control regrowth are chemical constituents and physical parameters.

1. *Chemical Constituents*

a. Nutrients

The growth of heterotrophic and coliform bacteria requires the presence of various nutrients, including carbon, nitrogen, and phosphorous. A general rule of thumb is that these substances must be present in a ratio of approximately 100:10:1, respectively, for balanced growth. Carbon-containing compounds are the source of energy (electron donor) and carbon (for anabolic processes) for these bacteria, while nitrogen and phosphorous are required for biosynthesis. Other growth factors such as trace metals and organic cofactors may also be necessary for bacterial growth, although these are generally presumed to be nonlimiting in a mixed microbial population. Heterotrophs also require a terminal electron acceptor, which in these systems is typically oxygen. It is possible, however, for the fermentative organisms such as coliforms to use an internal organic compound as their terminal electron acceptor in the absence of oxygen.

Because carbon must be present in significantly greater quantities than the other nutrients, a major focal point in recent investigations on regrowth has been the concentration of the utilizable organic carbon fraction in drinking water, which ranges from 3 to $>1000\ \mu g\ l^{-1}$.[25,35,36] This is an operational parameter which could be controlled at the treatment plant, particularly through the use of biological treatment. Van der Kooij[37] has found a significant correlation between AOC concentrations and the regrowth of HPC in Dutch distribution systems. Growth at levels $<10\ \mu g\ C\ l^{-1}$ was limited. In an investigation at a New Jersey facility, LeChevallier et al.[25] determined that the occurrence of coliform bacteria in the distribution system could be related to a number of factors, including AOC concentrations in excess of $50\ \mu g\ C\ l^{-1}$. In an earlier study[14] it was shown that AOC concentrations decreased with travel time of the water from the treatment plant. It was hypothesized that this decrease was due to utilization of the carbon by biofilm organisms. The loss of utilizable organic material with travel time from the plant has also been demonstrated with the BDOC procedure.[38] In contrast, Opheim and

Smith[39] demonstrated no correlation between AOC concentrations and the presence of coliforms in their distribution system.

Nitrogen is present in water in the form of organic nitrogen, ammonia, nitrate, and nitrite. All forms are utilizable by different microorganisms. Studies of nitrogen concentrations in distribution systems as related to microbial regrowth events are limited. Donlan and Pipes[21] indicated that there was no correlation between organic nitrogen, ammonia, nitrate, or nitrite with attached microbial population density. In another investigation, concentrations of nitrogen in water, which were adequate for balanced microbial growth, did not change with travel through a distribution system, whereas the AOC concentration decreased.[14,25] It is generally assumed that nitrogen is not limiting in distribution systems due to turnover of existing cellular nitrogen and the low concentrations which are required for growth and maintenance.

Similar to nitrogen, phosphorous is required in very low concentrations for microbial growth. It is also not presumed to be limiting in drinking water systems,[21] although one study has shown that low phosphorous concentrations may have restricted microbial growth in a section of distribution main.[40]

b. Disinfectants

It is generally recognized that simply raising the chlorine concentration in drinking water often does not control biofilms or the presence of indicator organisms.[14,17,18,23,41–43] In extreme cases chlorine doses up to 12 mg l^{-1}, which are far in excess of usual concentrations, were inadequate in attempts to control coliform regrowth.[44,45] When distribution-system biofilms were examined, no correlation was found between free chlorine residuals and the number of HPC organisms per unit surface area.[46]

Chloramines have become increasingly popular as a drinking water disinfectant. In a recent distribution-system comparison, chloramines were more effective at reducing the number of biofilm total coliforms and HPC than chlorine.[47] In another study where statistical evaluation of the influence of chloramine concentration on attached microbial populations in a distribution system was made, an inverse relationship was established.[21]

The reasons for the differences in disinfectant performance cannot be assessed in drinking water distribution systems due to the multitude of variables which influence disinfectant efficacy. However, these variables can be controlled in pilot and/or laboratory experiments, and their importance will be addressed in the following sections on pilot plant studies and laboratory investigations.

2. *Physical Attributes*

a. Temperature

As an interdependent variable, temperature has been implicated as contributing to increased microbial populations in distribution systems. Most, but certainly not all, regrowth events occur in summer months when water temperatures are highest. LeChevallier et al.[25] demonstrated an association between coliform regrowth and water temperatures above 15°C. In London, it was reported that coliform occurrences increased when the water temperature exceeded 20°C and declined when the temperature was less than 14°C.[48] Likewise, Donlan and Pipes[21] reported a direct relationship between water temperature and the attached microbial population density.

b. Precipitation

Increases in nutrients associated with runoff from rainfall have been proposed by some investigators as a contributing factor to regrowth events.[25,45] The nutrients are collected in surface water sources, pass through the treatment plant, and are then available for microbial growth in the distribution system.

c. Pipe Materials

Various pipe materials are used in distribution systems, including cement lined, asphaltic coated, mild steel, ductile iron, and PVC. When examining steel and iron surfaces, the interaction between the ferrous metal and disinfectants must be considered, as well as the production of corrosion products and tubercles. As stated before, coliforms have been found in association with tubercles in some distribution systems,[14,15] while not in others. When pipe materials have been compared, the number of bacteria on an unlined cast-iron pipe were the highest, while PVC-pipe biofilm densities were the lowest.[47] The relative importance of corrodible pipe materials in regrowth events has been shown indirectly by the reduction or control of coliform bacteria by application of corrosion inhibitors,[17,45] although this is not universally true.[14]

d. Hydraulic Regime

Hydraulic regime in distribution systems is controlled by original design and layout, but is also heavily influenced by customer water demand. Large daily fluctuations can occur between peak use periods (mornings, evenings) and night. Low flow conditions can lead to decreased disinfectant concentrations and nutrient transport. Higher flow rates result in increased nutrient transport to attached organisms, but also increases disinfectant flux and shear stresses at the surface. Associations between flow rates in distribution systems and regrowth events are therefore difficult to establish, although one study has shown an inverse relationship between flow rates and biofilm density.[21]

Residence time of the water in a system may also be an important factor in determining where organisms are most likely to occur on pipe surfaces, since AOC and BDOC concentrations have been shown to decrease as water moves through a distribution system.[14,25,37,38]

IV. PILOT PLANT STUDIES

A pilot system provides flexibility in conducting controlled experiments and accessing data under conditions which are far more controlled than those found in an operating system of water mains. Pilot systems must be carefully designed to simulate critical parameters in operating drinking water distribution systems and maximize separation of variables. These systems provide an important "proving ground" for relevancy of laboratory findings or the clarification of presumed phenomena in the field. Unfortunately, the presence of well-constructed and operated pilot facilities for research on regrowth are limited due to expense of construction and operation.

Two basic types of pilot pipe systems are found in the literature: (1) once-through and (2) recirculating. Examples in a review on pipe-loop systems for investigating corrosion by Levin and Schock[49] and others[50–55] indicate a majority of pilot systems have once-through flow. Characklis[2] has developed a staged pilot water-distribution system using RotoTorques with controllable residence time to represent plug-flow conditions and residence times up to several days, as found in the field. Once-through pilot systems lack this important capability. Haudidier et al.[5] describe the use of six pilot pipeline recirculation loops, configured in series, in a pilot system developed in Nancy, France, to conduct studies of attached microbial films in water-distribution systems. The French system is designed so that each loop can be operated as a perfectly mixed reactor and, when operated in series, can simulate the plug flow of a pipeline with a high axial-dispersion coefficient. An additional benefit of the latter system is the ability to separate residence-time effects from those of shear stress.

A. Cause-and-Effect Investigations

Few published papers are available on the use of pilot systems to investigate cause-and-effect relationships between drinking water quality parameters and biofilm accumulation. Using RotoTorque systems, it was demonstrated that the amount of biofilm which accumulated in sequential reactors depended upon chlorine concentration.[2,16] When no chlorine was present, most biofilm accumulation occurred at the entrance to the system. It was hypothesized that this occurred due to rapid consumption of available nutrients. At low chlorine concentrations, biofilm accumulation was reduced at the entrance to the system where chlorine concentrations were higher; as chlorine was consumed at greater residence times in the reactor, biofilm accumulation increased at the expense of the nutrients. In similar experiments with the same system, it was shown that a chlorinated backwash of the upstream pilot mixed-media (anthracite/sand) filter decreased the amount of biomass on the filter media with a subsequent increase in the amount of total organic carbon in the filter effluent. There was a concomitant increase in the amount of biofilm within the RotoTorques, indicating that the amount of available carbon influenced biofilm accumulation.[2] Using a pipe loop system, Levi and Joret[38] report that virtually all BDOC was consumed within 24 h. Both groups of investigators, although using different systems, concluded that the numbers of planktonic organisms in the water can originate only from detached biofilm organisms and not growth in the bulk phase. In a study using the French recirculating pipe-loop system, biofilm accumulation and AOC depletion were greatest in the first loop (40-h residence time).[5] LeChevallier et al.[50] utilized a model pipe system to determine the influence of chlorine and monochloramine disinfectant efficacy on biofilms which had accumulated on several pipe materials. Biofilms on galvanized, copper, or PVC surfaces were readily disinfected by free chlorine or monochloramine (1 mg l^{-1}). Iron-pipe surface-associated bacteria were more susceptible to monochloramine than free chlorine, but higher monochloramine concentrations (4 mg l^{-1}) were required than on the other surfaces.

V. LABORATORY INVESTIGATIONS

Although an abundance of basic laboratory research has been performed on biofilms,[7] the focus of this section will be restricted to those directly related to drinking water research. In well-designed laboratory investigations, proper control of variables is possible. However, the scope of these studies is often so narrow that applicability of the results to actual distribution systems is difficult. Proper design of experiments which take into account features relevant to pipe systems is critical.

Determining the response of bacteria to potential nutrients has been of key importance, and has been described in the section on AOC, BDOC, and regrowth-potential methods. This has not been the only area of interest. The ability of coliforms isolated from drinking water to grow on extremely low carbon concentrations and temperatures has been established.[56] The source of growth-stimulating compounds has also been investigated. Fedorak and Huck[57] demonstrated that cyanobacterial products, which may be a source of nutrients in drinking water, were readily mineralized by HPC bacteria. The importance of crushed tubercle material in enhancing coliform growth has been substantiated by several researchers[8,17,58,59] and refuted by one.[56]

Much of the fundamental research on efficacy of disinfectants against biofilms has been conducted in laboratory studies. Initially, it was shown that attachment of organisms to surfaces resulted in decreased disinfection by chlorine.[60–63] In more elaborate studies using biofilms on relevant surfaces, Berman et al.[63] reported that monochloramine was no more effective than chlorine. In another study, LeChevallier et al.[64] showed that there was decreased sensitivity to chlorine conveyed to organisms simply by being attached to a surface and that this effect was enhanced in older biofilms. When these biofilms were treated with monochloramine, more disinfection was observed than with free chlorine. In a related study, the interaction of oxidizing disinfectants, biofilms, and various pipe surfaces was investigated.[65] Monochloramine was less effective than free chlorine against suspended organisms, but the reverse effect was seen with biofilm bacteria. The same effect was seen in analyses where the CT values (concentration-time of biocide exposure for 99% inactivation) of biofilms and suspended (dispersed biofilm) cells of *Pseudomonas aeruginosa* exposed to chlorine and monochloramine were calculated.[66] A laboratory system for assaying the efficacy of monochloramine against biofilms containing coliforms has also been developed.[48] This device supports a stable heterotrophic bacteriological population representative of that found in their distribution system. When challenged with a coliform, it became incorporated into the biofilm and remained as a component. Monochloramine concentrations of 0.6 mg l^{-1} caused a loss of the coliform; at lower disinfectant concentrations it persisted.

VI. INTEGRATING SCALES OF OBSERVATION

Developing relevant information on distribution-system microbial growth which can be effectively applied by the water industry requires an integrated research effort. Microbial and chemical processes occur in the water-distribution system as a whole, within individual pipes, and in biofilms at the surface of the pipe. To investigate the problem, research must be defined in terms of scales of observation. Microscale research, e.g., millimeters in size ($<10^{-3}$ m), focuses on measurements and interactions at the cellular level within the biofilm and is represented in the above material by laboratory studies. At the mesoscale (1 mm to 10 m), fluid dynamics and system geometries become important, since the activity of the biofilm is dependent on the transport and interfacial-transfer phenomena created by these conditions (within a pipe). Well-designed pilot scale research falls within this category. The macroscale ($>$10 m), or field system, is influenced by system-operating parameters and environmental variables, such as influent water quality and water demand, as well as regulations regarding water quality. Experiments at the micro- and mesoscale must be carefully designed so the results can be "scaled up" to the macroscale. For example, microscale observations in a biofilm must be done under conditions relevant to the fluid dynamics in a pipe system (mesoscale) in order to model an operating distribution system (macroscale).

While the overview presented in this paper touches on the relevance of phenomena, parameters, and processes at each scale of observation, it is clear that integration of scales has yet to be accomplished. Nevertheless, some complimentary conclusions appear by comparing results at the various scales:

- Organic carbon is probably the most important nutrient in a distribution system, as it has been shown to support coliform growth in the laboratory as well as being correlated with regrowth in pilot and full-scale systems.
- While little work on the effect of temperature has been done at the micro- or mesoscale, and macroscale phenomena are often linked with other variable variations (e.g., nutrients, turbidity), it is reasonable to expect higher coliform proliferation with increased temperature due to increased enzyme kinetics. These organisms often have optimum growth temperatures well above those found in distribution systems and are quite sensitive to changes in the lower temperature range.
- Disinfection efficacy appears to vary widely with system parameters and variables as well as disinfectant type. It is tempting to conclude that slower-acting biocides (e.g., monochloramine) are superior to highly reactive compounds (e.g., free chlorine) in treating biofilms. However, there is insufficient supporting evidence at this time. Nevertheless, it is also clear that disinfectant alone may not be capable of controlling regrowth events.

REFERENCES

1. **Smith, D. B., Hess, A. F, and Hubbs, S. A.,** Survey of distribution system coliform occurrence in the United States, in *Proc. Water Quality Technology Conf., San Diego, CA,* American Water Works Association Research Association, Denver, 1103, 1990.
2. **Characklis, W. G.,** *Bacterial Regrowth in Distribution Systems,* American Water Works Association Research Foundation, Denver, 1988.
3. **van der Wende, E., Characklis, W. G., and Smith, D. B.,** Biofilms and bacterial drinking water quality, *Water Res.,* 23, 1313, 1989.
4. **van der Wende, E. and Characklis, W. G.,** Biofilms in potable water distribution systems, in *Drinking Water Microbiology,* McFeters, G. A., Ed., Springer-Verlag, New York, 1990, chap. 12.
5. **Haudidier, K, Paquin, J. L., Francais, T., Hartemann, P., Grapin, G., Colin, F., Jourdain, M. J., Block, J. C., Cheron, J, Pascal, O., Levi, Y., and Miazga, J.,** Biofilm growth in drinking water network: a preliminary industrial pilot plant experiment, *Water Sci. Technol.,* 20, 109, 1988.
6. **Olson, B. H.,** Assessment and Implications of Bacterial Regrowth in Water Distribution Systems, EPA-6001/52–82–072, U.S. Environmental Protection Agency, Cincinnati, 1982.
7. **Characklis, W. G. and Marshall, K. S.,** Eds., *Biofilms,* John Wiley & Sons, New York, 1990.
8. **Allen, M. J., Geldreich, E. E., and Taylor, R. H.,** The occurrence of microorganisms in water main encrustations, *J. Am. Water Works Assoc.,* 72, 614, 1980.
9. **Ridgway, H. F. and Olson, B. H.,** Scanning electron microscope evidence for bacterial colonization of a drinking water distribution system, *Appl. Environ. Microbiol.,* 41, 274, 1981.
10. **Ammann, R. I., Stromley, J., Deveraux, R., Key, R., and Stahl, D. A.,** Molecular and microscopic identification of sulfate-reducing bacteria in multispecies biofilms, *Appl. Environ. Microbiol.,* 58, 614, 1992.
11. **Rogers, J. and Keevil, C. W.,** Immunogold and fluorescein immunolabelling of Legionella pneumophila within an aquatic biofilm visualized by using episcopic differential interference contrast microscopy, *Appl. Environ. Microbiol.,* 58, 2326, 1992.
12. **Siebel, M. A. and Characklis, W. G.,** Observations of binary population biofilms, *Biotechnol. Bioeng.,* 37, 778, 1991.
13. **Zambon, J. J., Huber, P. S., Meyer, A. E., Slots, J., Fornalik, M. S., and Baier, R. E.,** *In situ* identification of bacterial species in marine microfouling films by using an immunofluorescence technique, *Appl. Environ. Microbiol.,* 48, 1214, 1984.
14. **LeChevallier, M. W., Babcock, T. M., and Lee, R. G.,** Examination and characterization of distribution system biofilms, *Appl. Environ. Microbiol.,* 53, 2714, 1987.
15. **Opheim, D., Growchowski, J., and Smith, D.,** Isolation of coliforms from water main tubercles, *Abstr. Annu. Meet. Am. Soc. Microbiol.,* 1988.
16. **van der Wende, E., Characklis, W. G., and Grochowski, J.,** Bacterial growth in water distribution systems, *Water Sci. Technol.,* 20, 277, 1988.
17. **Martin, R. S., Gates, W. H., Tobin, R. S., Sumarah, R., Wolfe, P., and Forestall, P.,** Factors affecting coliform bacterial growth in distribution systems, *J. Am. Water Works Assoc.,* 74, 34, 1982.
18. **Reilly, K. J. and Kippen, J. S.,** Relationship of bacterial counts with turbidity and free chlorine in two distribution systems, *J. Am. Water Works Assoc.,* 75, 309, 1983.
19. **Clark, T. F.,** Chlorine tolerant bacteria in a water distribution system, *Public Works,* 115, 65, 1984.
20. **Wierenga, J. T.,** Recovery of coliforms in the presence of a free chlorine residual, *J. Am. Water Works Assoc.,* 77, 83, 1985.

21. **Donlan, R. M. and Pipes, W. O.,** Selected drinking water characteristics and attached microbial population density, *J. Am. Water Works Assoc.,* 80, 70, 1988.
22. **Fransolet, G., Villers, G., and Masschelein, W. J.,** Influence of temperature on bacterial development in waters, *J. Int. Ozone Assoc.,* 7, 205, 1985.
23. **Oliveri, V. P., Bakalian, A. E., Bossung, K. W., and Lowther, E. D.,** Recurrent coliforms in water distribution systems in the presence of free residual chlorine, in *Water Chlorination, Chemistry, Environmental Impact, and Health Effects,* Jolley, R. L., Bull, R. J., Davis, W. P., Katz, S., Roberts, M. H., and Jacobs, V. A., Eds., Lewis Publishers, Chelsea, Michigan, 1985, 651.
24. **LeChevallier, M. W., Olson, B. H., and McFeters, G. A.,** *Assessing and Controlling Bacterial Regrowth in Distribution Systems,* American Water Works Association Research Foundation, Denver, 1990.
25. **LeChevallier, M. W., Schulz, W., and Lee, R. W.,** Bacterial nutrients in drinking water, *Appl. Environ. Microbiol.,* 57, 857, 1991.
26. **LeChevallier, M. W.,** Coliform regrowth in drinking water: a review, *J. Am. Water Works Assoc.,* 82, 74, 1990.
27. **Block, J. C.,** Biofilms in drinking water distribution systems, in *Biofilms — Science and Technology,* Bott, T. R., Melo, L., Fletcher, M., and Capdeville, B., Eds., NATO Advanced Study Institute, Portugal, Kluwer Publishers, The Netherlands, 1992, 469.
28. **Huck, P.,** Measurement of biodegradable organic matter and bacterial growth potential in drinking water, *J. Am. Water Works Assoc.,* 82, 78, 1990.
29. **Tuovinen, O. H., Button, K. S., Vuorinen, A., Carlson, L., Mair, D., and Yut, L. A.,** Bacterial, chemical, and mineralogical characteristics of tubercles in distribution pipelines, *J. Am. Water Works Assoc.,* 72, 626, 1980.
30. **Tuovinen, O. H. and Hsu, J. C.,** Aerobic and anaerobic microorganisms in tubercles of the Columbus, Ohio water distribution system, *Appl. Environ. Microbiol.,* 44, 761, 1982.
31. **Ridgway, H. F., Means, E. G., and Olson, B. H.,** Iron bacteria in drinking-water distribution systems: elemental analysis of *Gallionella* stalks using X-ray energy-dispersive microanalysis, *Appl. Environ. Microbiol.,* 41, 288, 1981.
32. **Bucklin, K. E., McFeters, G. A., and Amirtharajah, A.,** Penetration of coliforms through municipal drinking water filters, *Water Res.,* 25, 1013, 1991.
33. **McFeters, G. A., Kippin, J. S., and LeChevallier, M. W.,** Injured coliforms in drinking water, *Appl. Environ. Microbiol.,* 51, 1, 1986.
34. **McFeters, G. A.,** Injured indicator bacteria in drinking water, in *Drinking Water Microbiology,* McFeters, G. A., Ed., Springer-Verlag, New York, 1990, chap. 23.
35. **Bouwer, E. J. and Crowe, P. B.,** Biological processes in drinking water, *J. Am. Water Works Assoc.,* 80, 82, 1988.
36. **Servais, P. G., Billen, G., and Hascoet, M. C.,** Determination of the biodegradable fraction of dissolved organic matter in water, *Water Res.,* 21, 445, 1987.
37. **van der Kooij, D.,** Assimilable organic carbon as an indicator of bacterial regrowth, *J. Am. Water Works Assoc.,* 84, 57, 1992.
38. **Levi, Y. and Joret, J. C.,** Importance of bioeliminable dissolved organic carbon (BDOC) control in strategies for maintaining the quality of drinking water during distribution, in *Proc. Water Quality Technol. Conf., San Diego, CA,* American Water Works Association Research Foundation, Denver, 1990, 1267.
39. **Opheim, D. and Smith, D.,** The relationship of *Enterobacter cloacae,* AOC, and ADOC in raw and treated water to coliform episodes, in *Proc. Water Qual. Technol. Conf., San Diego, CA,* American Water Works Association Research Foundation, Denver, 1990, 1237.
40. **Herson, D. S., Marshall, D. R., and Victoreen, H. T.,** Bacterial persistence in the distribution system, in *Proc. Water Quality Technology Conf., Denver, CO,* American Water Works Association Research Foundation, Denver, 1984, 309.

41. **Ludwig, F.,** The occurrence of coliforms in the Regional Water Authority water supply system, Rept. submitted by the South Central Conn. Reg. Water Auth., New Haven, 1985.
42. **Centers for Disease Control,** Detection of elevated levels of coliform bacteria in a public water supply, *Morbid. Mortal. Weekly Rep.*, 34, 142, 1985.
43. **Hudson, L. D., Hankins, J. W., and Battaglia, M.,** Coliforms in a water distribution system: a remedial approach, *J. Am. Water Works Assoc.*, 75, 564, 1983.
44. **Earnhardt, K. B.,** Chlorine resistant coliforms — the Muncie, Indiana experience, in *Proc. Water Quality Technology Conf., Miami Beach, Fla.*, American Water Works Association Research Foundation, Denver, 1980, 371.
45. **Lowther, E. D. and Moser, R. H.,** Detecting and eliminating coliform regrowth, in *Proc. Water Quality Technology Conf., Denver, CO,* American Water Works Association Research Foundation, Denver, 1984, 323.
46. **Nagy, L. A. and Olson, B. H.,** Occurrence and significance of bacteria, fungi, and yeasts associated with distribution pipe surfaces, in *Proc. Water Quality Technol. Conf., Houston, TX,* American Water Works Association Research Foundation, Denver, 1985, 213.
47. **Neden, D. G., Jones, R. J., Smith, J. R., Kirmeyer, G. J., and Foust, G. W.,** Comparing chlorination and chloramination for controlling bacterial regrowth, *J. Am. Water Works Assoc.*, 84, 80, 1992.
48. **Colbourne, J. S., Dennis, P. J., Kevil W., and Nackerness, C.,** The operational impact of growth of coliforms in London's distribution system, in *Proc. Water Qual. Technol. Conf., Orlando, Fla.*, American Water Works Association Research Foundation, Denver, 1991, 799.
49. **Levin, R. and Schock, M. R.,** The use of pipe loop tests for corrosion control diagnostics, in *Proc. Water Quality Technol. Conf., San Diego, CA,* American Water Works Association Research Foundation, 697, 1991.
50. **LeChevallier, M. W., Lowry, C. D., and Lee, R. G.,** Disinfecting biofilms in a model distribution system, *J. Am. Water Works Assoc.*, 82, 87, 1990.
51. **Gardels, M. C. and Sorg, T. J.,** A laboratory study of the leaching of lead from water faucets, *J. Am. Water Works Assoc.*, 81, 101, 1989.
52. **Heumann, D. W.,** Solid lead gooseneck slug dispersion in consumer plumbing systems, in *Proc. Water Quality Technol. Conf., Philadelphia, PA,* American Water Works Association Research Foundation, Denver, 1989, 525.
53. **Birden, H. H., Jr., Calabrese, E. J., and Stoddard, A.,** Lead dissolution from soldered joints, *J. Am. Water Works Assoc.*, 77, 66, 1985.
54. **Treweek, G. P., Glicker, J., Chow, B., and Sprinker, M.,** Pilot-plant simulation of corrosion in domestic pipe materials, *J. Am. Water Works Assoc.*, 77, 74, 1985.
55. **American Water Works Association Research Foundation — DVGW — Forschungsstelle,** *Internal Corrosion of Water Distribution Systems, Research Report,* American Water Works Association Research Foundation, Denver, 1985.
56. **Camper, A. K., McFeters, G. A., Characklis, W. G., and Jones, W. L.,** Growth kinetics of coliform bacteria under conditions relevant to drinking water distribution systems, *Appl. Environ. Microbiol.*, 57, 2233, 1991.
57. **Fedorak, P. M. and Huck, P. M.,** Microbial metabolism of cyanobacterial products: batch culture studies with applications to drinking water treatment, *Water Res.*, 22, 1267, 1988.
58. **Victoreen, H. T.,** The stimulation of coliform growth by hard and soft water main deposits, in *Proc. Water Qual. Technol. Conf., Miami, Fla.*, American Water Works Association Research Foundation, Denver, 11, 1980.
59. **Victoreen, H. T.,** The role of rust in coliform regrowth, in *Proc. Water Qual. Technol. Conf., Denver, CO,* American Water Works Association Research Foundation, Denver, 1984, 253.

60. **LeChevallier, M. W., Hassenauer, T. S., Camper, A. K., and McFeters, G. A.,** Disinfection of bacteria attached to granular activated carbon, *Appl. Environ. Microbiol.,* 48, 918, 1984.
61. **Ridgway, H. F. and Olson, B. H.,** Chlorine resistance patterns of bacteria from two drinking water distribution systems, *Appl. Environ. Microbiol.,* 44, 972, 1982.
62. **Herson, D. S., McGonigle, B., Payer, M. A., and Baker, K. H.,** Attachment as a factor in the protection of *Enterobacter cloacae* from chlorination, *Appl. Environ. Microbiol.,* 53, 1178, 1987.
63. **Berman, D., Rice, E. W., and Hoff, J. C.,** Inactivation of particle-associated coliforms by chlorine and monochloramine, *Appl. Environ. Microbiol.,* 55, 507, 1988.
64. **LeChevallier, M. W., Cawthon, C. D., and Lee, R. G.,** Factors promoting survival of bacteria in chlorinated water supplies, *Appl. Environ. Microbiol.,* 54, 649, 1988.
65. **LeChevallier, M. W., Cawthon, C. D., and Lee, R. G.,** Inactivation of biofilm bacteria, *Appl. Environ. Microbiol.,* 54, 2492, 1988.
66. **Griebe, T., Chen, C.-I., Srinivasan, R., and Stewart, P. S.,** Analysis of biofilm disinfection by monochloramine and free chlorine, Chapter 9 in this book.

Section II
BIOFOULING

7

Effect of Mixed Sulfate-Reducing Bacterial Communities on Coatings

Joanne Jones-Meehan, Marianne Walch, Brenda J. Little, Richard I. Ray, and Florian B. Mansfeld

I. INTRODUCTION

Epoxy coatings are widely used as heavy-duty moisture- and chemical-resistant coatings and linings in immersion and atmospheric marine environments to protect underlying metals against corrosion.[1,2] Nylon coatings are used because of their strength, low coefficient of friction and wear resistance.[3] Polyurethanes are used as protective coatings to prevent environmental weathering of materials.[4] These types of coatings are used routinely in marine immersion service by the U.S. Navy and others. However, for protective coatings to be effective, they must be resistant to microbial degradation.

Corrosion is often extremely rapid at small discontinuities in coatings, and breaks or blisters in coatings may allow access of corrosion-inducing microbes, such as sulfate-reducing bacteria (SRB), to the metal beneath.[5] SRB are a diverse group of anaerobic bacteria that can be isolated from many anaerobic environments, but their principal habitat is the marine environment where the concentration of sulfate in seawater is high and fairly constant.[6] Jones et al.,[7] using scanning electron microscopy (SEM) coupled with energy dispersive X-ray spectrometry (EDS) and electrochemical impedance spectroscopy (EIS), described the microbial attack of epoxy- and nylon-coated 4140 steel by mixed

0-87371-928-X/94/$0.00+$.50

communities of strict anaerobes (SRB) and facultative anaerobes isolated from corroding, in-service naval materials.

Using SEM, Terry and Edyvean[8] demonstrated that fouling organisms attached to freely corroding steel, cathodically protected steel, and to steel surfaces coated with a nontoxic penetrating the epoxy coal-tar paint. An SEM micrograph showed a pennate diatom epoxy coal-tar paint surface from a static laboratory culture after 100 d.[8] Polyester-containing polyurethanes have been found to be more susceptible to microbial attack and experience more in-service failure than those based on polyether.[4] Recent studies suggest that microbes may be involved in opening the aromatic rings of the di-isocyanate and in direct hydrolysis of the urethane, amide, and urea groups.[4]

Stranger-Johannessen[9] has described blistering and debonding of corrosion-protective coatings by marine and soil microbes, as well as fungal deterioration of coatings. Stranger-Johannessen and Norgaard[10] reported on observations that indicate microbial attack at the coating surface, thereby changing its chemical and physical properties.

In 1981, Kinoshita et al.[11] described a *Flavobacterium* which grew in a medium containing 6-amino-hexanoic acid cyclic dimer (a by-product of the nylon-6 industry) as the sole carbon and nitrogen source. *Flavobacterium* and *Pseudomonas* species were isolated from the wastewater of a nylon factory in Japan, both were able to degrade a nylon oligomer, and two enzymes were found to be responsible for the degradation (6-aminohexanoate-cyclic-dimer hydrolase and 6-aminohexanoate-dimer hydrolase).[12–16]

Coating deterioration in the presence of an adherent biofilm (i.e., mixed communities containing SRB; microbial metabolites, polymers, and enzyme activities; inorganic ion deposits) is of interest to our laboratories. In this report, we describe an investigation of microbial attack on protective coatings using mixed communities consisting of strict (SRB) and facultative anaerobes (non-sulfate reducers). Coating deterioration was assessed by visual observations, SEM/EDS, environmental scanning electron microscopy (ESEM)/EDS, and EIS.

II. MATERIALS AND METHODS

A. Steel Alloy and Coatings

Corrosion coupons (3″ × 3″ and 3/4″ × 3/4″) were prepared from steel alloy 4140 with the following chemical composition (wt%): 0.38 to 0.43% C, 0.75 to 1.00% Mn, 0.035% P (max), 0.040% S, 0.15 to 0.30% Si, 0.80 to 1.10% Cr, and 0.15 to 0.25% Mo. The density was 7.84 g/cm^3. Coupons in categories I to V (see Table 1) were glass-peened, degreased, primed, and coated with light gray 2770 epoxy, dark gray 17104 epoxy, or black

TABLE 1. Coatings Applied to 4140 Steel for Use in EDS and EIS Analyses

Category[a]	Primer	Topcoat(s)
1	Zinc[b]	epoxy[c] + polyurethane[d]
2	Phosphate Coat[e]	Epoxy[c]
3	None (abrasive blast 100- or 120-grit aluminum oxide at 85 to 100 PSI)	Epoxy[c]
4	IVD aluminum[f]	Epoxy[c]
5	IVD aluminum[f]	Epoxy[c] + polyurethane[d]
6	Phosphate coat[e]	Epoxy[c] + polyurethane[d]
7	Zinc[b]	Epoxy[c]
8	None (abrasive blast 100- or 120-grit aluminum oxide at 85 to 100 PSI)	Epoxy[c] + polyurethane[d]
I	Five-step iron phosphate[g]	Nylon
II	IVD aluminum[h]	Epoxy
III	Luberite[i]	Epoxy
IV	Zinc[j]	Epoxy
V	Five-step iron phosphate[k]	Epoxy

[a] Categories 1 to 8 were coated at AAA Plating in Compton, CA, following MIL specs. Categories I to V were glass-peened, degreased, primed, and coated (light grey 2770 epoxy, dark grey 17104 epoxy, or black 3147 nylon; topcoats were produced by the Polymer Corporation, Reading, PA) at NAVSWC/Code R35 (Dahlgren VA) following MIL specs.

[b] Zinc plate per QQ-Z-325 Rev. C (Type II, Class 2).

[c] Two coats of epoxy polyamide primer MIL-P — 23377 Rev. E (Type I, Class 3) per MIL-F-18264.

[d] One coat polyurethane #36375 lusterless gray FED-STD-595 per MIL-F-18264.

[e] Phosphate coat DOD-P-16232 Rev. F (Type Z, Class 3); hydrogen embrittlement relieved for 8 h at 210 to 225 degrees.

[f] IVD aluminum per MIL-C-83488-C (Notice 1, Type II, Class 1).

[g] Five-step iron phosphate was applied according to TT-C-490C.

[h] IVD aluminum was applied according to MIL-C-83488, Class a, Type II.

[i] Parco Luberite was applied according to Parker specification #27.

[j] Zinc plate was applied according to ASTM-B633–78, Type II.

[k] Five-step iron phosphate was applied according to TT-C-490C.

From Jones, J. M., Walch, M., and Mansfeld, F. B., Corrosion/91, Paper No. 108, NACE, Houston, TX, 1991. With permission.

3147 nylon at NAVSWC (Dahlgren, VA). These polymer coatings were electrostatically applied as powders to the primer coating (average coating thickness was 3.7 to 4.6 mils). The integrity of the applied coatings was assessed by impedance measurements using a Solartron® 1250 frequency-response analyzer, a Solartron® 1286 potentiostat, Z-plot software, and IBM® XT computer. For laboratory microbiologically influence corrosion (MIC) testing, all coupons in categories I to V (see Table 1) were sterilized in a dry heat oven at 150°C for 4 to 5 h. Impedance studies showed no deterioration of the coatings from the sterilization procedure.

The following six primer/coating treatments were used for the EDS analyses: (1) five-step iron phosphate primer + epoxy topcoat; (2) zinc primer + epoxy topcoat; (3) ion vapor deposited (IVD) aluminum primer + epoxy topcoat; (4) bare (no primer) + nylon topcoat; (5) five-step iron phosphate + nylon topcoat; and (6) zinc primer + nylon topcoat.

Coupons in categories 1 to 8 (see Table 1) were prepared according to MIL specs by AAA Plating (Compton, CA). The coated steel coupons shown in Table 1 were used in EIS analyses.

B. Microbial Sampling and Maintenance of Cultures

The isolation, maintenance, and characterization of the mixed communities containing SRB have been described previously.[7] All cultures were grown anaerobically at room temperature in sealed bottles.

Table 2 shows the material sampled for the isolation of the mixed, microbial cultures (obligate and facultative anaerobes; SRB and nonsulfate reducers). The SRB mixed cultures were tested for hydrogenase activity using Caproco Hydrogenase Test Kits (Caproco Limited, Edmonton, Alberta, Canada). Color development was rated after 4 h, but to detect weak enzyme activities color development was allowed to occur for 24 h (see Table 2).

C. SEM/EDS Studies

At the end of the exposure period of 2 months, 1 cm^2 corrosion coupons were fixed overnight in 2.5% glutaraldehyde in filtered (0.2-μm pore size) 0.1 *M* cacodylate-buffered seawater, pH 7.2. The coupons were taken through a dehydration series with acetone and critical point dried with CO_2. They were then sputter coated with carbon and examined using an Amray Model 1000A SEM. EDS analyses were performed using an EDS 9100 Energy Dispersive X-ray Spectrometer. At least six areas of each coupon were analyzed by EDS, and at least eight spectra were obtained for each area. Coupons were coated with gold/palladium for scanning electron micrographs.

D. ESEM/EDS Studies

At the end of the exposure period of 8 months, surface topography and chemistry were documented using an Electroscan® Model E-30 ESEM and a Tracor® Northern Model 5502 EDS. Coupons were removed from the culture medium, carried through a series of salt water/distilled water washes, and examined directly from distilled water.

TABLE 2. Hydrogenase Activity in the Mixed, Microbial Populations Containing Sulfate-Reducing Bacteria (SRB)

Culture[a]	Isolated originally on: Material & Topcoat(s)	Hydrogenase activity[b]		
		2½ h	4 h	24 h
I	4140 Steel + five-step iron phosphate	0	+1	+2
II	4140 Steel + five-step iron phosphate + epoxy	+1	+1	+2
III	4140 Steel + zinc plate	+2	+2	+2
IV	4140 Steel + IVD aluminum + nylon	+2	+2	+2
V	4140 Steel + five-step iron phosphate + nylon	0	0	0
VI	Carbon steel + proprietary primer and topcoats	0	0	0
VII	Aluminum alloy + epoxy + polyurethane	0	0	+2

[a] Cultures 1 to V were isolated from a constant-immersion flume tank at NAVSWC/Ft. Lauderdale, FL. Culture VI was isolated from the seawater-piping system of a surface ship at Long Beach Naval Station, Long Beach, CA. SRB in cultures I to VI require NaCl for growth (halophilic). Culture VII was isolated from moisture trapped under the cargo ramp of a C-130 transport plane at the NADEP in Cherry Point, NC. SRB in culture VII are facultative halophiles (do not need NaCl for growth, but can grow in seawater concentrations of NaCl).

[b] Activity of the hydrogenase enzyme is measured with a Caproco Hydrogenase Test Kit. The rating system is from 0 to 3. A negative reaction is rated as 0, weak reaction is +1 (0 to 0.5 nmol of hydrogen uptake/min), moderate reaction is +2 (0.05 to 5 nmol of hydrogen uptake/min), and a strong reaction is +3 (5 to 5000 nmol of hydrogen uptake/min).

E. EIS Analyses

Coated coupons were exposed to the sterile medium and to mixed, microbial cultures for 2 months in the laboratory. EIS analyses were performed after this exposure period using a Solartron® potentiostat Model 1286 and a Solartron® Frequency Analyzer Model 1250. The impedance data were collected and analyzed using software developed at the Corrosion and Environmental Effects Laboratory (CEEL) at the University of Southern California (USC) (Los Angeles). The EIS data were collected at the applied corrosion potential after 2 h exposure to 0.5 *N* NaCl. The tested area was 20 cm². A Pt wire pseudoreference electrode coupled capacitively to a Standard Calomel Electrode (SCE) reference electrode was used to reduce the phase shift at high frequencies. The frequency range of the applied signal was 65 KHz to 10 mHz.

III. RESULTS

A. SEM and ESEM Studies

Scanning electron micrographs of nylon- and epoxy-coated 4140 steel showed heavy microbial colonization of the coupon surfaces after exposure to the mixed, bacterial cultures. Figure 1 shows a scanning electron micrograph of nylon- and epoxy-coated steel surfaces exposed to mixed Culture I. A variety of cell morphologies (including short rods, long rods, vibrio, spirillum, and cocci) were seen in the mixed communities attached to the coupon surfaces. Figure 2 shows an environmental scanning electron micrograph of 4140 steel coupon + phosphate coat + epoxy polyamide topcoat exposed to marine, mixed culture II.

Coupons in groups I to V (see Table 1) were sterilized by dry heat while coupons in groups 1 to 8 were not sterilized; therefore, microbes present on the coupons such as bacilli, molds, and fungi grew in these control samples. In the control samples (no SRB present), breaching of the coatings was not detected by SEM. However, biodegradation of some coatings exposed to SRB mixed communities was detected using SEM (approximately 5- to 10-μm holes were observed with bacteria). ESEM detected breaches in some of the coatings tested and was used to characterize the topography of wet biofilms on the coated 4140 steel surfaces. Bacteria were attached to the coated surface between layers of corrosion product, as well as in and around holidays/breaches in the coatings. The microbes were distributed throughout the biofilm matrix of microbial exopolymers and corrosion products. The chemical composition of the biofilm/corrosion layers was characterized using EDS analyses.

B. EDS Analyses

EDS analyses were performed on the coated coupons in categories I to V listed in Table 1. The EDS spectra for zinc primer + epoxy-coated 4140 steel coupons exposed to marine, mixed cultures I, II, and III are shown in Figure 3. Breaching of the epoxy coating is evident in the samples exposed to the bacterial cultures which included SRB, as seen by the appearance of zinc peaks (from the primer) and iron peaks (from the steel) in the EDS spectra. Neither zinc nor iron is seen in the control EDS spectrum. Epoxy- and nylon-coated coupons were also exposed to the facultative halophilic, mixed culture VII (no chloride ions added to the culture medium). The EDS spectra for 4140 steel coupons with zinc primer and an epoxy topcoat exposed to sterile growth medium and to mixed culture IV are shown in Figure 4. Breaching of the epoxy topcoat was evident, with both zinc and iron peaks appearing in the spectra of coupons exposed to the bacteria.

The nylon-coated coupons showed no visible signs of coating deterioration after a 1-month exposure to the mixed cultures. Breaching of

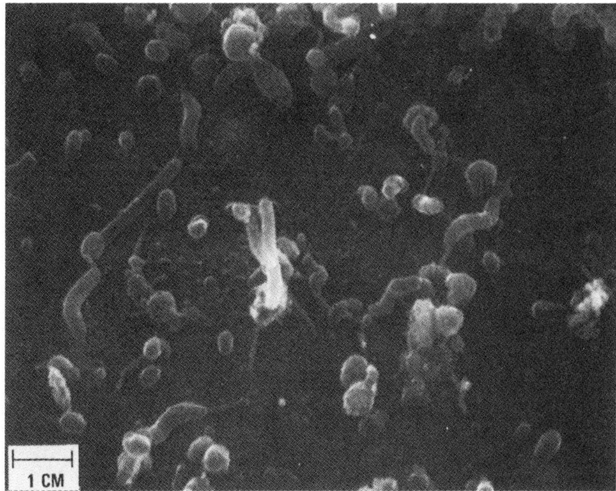

(A) NYLON COATED 4140 STEEL COUPON.
1 cm BAR INDICATES 2 μm.

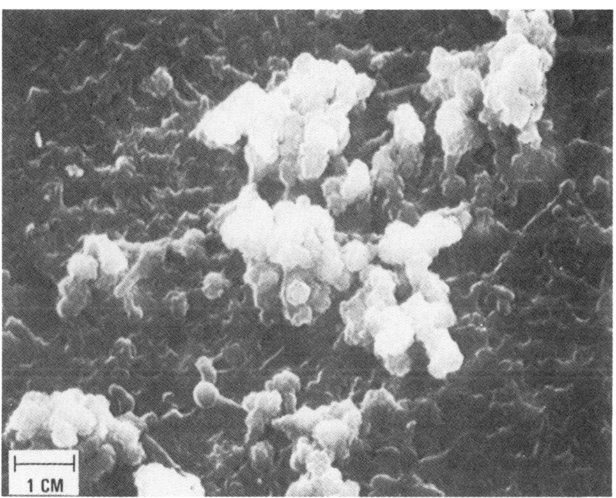

(B) EPOXY COATED 4140 STEEL COUPON.
1 cm BAR INDICATES 2 μm.

FIGURE 1. Scanning electron micrographs of nylon- and epoxy-coated 4140 steel exposed to marine, mixed culture I containing strict anaerobes (sulfate-reducing bacteria; SRB) and facultative anaerobes (non-sulfate reducers). (From Jones, J. M., Walch, M., and Mansfeld, F. B., Corrosion/91, Paper No. 108, NACE, Houston, TX, 1991. With permission.)

(A)

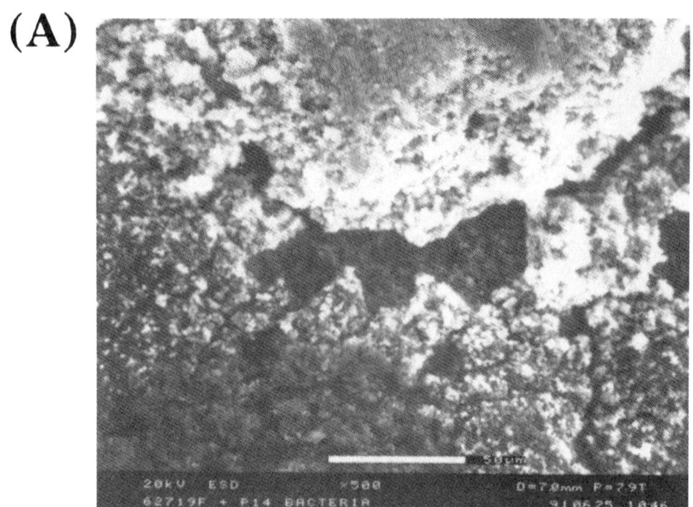

(B)

FIGURE 2. ESEM of phosphate coat with hydrogen embrittlement relief + epoxy-coated 4140 steel (category 2 in Table 1) exposed to marine, mixed culture II containing SRB and non-sulfate reducers. (From Jones, J. M., Walch, M., and Mansfeld, F. B., Corrosion/91, Paper No. 108, NACE, Houston, TX, 1991. With permission.)

(C)

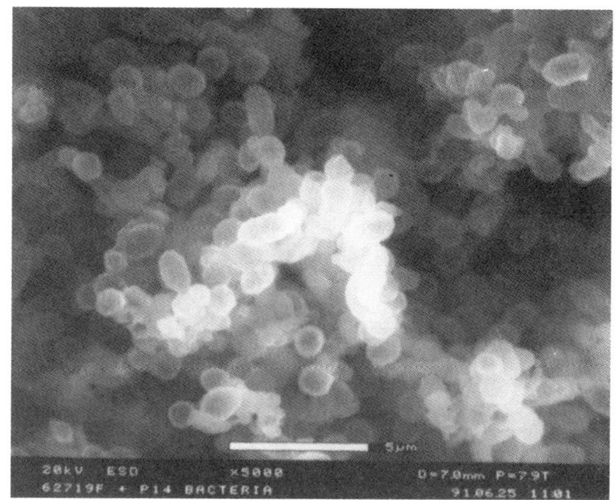

FIGURE 2 (continued)

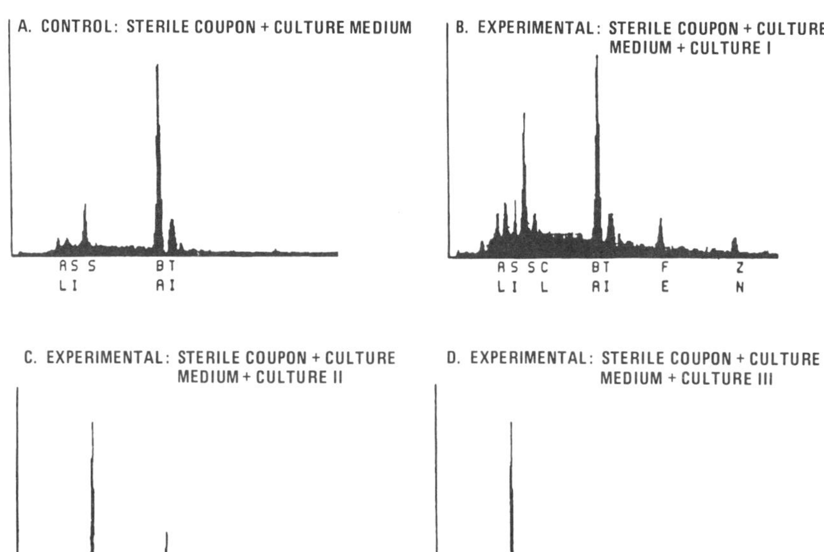

FIGURE 3. EDS spectra of zinc primer + epoxy-coated 4140 steel (category IV in Table 1) exposed to sterile culture medium with 2.5% NaCl (control) or to marine, mixed cultures I, II, and III in growth medium with 2.5% NaCl.

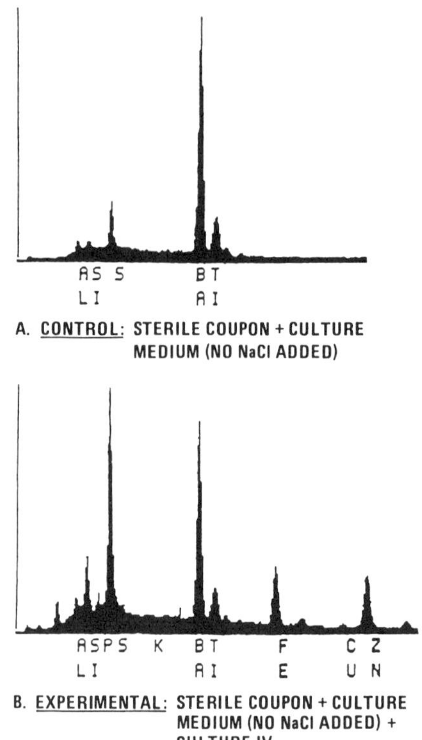

FIGURE 4. EDS spectra of zinc + epoxy-coated 4140 steel (category IV in Table 1) exposed to sterile culture medium without NaCl (control) or to marine, mixed culture IV in growth medium without NaCl.

the nylon coatings was not detected in the controls (coated coupon exposed to sterile medium). When SRB mixed communities were present, breaching of some of the nylon coatings was detected in EDS spectra which showed peaks from the primers (i.e., zinc, aluminum, or iron) and from the substrate steel (data not shown). A summary of the coating deterioration as detected by EDS analyses from SEM studies of coated coupons is shown in Table 3. A summary of the EDS spectra analyses from ESEM studies of coated coupons is shown in Table 4.

C. Visual Observations of Coupons Used for EIS Analyses

Some epoxy-coated coupons showed visible signs of coating deterioration (i.e., small holes under deposits, blistering without rupturing of the coating, ruptures from underneath the coating, and peeling along the coupon edges) after a 1-month exposure to the bacterial communi-

TABLE 3. Summary of EDS Spectra Analyses from SEM on Coated 4140 Steel Coupons Exposed to Halophilic and Nonhalophilic, Mixed Microbial Cultures Containing SRB

Coupon description[a]	Breaching of the coating[b] (halophilic, mixed cultures)		
	Culture I	Culture II	Culture III
Phosphate + epoxy	Yes	N.D.[c]	N.D.
Zinc + epoxy	Yes	Yes	Yes
IVD aluminum + epoxy	Yes	No	Yes
No primer + nylon	No	Yes	No
Phosphate + nylon	Yes	No	No
Zinc + nylon	Yes	N.D.	N.D.
	Breaching of the coating[b] (nonhalophilic, mixed culture VII)		
Zinc + epoxy		Yes	
IVD aluminum + epoxy		Yes	
No primer + nylon		No	
Phosphate + nylon		No	

[a] All coupons were 4140 steel; the primer and then the topcoat are listed.

[b] Breaching of the nylon or epoxy coatings was determined by EDS analyses showing the presence of the primer (zinc, phosphate, or aluminum) and the steel substrate (iron) in the spectra of the experimental samples, but not in the controls (culture medium only; no bacteria added).

[c] N.D. = not determined.

From Jones, J. M., Walch, M., and Mansfeld, F. B., Corrosion/91, Paper No. 108, NACE, Houston, TX, 1991. With permission.

ties. Blistering is an indication of microbial and/or inorganic ion processes occurring beneath the coating that may lead to failure of the coating. Examples of coating deterioration after a 2-month exposure in the laboratory to mixed, microbial cultures are shown in Figures 5 through 7. Figure 5 clearly shows microbial attack or microbial + chloride ion attack along the edges of the coupons, while Figures 6 and 7 show the corrosion attack over the surface of the coupon (under-deposit biocorrosion). Many of the epoxy-coated coupons showed attack in and around the suspension hole, at corners, and along edges of the coupons. An example of good coating integrity (by visual observation and by EIS data shown in Table 5) was with the zinc + epoxy + polyurethane-coated 4140 steel (category 1 in Table 1). Figure 8 shows the importance of having a primed

TABLE 4. Summary of EDS Spectra Analyses From ESEM Studies of Coated 4140 Steel Coupons Exposed to Sterile Growth Medium (No SRB) or to Mixed Culture II Containing SRB

Coupon description[a]	Breaching of coating[b]	
	Control (no SRB)[c]	Mixed culture II (SRB)
IVD aluminum + epoxy	(no Al detected)	+ + + (22.52 wt% Al)[d]
IVD aluminum + epoxy + polyurethane	(1.45 wt% Al)[e]	+ + + (71.72 wt% Al)
Phosphate + epoxy	(no P detected)	+ + (9.60 wt% P)
Phosphate + epoxy + polyurethane	(10.70 wt% P and 13.24 wt% Fe)	+ + (20.24 wt% P and 38.54 wt% Fe)
Zinc + epoxy	+ (3.02 wt% Zn)	+ + + (59.07 wt% Zn)
Epoxy	+ (14.37 wt% Fe)	+ + + (70.40 wt% Fe)
Epoxy + polyurethane	+ (17.60 wt% Fe)	+ + + (60.46 wt% Fe)

[a] All coupons were 4140 steel; the primer and then the topcoat(s) are listed.

[b] Breaching of the coatings was determined using ESEM/EDS analyses which showed the presence of the primer (aluminum, phosphate, or zinc) and often the steel substrate (iron) in the spectra of the experimental sample containing SRB, but not in the controls (no SRB).

[c] Coated coupons were not sterilized; controls contained sterile growth medium which allowed for growth of the non-sulfate-reducing microbes present on the coated coupons.

[d] The number in parentheses represents the element wt% from EDS spectral analyses that is relevant to breaching of the topcoat which allows the detection of the primer applied to the steel coupon.

[e] Visual observation of this sample did not show blisters, holes, or delamination, but the coating may have started to degrade, resulting in detection of the primer by EDS.

surface for coating applications. The epoxy-coated 4140 steel coupon shows many pinpoint breaches in the coating when exposed to sterile medium with NaCl and the pinpoint holes along with 20 areas where blisters have ruptured with intense corrosion of the exposed steel.

D. EIS Analyses

For the recording of impedance spectra, the side of the sample which showed the least corrosion damage was used. For the majority of samples, both sides were similar. The test cell defined an area of 20 cm^2 for the measurement of the EIS data. The impedance data were analyzed using the ANALEIS-COATFIT software which was developed at USC and is based on the model shown in Figure 9, where C_C is the capacitance of the polymer coating, R_{po} is the pore resistance which is related to the conductivity of the coating, R_p is the polarization resistance of the area under the coating at which corrosion occurs, and C_{dl} is the corresponding

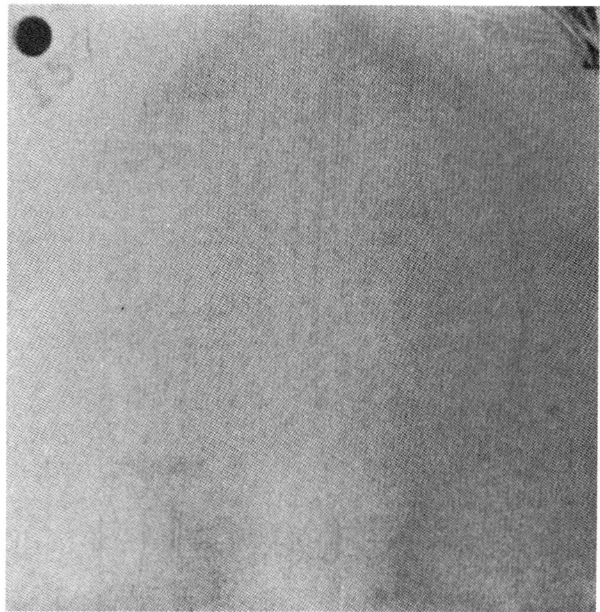

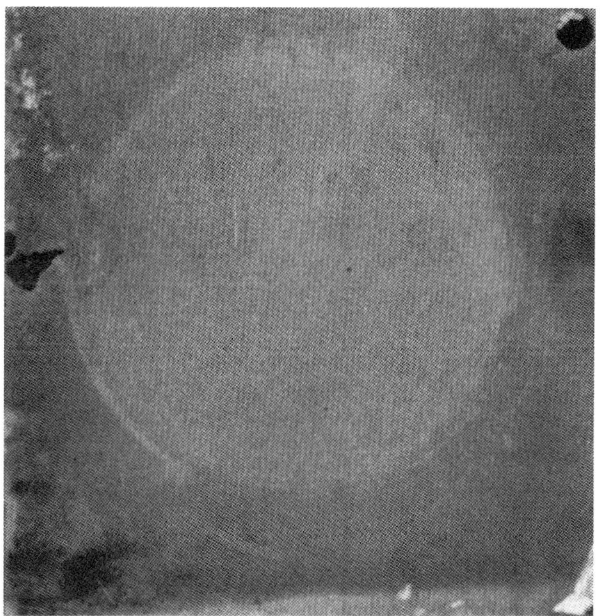

FIGURE 5. IVD aluminum + epoxy coated 4140 steel (category II in Table 1) exposed for 2 months in the laboratory to sterile medium with 2.5% NaCl (control coupon on the left) or to the same medium with marine, mixed culture III (coupon on the right). (From Jones, J. M., Walch, M., and Mansfeld, F. B., Corrosion/91, Paper No. 108, NACE, Houston, TX, 1991. With permission.)

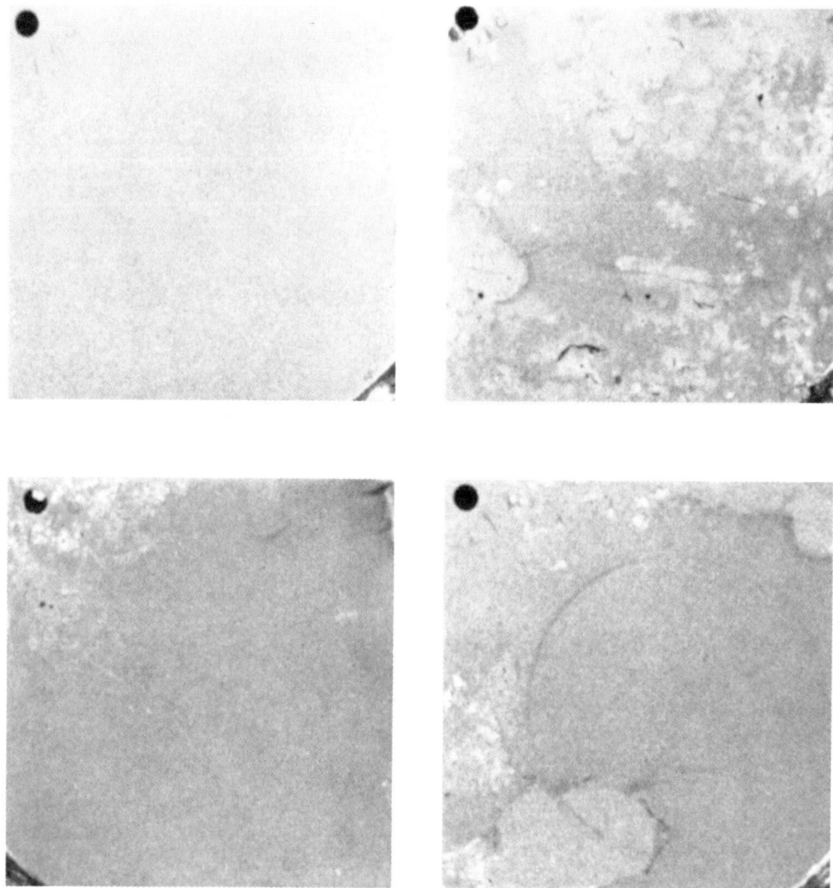

FIGURE 6. IVD aluminum + epoxy + polyurethane-coated 4140 steel (category 5 in Table 1) exposed for 2 months in the laboratory as follows: sterile medium with 2.5% Na Cl (control top left coupon), same medium with mixed culture I (top right coupon), mixed culture II (bottom left coupon), or mixed culture III (bottom right coupon). (From Jones, J. M., Walch, M., and Mansfeld, F. B., Corrosion/91, Paper No. 108, NACE, Houston, TX, 1991. With permission.)

capacitance.[17,18] R_s is the solution resistance between the reference electrode and the sample surface. The impedance spectra are presented in the following as Bode plots in which the logarithm of the modulus /Z/ of the impedance and the phase angle are plotted vs. the logarithm of the frequency f of the applied AC signal.[17,18] The analysis of the impedance spectra results in the values of the elements in the equivalent circuit of Figure 9 (Model I). The coating capacitance C_c is determined

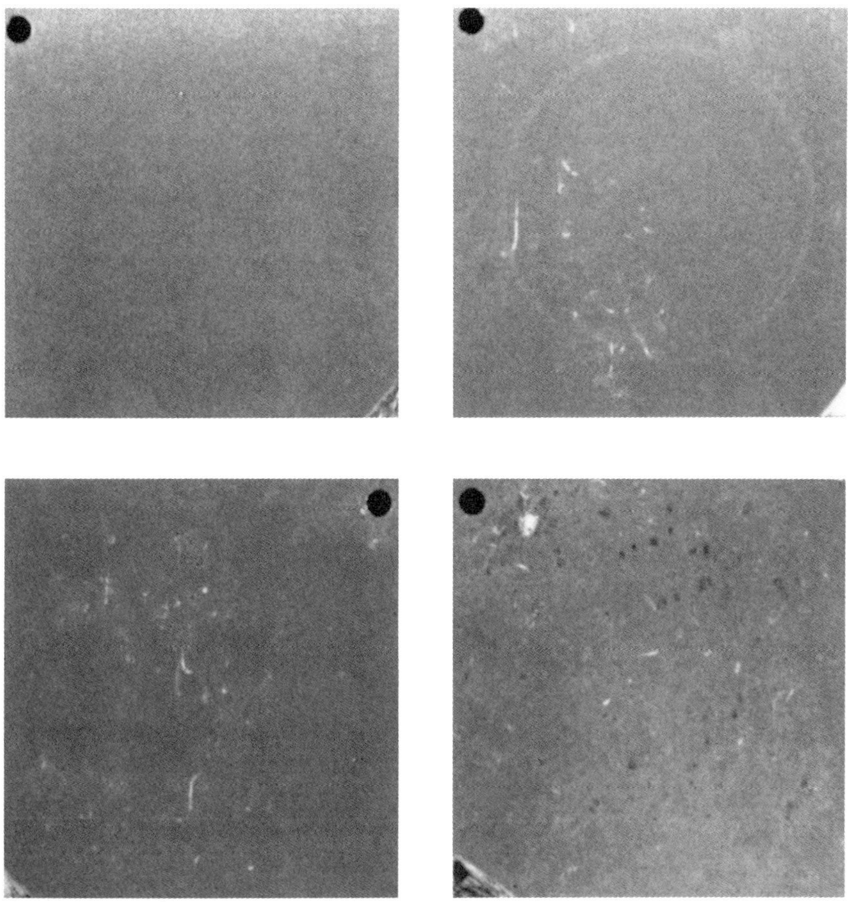

FIGURE 7. Zinc + epoxy-coated 4140 steel (category IV in Table 1) exposed for 2 months in the laboratory as follows: sterile medium with 2.5% NaCl (control top left coupon), same medium with mixed culture I (top right coupon), mixed culture II (bottom left coupon), or mixed culture III (bottom right coupon). (From Jones, J. M., Walch, M., and Mansfeld, F. B., Corrosion/91, Paper No. 108, NACE, Houston, TX, 1991. With permission.)

by the ratio of the dielectric constant ϵ to the coating thickness d:

$$C_c = \epsilon\epsilon_o/d \tag{1}$$

C_c can increase due to water uptake of the coating. R_{po} has been used by Mansfeld and Kendig[18] as an indication of damage to the coating as a result of corrosion at the metal-coating interface. Mechanical action of corrosion products against the coating can lead to the formation of con-

TABLE 5. Results of Analysis of EIS Data

Coating category	Microbial culture	E_{corr}(V) (vs. SCE)	C_c(nF)	C_{dl}(F)	R_{po}(ohm)	R_p(ohm)	D	A_{corr}(cm^2)
1 (Zinc + epoxy + polyurethane)	None	−0.580	3.7	1.6×10^{-8}	3.4×10^{6}	1.9×10^{8}	2.77	8.0×10^{-4}
	I	−0.700	3.9	1.5×10^{-8}	3.3×10^{6}	6.2×10^{7}	2.68	7.5×10^{-4}
	II	−0.696	4.4	4.1×10^{-8}	3.4×10^{5}	1.1×10^{8}	2.73	2.1×10^{-3}
	III	−0.705	4.5	2.3×10^{-8}	6.0×10^{5}	2.9×10^{8}	2.65	1.2×10^{-3}
2 (Phosphate coat + epoxy)	None	−0.374	5.6	1.7×10^{-8}	1.6×10^{5}	6.8×10^{5}	1.80	8.5×10^{-2}
	I	−0.304	2.9	3.5×10^{-4}	3.5×10^{3}	2.1×10^{5}	0.27	1.7×10
	II	−0.323	3.0	5.0×10^{-5}	7.4×10^{5}	3.7×10^{7}	2.35	2.5
	III	−0.370	3.2	1.9×10^{-8}	2.5×10^{6}	5.9×10^{7}	2.78	9.5×10^{-4}
3 (Epoxy)	None	−0.464	8.9	4.0×10^{-3}	1.3×10^{3}	2.0×10^{4}	0.21	2.0×10^{2}
	I	−0.347	3.4	7.9×10^{-5}	4.7×10^{2}	1.9×10^{5}	0.76	4.0
	II	−0.456	N.D.	3.7×10^{-4}	2.3×10^{2}	7.0×10^{4}	0.47	1.9×10
	III	−0.398	3.4	7.0×10^{-5}	3.2×10^{2}	3.5×10^{5}	0.93	3.5
4 (IVD Al + epoxy)	None	−0.760	2.3	3.7×10^{-5}	3.4×10^{4}	7.3×10^{5}	1.94	1.9
	I	−0.690	3.4	2.8×10^{-8}	3.5×10^{6}	3.0×10^{8}	2.78	1.4×10^{-3}
	II	−0.760	3.1	2.7×10^{-6}	1.3×10^{4}	1.9×10^{6}	1.33	1.4×10^{-1}
	III	−0.658	3.9	2.6×10^{-7}	2.5×10^{6}	1.1×10^{8}	2.63	1.3×10^{-2}
5 (IVD Al + epoxy +polyurethane)	None	−0.728	5.2	4.2×10^{-8}	4.4×10^{5}	1.0×10^{8}	2.54	2.1×10^{-3}
	I	−0.600	5.6	3.7×10^{-8}	7.6×10^{5}	1.3×10^{8}	2.56	1.9×10^{-3}
	II	−0.565	9.4	7.7×10^{-8}	3.5×10^{5}	6.4×10^{7}	2.44	3.9×10^{-3}
	III	−0.525	5.3	2.3×10^{-6}	6.3×10^{5}	8.0×10^{7}	2.59	1.2×10^{-1}
6 (Phosphate coat + epoxy + polyurethane)	None	−0.480	5.1	1.4×10^{-8}	5.7×10^{5}	9.5×10^{6}	2.58	7.0×10^{-4}
	I	−0.300	3.9	9.1×10^{-9}	2.7×10^{7}	4.0×10^{7}	2.69	4.6×10^{-4}
	II	−0.613	5.9	2.5×10^{-8}	3.5×10^{6}	2.4×10^{8}	2.62	1.3×10^{-3}
	III	−0.200	5.5	3.3×10^{-8}	1.1×10^{6}	5.7×10^{7}	2.60	1.7×10^{-3}
7 (Zinc + epoxy)	None	−0.800	4.4	3.9×10^{-8}	1.5×10^{5}	9.1×10^{8}	2.32	2.0×10^{-3}
	I	−0.995	2.2	2.9×10^{-8}	1.7×10^{6}	1.0×10^{8}	2.73	1.5×10^{-3}
	II	−0.876	2.4	7.3×10^{-9}	3.3×10^{6}	1.8×10^{8}	2.74	3.7×10^{-4}
	III	−0.980	2.8	1.2×10^{-8}	2.4×10^{8}	8.5×10^{6}	2.62	6.0×10^{-4}

8 (Epoxy + polyurethane)	None	−0.430	6.8	7.6×10^{-8}	4.5×10^{5}	1.7×10^{7}	2.49	3.8×10^{-3}
	I	−0.290	6.7	4.9×10^{-8}	1.1×10^{6}	1.1×10^{8}	2.52	2.5×10^{-3}
	II	−0.232	4.3	7.7×10^{-8}	2.1×10^{5}	9.5×10^{8}	2.54	3.9×10^{-4}
	III	−0.260	5.9	7.7×10^{-8}	6.0×10^{5}	6.0×10^{7}	2.57	3.9×10^{-3}
I (Five-step iron phosphate + nylon)	None	−0.200	2.2			1.1×10^{8}	2.70	
	IV	−0.200	1.2	2.0×10^{-9}	3.3×10^{6}	5.0×10^{8}	2.65	1.0×10^{-4}
	V	−0.600	1.5	1.8×10^{-9}	3.2×10^{6}	1.2×10^{8}	2.66	9.0×10^{-5}
II (IVD Al + epoxy)	None	−0.400	5.3			7.5×10^{8}	2.89	
	I	−0.500	3.3			5.7×10^{8}	2.90	
	II	−0.340	5.7			4.3×10^{8}	2.89	
	III	−0.600	6.4			3.7×10^{8}	2.89	
	None (−NaCl)	−0.200	5.0			3.7×10^{8}	2.80	
	VII	−0.500	4.1			6.5×10^{8}	2.90	
III (Luberite + epoxy)	None	−0.180	9.2			3.5×10^{8}	2.92	
	I	−0.200	5.6	6.5×10^{-9}	2.3×10^{5}	4.0×10^{7}	2.67	3.3×10^{-4}
	II	−0.310	4.2	1.9×10^{-9}	1.0×10^{5}	3.5×10^{7}	2.86	1.0×10^{-4}
	None (−NaCl)	−0.620	2.4	0.9×10^{-9}	1.8×10^{6}	1.5×10^{7}	2.43	4.5×10^{-5}
	VII	−0.370	6.1	1.4×10^{-5}	4.6×10^{3}	1.3×10^{6}	0.69	7.0×10^{-1}
IV (Zinc + epoxy)	None	−0.900	4.5			3.9×10^{8}	2.88	
	I	−1.115	2.3	7.6×10^{-7}	1.1×10^{5}	1.6×10^{5}	1.33	3.8×10^{-2}
	II	−0.100	3.1			3.1×10^{8}	2.90	2.9×10^{-3}
	III	−1.030	2.3	5.8×10^{-8}	4.9×10^{4}	3.5×10^{5}	1.54	2.5×10^{-1}
	V	−1.050	3.3	5.0×10^{-6}	6.0×10^{3}	4.0×10^{3}	0.37	1.6×10^{-2}
	IV	−1.050	3.4	3.2×10^{-7}	3.3×10^{3}	1.9×10^{3}	0.24	
	VII	−1.050	4.0			4.6×10^{4}	1.04	
	None (−NaCl)	−0.950	7.3			1.1×10^{8}	2.90	
	VII	−1.020	4.1	8.7×10^{-7}	4.0×10^{4}	2.1×10^{5}	1.32	4.4×10^{-2}

TABLE 5. (continued) Results of Analysis of EIS Data

Coating category	Microbial culture	E_{corr}(V) (vs. SCE)	C_c(nF)	C_{dl}(F)	R_{po}(ohm)	R_p(ohm)	D	A_{corr}(cm^2)
V (Five-step iron phosphate + epoxy)	None	+0.100	2.2			3.4×10^8	2.90	
	I	−0.020	2.9			2.2×10^8	2.93	
	II	−0.410	4.2	1.6×10^{-7}	2.5×10^4	5.7×10^5	1.73	8.0×10^{-3}
	III	+0.050	3.4			3.0×10^8	2.89	
	V	−0.120	3.4	9.3×10^{-8}	8.6×10^6	1.6×10^8	2.83	4.7×10^{-3}
	IV	−0.500	4.3			1.6×10^8	2.88	
	VI	−0.040	3.8			1.4×10^8	2.89	
	VII	−0.530	4.3			6.0×10^7	2.88	
	None (−NaCl)	−0.635	2.9	5.0×10^{-7}	1.4×10^4	1.9×10^5	1	
	VII	−0.020	3.1			2.3×10		

C_c Capacitance of the polymer coating.
C_{dl} Capacitance of the polymer coating where corrosion occurs.
R_{po} Pore resistance which is related to the conductivity of the coating.
R_p Polarization resistance of the area under the coating at which corrosion occurs.
D Damage function defined as $D = \log (Z_{10}/Z_{10,000})$, where Z_{10} is the impedance of f = 10 Hz and $Z_{10,000}$ is the impedance at f = 10,000Hz. For a coating for which the impedance is truly capacitive in this frequency range, D = 3.0. D will be smaller than 3.0 for coatings with spectra deviating from capacitive behavior due to damage during exposure to corrosive environments.
A_{corr} Corroding area.

From Jones, J. M., Walch, M., and Mansfeld, F. B., Corrosion/91, Paper No. 108, NACE, Houston, TX, 1991. With permission.

(A)

(B)

FIGURE 8. (A) 4140 steel + epoxy topcoat (no primer used) exposed to sterile medium with 2.5% NaCl; (B) 4140 steel + epoxy topcoat (no primer used) exposed to medium with 2.5% NaCl and to mixed culture II.

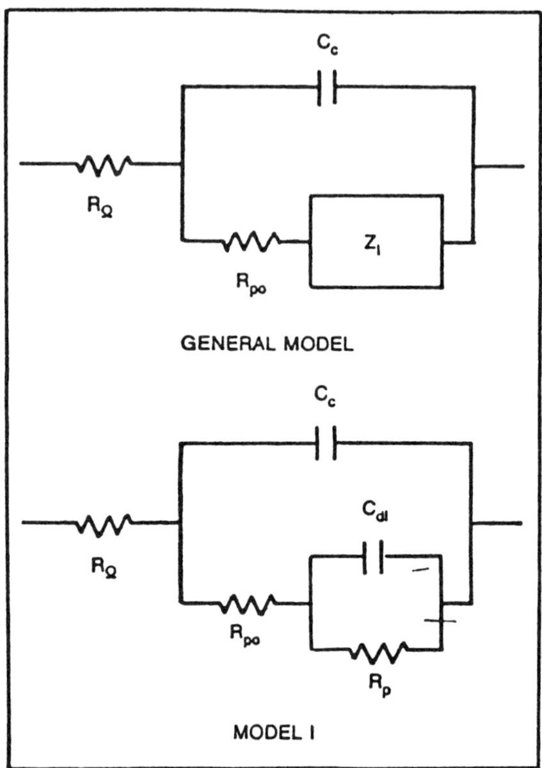

FIGURE 9. Equivalent circuits for the impedance of a polymer coating. (From Jones, J. M., Walch, M., and Mansfeld, F. B., Corrosion/91, Paper No. 108, NACE, Houston, TX, 1991. With permission.)

ductive paths and a lowering of R_{po} with exposure time. The experimental values of C_{dl} can be used to estimate the corroding area A_{corr} as

$$A_{corr} = (C_{dl}/C^{o}_{dl})A \tag{2}$$

based on the theoretical value of the double-layer capacitance $C^{o}_{dl} = 20$ $\mu F/cm^2$. The corrosion rate of the corroding area A_{corr} can be estimated from the value of $R_p \times A_{corr}$, which has the dimensions of Ohm $\times$ cm^2.

In general, it was observed that the impedance spectra for the samples within a given coating system (see Table 5) were very similar. Some systems showed mainly capacitive behavior, which is indicative of very little degradation, and some systems showed two time constants as predicted by Model I in Figure 9 for coatings with pronounced degradation. These results suggest that for the majority of the cases studied the impedance spectra reflect the inherent protective properties of the coating systems and the corrosion characteristics of the metal-coating interface.

Very good performance was found for the nylon-coated steel (category I in Table 5). Figure 10 is an example of a coating system (category 1 in Table 5; zinc plate + epoxy + polyurethane) for which little degradation had occurred during the 2-month exposure period. The spectra seemed to be dominated by C_c in the frequency range down to 0.1 Hz. However, closer analysis of the frequency dependence of the phase angle shows deviations in the intermediate frequency range from the 90° value expected for purely capacitive behavior of a perfect coating, which indicates that some interaction of the coating with the environment must have occurred. The corroding area A_{corr} is very small (less than 10^{-3} cm^2). Very different impedance spectra were obtained for samples which were covered with only an epoxy topcoat (category 3 in Table 5; see Figure 11). The impedance was very low, indicating the formation of conductive paths in the coating and the occurrence of corrosion and delamination at the metal-coating interface. The phase angle (Figure 11) has maxima at the highest frequencies (which are not too clear in the limited high-frequency range of these measurements) and at low frequencies.

An interesting case is the system of a zinc primer with an epoxy topcoat (category IV in Table 5; see Figure 12). The spectra for the control sample (curve 1) and one of the samples which had been exposed to mixed culture II (curve 3) are entirely capacitive as for intact coatings. The spectra for the other five samples are dominated by the corrosion reaction at the zinc surface, which is exposed at sites where the coating has been attacked and has become delaminated. This attack could be due to the action of microorganisms and has to be verified by SEM and Energy Dispensive Spectroscopy (EDS) analyses.

A summary of the analysis of the EIS data for all samples tested is given in Table 5. The corrosion potential E_{corr} was measured after 2 h exposure to NaCl. The values of C_c, C_{dl}, R_{po}, and R_p are given for all samples except for those with virtually unchanged coating properties which showed capacitive spectra from which only C_c and R_p can be determined. The average value of C_c = 5nF (2.5×10^{-10} F/cm^2) corresponds to a coating thickness of 14 μm. The nylon coating was about twice as thick. The values of A_{corr} were calculated as defined in Equation 2. In addition, the damage function D is listed in Table 5, which is defined as

$$D = \log(Z_{10}/Z_{10,000}) \tag{3}$$

where Z_{10} is the impedance at f = 10 Hz and $Z_{10,000}$ is the impedance at f = 10,000 Hz. For a coating for which the impedance is truly capacitive

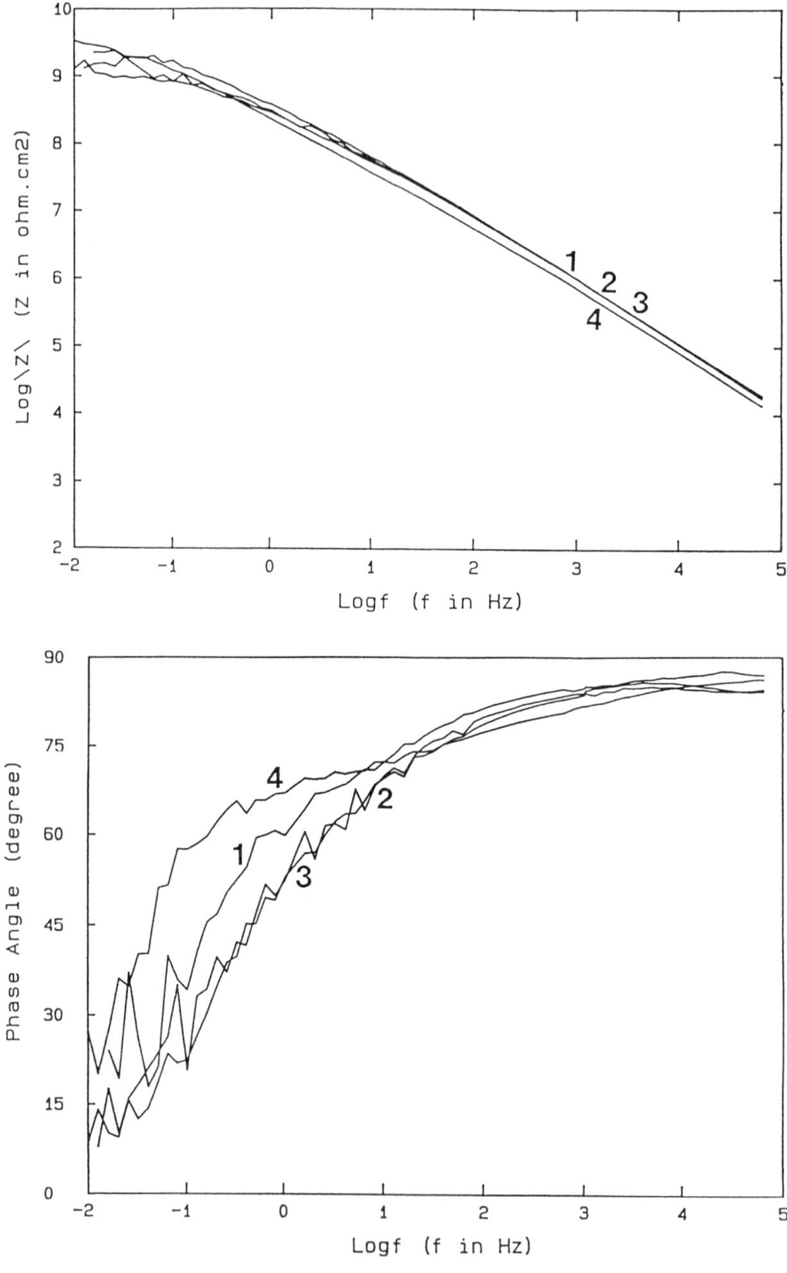

FIGURE 10. Bode plots for zinc plate + epoxy primer + polyurethane topcoat on 4140 steel (category 1 in Table 1) exposed for 2 months to sterile medium with 2.5% NaCl or to the same medium with mixed microbial cultures. 1, control (no SRB); 2, mixed culture I; 3, mixed culture II; and 4, mixed culture III. (From Jones, J. M., Walch, M., and Mansfeld, F. B., Corrosion/91, Paper No. 108, NACE, Houston, TX, 1991. With permission.)

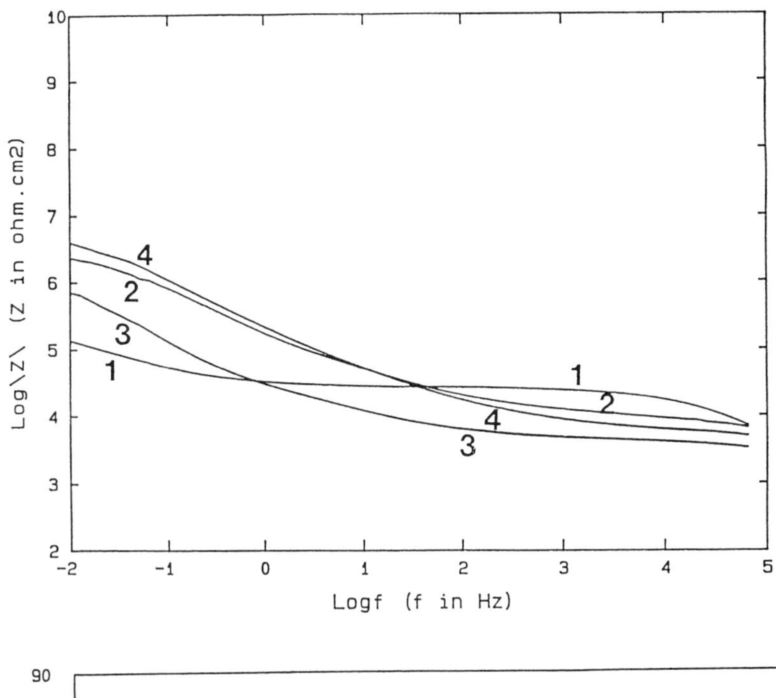

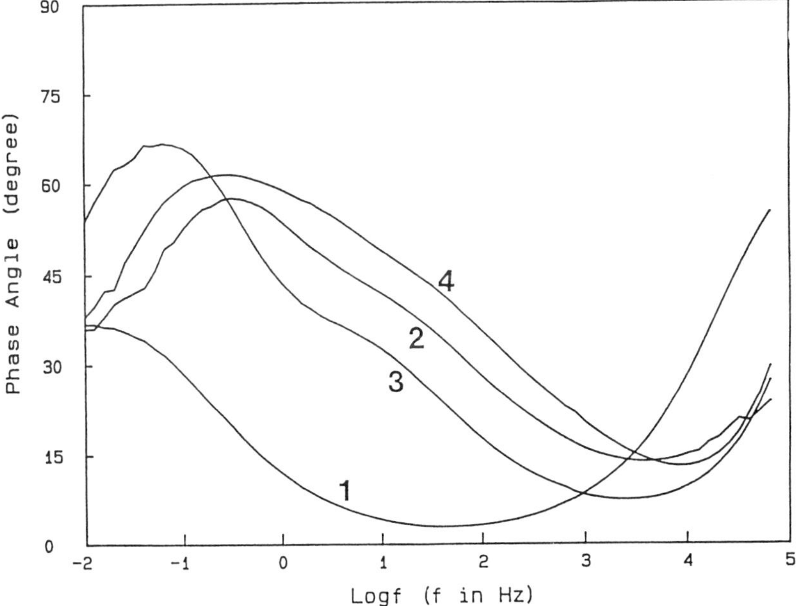

FIGURE 11. Bode plots for epoxy topcoat on 4140 steel (category 3 in Table 1) exposed for 2 months to sterile medium with 2.5% NaCl or to the same medium with mixed microbial cultures. (From Jones, J. M., Walch, M., and Mansfeld, F. B., Corrosion/91, Paper No. 108, NACE, Houston, TX, 1991. With permission.)

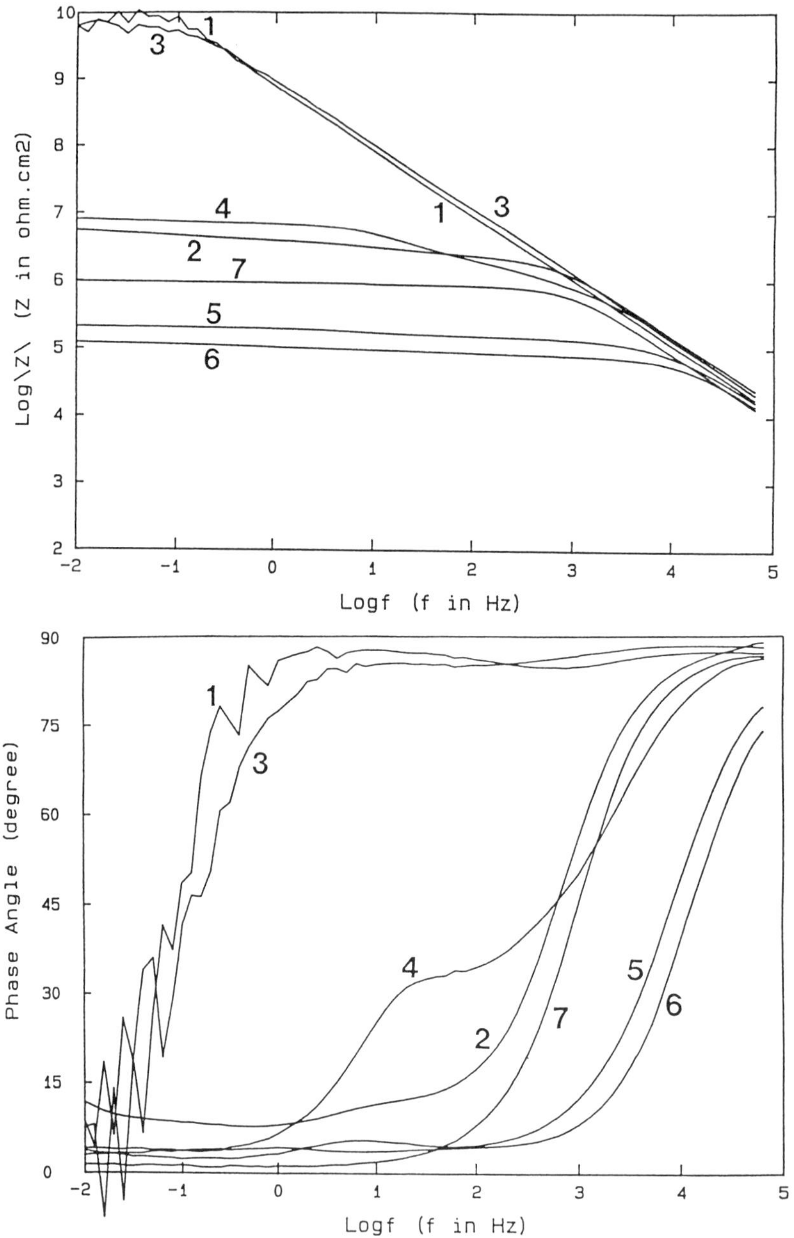

FIGURE 12. Bode plots for zinc + epoxy topcoat on 4140 steel (category IV in Table 1) exposed for 2 months to sterile medium with 2.5% NaCl or to the same medium with mixed microbial cultures. 1, control (no SRB); 2, mixed culture I; 3, mixed culture II; 4, mixed culture III; 5, mixed culture V; 6, mixed culture IV; and 7, mixed culture VII. (From Jones, J. M., Walch, M., and Mansfeld, F. B., Corrosion/91, Paper No. 108, NACE, Houston, TX, 1991. With permission.)

in this frequency range, D = 3.0 (straight line with a slope of −1 in a log /Z/ − log f plot). For coatings in which the spectra deviate from capacitive behavior due to damage during exposure to corrosive environments, D will be smaller than 3.0. The damage function allows a qualitative grouping of the different coating systems and of individual samples in a given system in terms of corrosion damage.

A large number of systems (categories 1, 4 to 8, I, and II in Table 5) had D values exceeding 2.5 for all samples in the grouping. The best performance was exhibited by category II (IVD aluminum with epoxy topcoat) which for all samples had D values exceeding 2.8 and for which only capacitive behavior was observed. Only data for C_c and R_p can be determined in this case; a value of A_{corr} does not exist, since corrosion was not initiated. Very good performance was also found for the nylon-coated steel (category I in Table 5).

Systems with very low D values were those coupons with a phosphate primer + epoxy topcoat (category 2 in Table 5) and coupons with only an epoxy topcoat (category 3 in Table 5). For category III (luberite + epoxy topcoat), all but one sample had high D values, while for category V (five-step iron phosphate + epoxy topcoat) one of the two control samples had the lowest D value of all coupons tested. Category IV (zinc + epoxy topcoat) had high D values for the two control samples, but low D values for some of the samples exposed to microbes. In most cases listed in Table 5, A_{corr} was extremely small, indicating that these coating systems were still providing adequate corrosion protection in the area tested.

In summarizing the results of the analysis of the EIS data for the different coatings, it can be stated that evidence for accelerated attack of polymer coatings by microorganisms can only be found for the zinc + epoxy topcoat coupons (category IV in Tables 1 and 5). The fact that the impedance was capacitive for the control samples, but decreased significantly for most coupons which had been exposed to microorganisms, suggests that under these laboratory testing conditions conductive paths had been formed which reached the steel surface and caused corrosion and coating delamination. However, it must be kept in mind that although EIS is a very sensitive tool for the detection of coating damage, especially in its initial stages, it does not identify the factors which have caused this damage. Other tools such as SEM, ESEM, and EDS analyses have been used to confirm that the observed degradation was due to the action of microorganisms and not to factors such as variations in the quality of the coating application.

IV. DISCUSSION

Each of the mixed cultures contains a variety of different microorganisms as evidenced by SEM and ESEM. Water vapor was used as the specimen environment with the ESEM; therefore, wet samples were examined directly from the culture medium and the EDS spectra collected at the same time as the sample morphology and topography were photographed. Bacteria were attached to the coated surface and were distributed throughout the biofilm layers of corrosion products, bacterial exopolymers, and inorganic ion deposits.

The results of experiments described here show that microorganisms exist in both marine and nonmarine environments whose growth can lead to deterioration of epoxy coatings, nylon coatings, and polyurethane coatings applied to 4140 steel. Deterioration of nylon coatings appeared to proceed at a slower rate than that of the epoxy coatings and very little damage was visible in scanning electron micrographs. However, none of the cultures used in these experiments were isolated from a nylon-coated material. Halophilic, mixed cultures have been isolated from nylon-coated 4140 steel and these mixed cultures will be tested for their effect on the nylon coatings. It is expected that the bacteria isolated from a nylon substratum will have a greater potential for degradation of the nylon coatings.

The microbial attack on coatings has important implications because such coatings must remain intact for effective corrosion protection. Breaches in protective coatings can lead to rapid localized corrosion and may allow access of other bacteria to the underlying metal, perhaps further exacerbating the corrosion problem.[5] Many other protective coatings may also be subject to biodeterioration, given that bacteria are capable of degrading an incredible variety of organic polymers and materials. It is not yet clear whether the coating deterioration observed here resulted from a specific enzymatic degradation or from corrosive metabolites.

SRB hydrogenase activity may influence the biocorrosion process and may be repressed or induced, depending on the substrate and the bacterial consortia present in the biofilm.[19] The microbial composition of a mixed population which included SRB and the presence or absence of hydrogenase enzyme were two variables that were found to influence the biocorrosion process on SAE 1020 steel.[19] All the mixed SRB populations used in this study were hydrogenase positive. To our knowledge, no one has looked at hydrogenase-negative and hydrogenase-positive mixed SRB populations to determine whether the hydrogenase enzyme plays a role in the microbial degradation of coated steel.

We have isolated pure cultures of SRB from the mixed communities. Many of the SRB pure cultures are desulfoviridan positive, which is indicative of *Desulfovibrio*. Several hydrogenases, which differ in cellular localization, in the composition of the active site and with respect to hydrogen production and oxidation have been isolated from *Desulfovibrio* spp.[20] Three different hydrogenases have been found in *D. vulgaris* (Hildenborough).[21] Two of the hydrogenase enzymes are located in the cytoplasmic membrane and contain nickel as well as iron-sulfur clusters,[22] while the third hydrogenase is soluble and periplasmic[23] and has only iron-sulfur clusters. This periplasmic, Fe-only hydrogenase is present in *D. vulgaris* (Hildenborough) under all growth conditions tested (lactate, pyruvate, glycerol, or H^2 as an energy source; sulfate or sulfite as an electron donor).[24]

Further MIC and coating degradation studies using pure and mixed SRB cultures (including hydrogenase-negative and hydrogenase-positive SRB) will provide more information about which bacterial species are responsible for the corrosive attack of the coatings and whether SRB hydrogenase plays a role in coating degradation.

EIS methods provide a capability for assessing the degree of coating deterioration[17,18,25–27] and for detecting the initiation of corrosion at the metal-coating interface provided that appropriate analysis of the impedance parameters is used. EIS has been shown to have some utility in following epoxy coating degradation in aerated American Society for Testing Materials (ASTM) seawater.[28] Further testing will be necessary to determine whether EIS can be successful in separating coating deterioration due to microorganisms (mixed cultures that include SRB) from that caused by corrosion occurring in similar environments, but in the absence of microorganisms.

ESEM offers a nondestructive method for observing the activity of microbial biofilms. The applicability and advantages of ESEM/EDS in studies of MIC have been discussed by Little et al.[29] and Wagner et al.[30] and have been used in this study to show microbial degradation of epoxy and epoxy + polyurethane coatings.

ACKNOWLEDGMENTS

Work at NAVSWC was supported by Capt. Steve Snyder, ONR/Code 1141MB and by the AEGIS Combat Systems Program Office NO5 (NAVSWC/Dahlgren, VA). The work at the Center of Marine Biotechnology and USC was supported by the AEGIS Combat Systems Program Office NO5 (NAVSWC/Dahlgren, VA). The EIS data were obtained by Dr. S. Liu (USC).

REFERENCES

1. **Brady, R. F.,** Epoxy resins for protective coatings and linings, *J. Protect. Coat. Linings,* 2, 24, 1985.
2. **Lewis, C. E.,** Colorful protection with powder coatings, *Mater. Eng.,* 104, 37, 1987.
3. **Baker, R. G., Kitchen, G. H., and Wells, R. R.,** Nylon 11 as a lubricant and corrosion preventative coating, *Natl. Bur. Stand. (U.S.), Spec. Publ.,* 452, 25, 1976.
4. **Allsopp, D. and Seal, K. J.,** Biodeterioration of refined and processed materials, in *Introduction to Biodeterioration,* Edward Arnold, London, 1986, chap. 3.
5. **Miller, J. D. A.,** *Microbial Aspects of Metallurgy,* American Elsevier, New York, 1970, 31 and 97.
6. **Battersby, N. S.,** 1988, Sulphate-reducing bacteria, in *Methods in Aquatic Bacteriology,* Austin, B., Ed., John Wiley & Sons, New York, 1988, 269.
7. **Jones, J. M., Walch, M., and Mansfeld, F. B.,** Microbial and electrochemical studies of coated steel exposed to mixed microbial communities, Corrosion/91, Paper No. 108, NACE, Houston, TX, 1991.
8. **Terry, L. A. and Edyvean, R. G. J.,** Microalgae and corrosion, *Bot. Mar.,* XXIV, 177, 1981.
9. **Stranger-Johannessen, M.,** The role of microorganisms in the blistering and debonding of corrosion protective organic coatings, in *Proceedings of the XVIII FATIPEC-Congress,* Venezia, 1986, Vol. 3, 1.
10. **Stranger-Johannessen, M. and Norgaard, E.,** Deterioration of anti-corrosive paints by extracellular microbial products, *Int. Biodeterior. Bull.,* 27, 157, 1991.
11. **Kinoshita, S., Terada, T., Taniguchi, T., Takene, Y., Masuda, S., Matsunaga, N., and Okada, H.,** Purification and characterization of 6-aminohexanoic-acid-oligomer hydrolase of *Flavobacterium* sp. KI72, *Eur. J. Biochem.,* 116, 547, 1981.
12. **Negoro, S., Taniguchi, T., Kanaoka, M., Kimura, H., and Okada, H.,** Plasmid-determined enzymatic degradation of nylon oligomers, *J. Bacteriol.,* 155, 22, 1983.
13. **Okada, H., Negoro, S., Kimura, H., and Nakamura, S.,** Evolutionary adaptation of plasmid-encoded enzymes for degrading nylon oligomers, *Nature,* 306, 203, 1983.
14. **Negoro, S., Nakamura, S., and Okada, H.,** DNA-DNA hybridization analysis of nylon oligomer-degradative plasmid pOAD2: Identification of the DNA region analogous to the nylon oligomer degradative gene, *J. Bacteriol.,* 158, 419, 1984.
15. **Kanagawa, K., Negoro, S., Takada, N., and Okada, H.,** Plasmid dependence of *Pseudomonas* sp. strain NK87 enzymes that degrade 6-aminohexanoate-cyclic dimer, *J. Bacteriol.,* 171, 3181, 1989.
16. **Tsuchiya, K., Fukuyama, S., Kanzaki, N., Kanagawa, K., Negoro, S., and Okada, H.,** High homology between 6-aminohexanoate-cyclic-dimer hydrolases of *Flavobacterium* and *Pseudomonas* strains, *J. Bacteriol.,* 171, 3187, 1989.
17. **Kendig, M. W., Allen, A. T., Jeanjaquet, S. L., and Mansfeld, F.,** The application of impedance spectroscopy to the evaluation of corrosion protection by inhibitors and polymer coatings, Corrosion/85, Paper No. 74, NACE, Houston, TX, 1985.
18. **Mansfeld, F. and Kendig, M. W.,** *Laboratory Corrosion Tests and Standards,* Haynes, G. S. and Baboian, R., Eds., ASTM-STP 866, Philadelphia, 1985, 122.
19. **Bryant, R. D., Jansen, W., Boivin, J., and Laishley, E. J.,** Effect of hydrogenase and mixed sulfate-reducing bacterial populations in the corrosion of steel, *Appl. Environ. Microbiol.,* 57, 2804, 1991.
20. **Fauque, G., Peck, H. D., Jr., Moura, J. J. G., Huynh, B. H., Berlier, Y., Dervartanian, D. V., Teixeira, M., Przybyla, A. E., Lespinat, P. A., Moura, I., and LeGall, J.,** The three classes of hydrogenase from sulfate-reducing bacteria of the genus *Desulfovibrio, FEMS Microbiol. Rev.,* 54, 299, 1988.
21. **Lissolo, T., Choi, E. S., LeGall, J., and Peck, H. D., Jr.,** The presence of multiple intrinsic membrane nickel containing hydrogenase in *Desulfovibrio vulgaris* (Hildenborough), *Biophys. Res. Commun.,* 139, 701, 1986.

22. **Peck, H. D., Jr. and Lissolo, T.,** Assimilatory and dissimilatory sulphate reduction: enzymology and bioenergetics, *Symp. Soc. Gen. Microbiol.*, 42, 99, 1988.
23. **Van der Westen, H., Mayhew, S. G., and Veeger, C.,** Separation of hydrogenase from intact cells of *Desulfovibrio vulgaris*. Purification and properties, *FEBS Lett.*, 86, 122, 1978.
24. **Van Den Berg, W. A. M., Van Dongen, W. M. A. M., and Veeger, C.,** Reduction of the amount of periplasmic hydrogenase in *Desulfovibrio vulgaris* (Hildenborough) with antisense RNA: direct evidence for an important role of this hydrogenase in lactate metabolism, *J. Bacteriol.*, 173, 3688, 1991.
25. **Scully, J. R.,** Evaluation of organic coating deterioration and substrate corrosion in seawater using electrochemical impedance measurements, Corrosion/86, Paper No. 222, NACE, Houston, TX, 1986.
26. **Scully, J. R.,** Electrochemical impedance of organic coated steel: correlation of impedance parameters with long term coating deterioration, DTRC/SME-86/108, Ship Materials Engineering Department, Research and Development Report, David Taylor Research Center, Bethesda, MD, 1988.
27. **Boxall, J.,** Advances in protective coatings: electrochemical testing, *Polym. Paint Col. J.*, 175, 597, 1985.
28. **Murray, J. N. and Hack, H. P.,** Long term testing of epoxy coated steel in ASTM sea water using E.I.S., Corrosion/90, Paper No. 140, NACE, Houston, TX, 1990.
29. **Little, B., Wagner, P., Ray, R., Pope, R., and Sheetz, R.,** Biofilms: an ESEM evaluation of artifacts introduced during SEM preparation, *J. Ind. Microbiol.*, 8, 213, 1991.
30. **Wagner, P. A., Little, B. J., Ray, R. I., and Jones-Meehan, J.,** Investigations of microbiologically influenced corrosion using environmental scanning electron microscopy, Corrosion/92, Paper No. 185, NACE, Houston, TX, 1992.

Section II
BIOFOULING

8

On-Line Side-Stream Monitoring of Biofouling

Frank L. Roe, Eric Wentland, Nick Zelver, Bryan Warwood, Ralf Waters, and William G. Characklis

I. INTRODUCTION

Bacteria will adhere to and grow on most surfaces in flowing aquatic environments. The deposits which accumulate are called biofilms.[1,2] Biofouling, which is the undesirable accumulation of biofilm, imposes a significant cost on society. These costs include increased energy needs in pumping systems, decreased efficiency in heat exchangers, early equipment failure due to biocorrosion, decreased life expectancy of equipment due to harsh biofouling control procedures, aesthetic degradation of products, and increased health risks.

Biofilms consist of cells growing in a matrix of a viscous permeable polysaccharide secretion or biopolymer and often include embedded inorganic particles such as sediments, scale deposits, or corrosion deposits. Biofilms continuously change in thickness, surface distribution, microbial populations, and chemical composition, and respond to changes in environmental factors such as water temperature, water chemistry, and surface conditions. This dynamic nature of biofilms makes it difficult to measure and predict the course of biofouling, and thus often reduces the effectiveness of treatment strategies. Site-specific continuous monitoring of biofouling rates and extent of fouling is essential to verify mathematical models and to fine tune procedures for efficient and

0-87371-928-X/94/$0.00+$.50

effective treatment of biofouling. A variety of different monitors for biofouling are reported in the literature.[3,4]

The purpose of this study was to evaluate on-line side-stream monitoring of biofilm development using an instrumented annular reactor and compare it with biofilm development in a tubular system. Continuously collected data — torque and light transmittance in the annular reactor, and heat transfer resistance (HTR) and pressure drop (PD) in the tubular system — were complemented by measuring attached mass for both systems at the end of each experiment. The goal was to show that a suitably instrumented annular reactor could predict biofouling development in an industrially relevant tubular system.

The annular reactor[5–8] is a side-stream reactor which can be used to monitor accumulation of biofilm in any fluid flow system. It accepts a side-stream portion of the process water and measures fouling potential at a surface employing a user-selected shear stress. It has commonly been used for laboratory studies of biofouling, but is presented here as a potential device for continuous monitoring in field situations. It is an ideal platform for mounting biofouling probes to assess biofilm accumulation; it is small, simple, versatile, and rugged. Sensors that have been suggested include torque, light transmittance, HTR, and a variety of specific ion probes. All of these can be monitored in real time using a computer data-acquisition system. We report here on use of a torque monitor and a light transmittance probe with the annular reactor for continuous monitoring of biofouling.

We chose torque to study in detail for a variety of reasons, the most compelling being that it was an average measure of fouling, rather than a localized measure. Localized variables such as light transmittance of the biofilm or near-surface dissolved-oxygen concentration would be expected to show significant spatial variability. This variability can be valuable in other contexts; however, we felt that there was less chance for false-positive or -negative indications of fouling with torque. Another advantage is that the torque monitor, once calibrated, maintains its accuracy over long periods. Chemical sensors have an added drawback in that the measured signals drift and the probes are difficult to calibrate *in situ*. Optical sensors, which have an initial limited range of sensitivity, are also sensitive to qualitative changes in the biofilm and process water, further reducing their usefulness.

II. EXPERIMENTAL

A. Equipment

1. Loop System

The simulated industrial system consisted of a tubular recycle loop shown in Figure 1. It was fabricated from 0.5-in.-diameter stainless steel tubing.

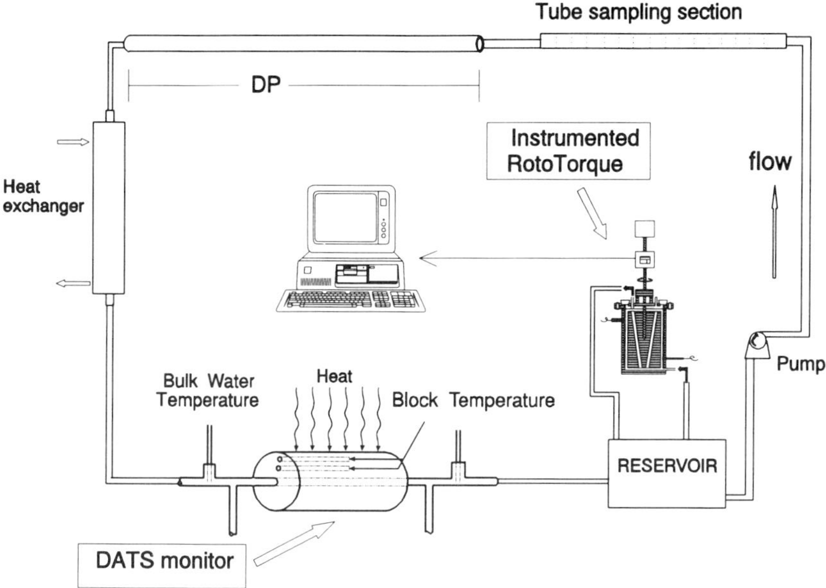

FIGURE 1. Instrumented annular reactor installed side-stream to a recycle loop system.

A centrifugal pump took water from the mixing tank and passed it through the loop at a flow rate controlled by a motorized valve. HTR was measured in a simulated heat exchanger located on one section of the loop. PD was measured across another section of the loop. From a third section of the loop, 5-in. sections of pipe were cut for mass accumulation measurements.

The mixing tank was operated as a continuous-flow stirred-tank reactor with a total volume of approximately 110 l. Dilution water (56.7 l/h) and nutrients (284 mg/h glucose plus 7 mg/h nitrogen and 14 mg/h phosphorous) were fed to the mixing tank using a tubing pump. Temperature in the mixing tank was controlled at 24 ± 2°C.

2. *Annular Reactors*

The annular reactors were fabricated from PVC (see Figure 2). Each consisted of a stationary outer drum and a rotating inner cylinder. A port located in the bottom housing allowed introduction of fluids, while removal of fluids took place at the top.

Water from the mixing tank was recirculated through two annular reactors operated independently and in parallel.

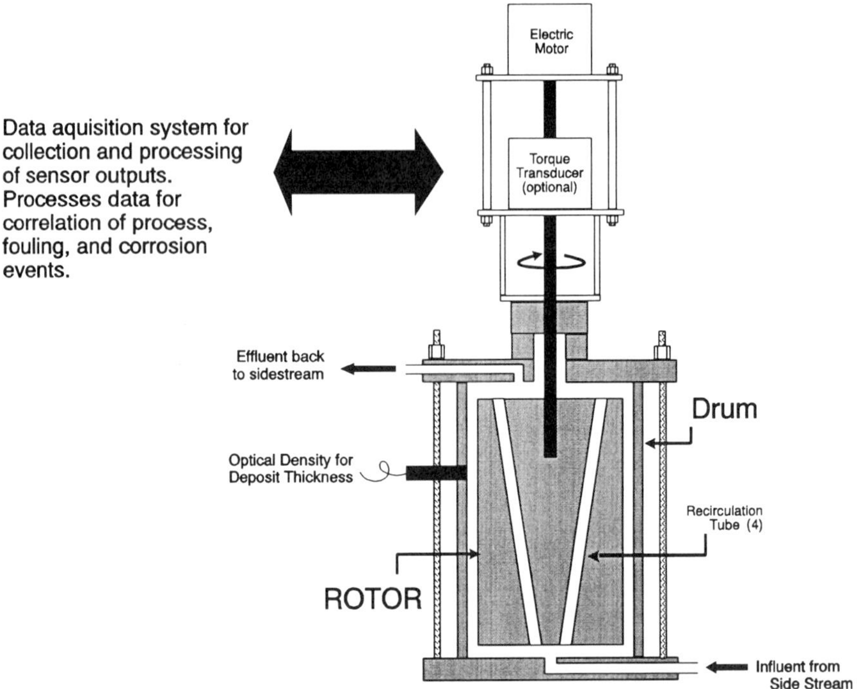

FIGURE 2. Annular reactor system.

Torque was measured using a General Thermodynamics torque monitor.

3. Data Acquisition

Computer data-acquisition systems were used to collect continuous torque, HTR, differential pressure (DP), loop flow rate, and loop temperature data.

For annular reactor torque, signal conditioning was accomplished using the circuit in Figure 3. A 5-V supply provided the driving voltage for the bridge circuit which consisted of the torque-proportional potentiometer in the torque monitor and a zeroing potentiometer in the signal conditioner. Output from the signal conditioner was fed to a Metrabyte® DAS-8 A/D board in a Zenith® 386 25-MHz microcomputer. LabTech® Notebook data-acquisition software was used to collect data from the torque monitors.

Instrumentation for the light transmittance probe is shown in Figure 4. Modulated light from an infrared light-emitting diode (IRLED) passes through a fiber optics cable embedded in the outer drum wall of the

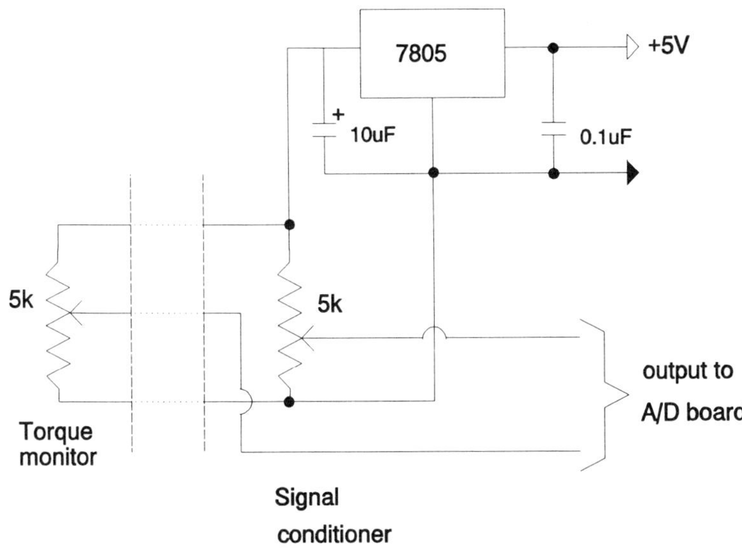

FIGURE 3. Torque monitor signal conditioner.

annular reactor, through the biofilm on the outer wall, through the biofilm on the inner wall, and is reflected back along the same path to the detector. The signal conditioner detects light of the same phase as that emitted by the IRLED, converts it to a frequency proportional to intensity, and then finally converts it to a voltage using a frequency-to-voltage converter.

For the tubular recycle system, HTR, PD, flow rate, and water temperature were automatically logged to a Bridger Scientific DATS Fouling Monitor System. Periodically, data was dumped from the DATS to the microcomputer using Bridger Scientific DATS-specific software.

All computer-collected data were eventually merged and analyzed using Borland's® Quattro Pro spreadsheet software.

4. *Manual Measurements*

Biofilm mass was manually measured at the end of each experiment. For the annular reactor, the inner rotor surface was allowed to drain for 5 min, and then was scraped with a rubber spatula into an aluminum boat. The residue was dried at 60°C for 24 h and weighed. For the tubular system, three 10-cm sections were cut from the loop, drained for 5 min, dried at 60°C for 24 h, and weighed. The tube sections were then cleaned with a bristle brush, dried as before, and reweighed. The difference between the two weights was the dry mass of the deposits on the tube sections.

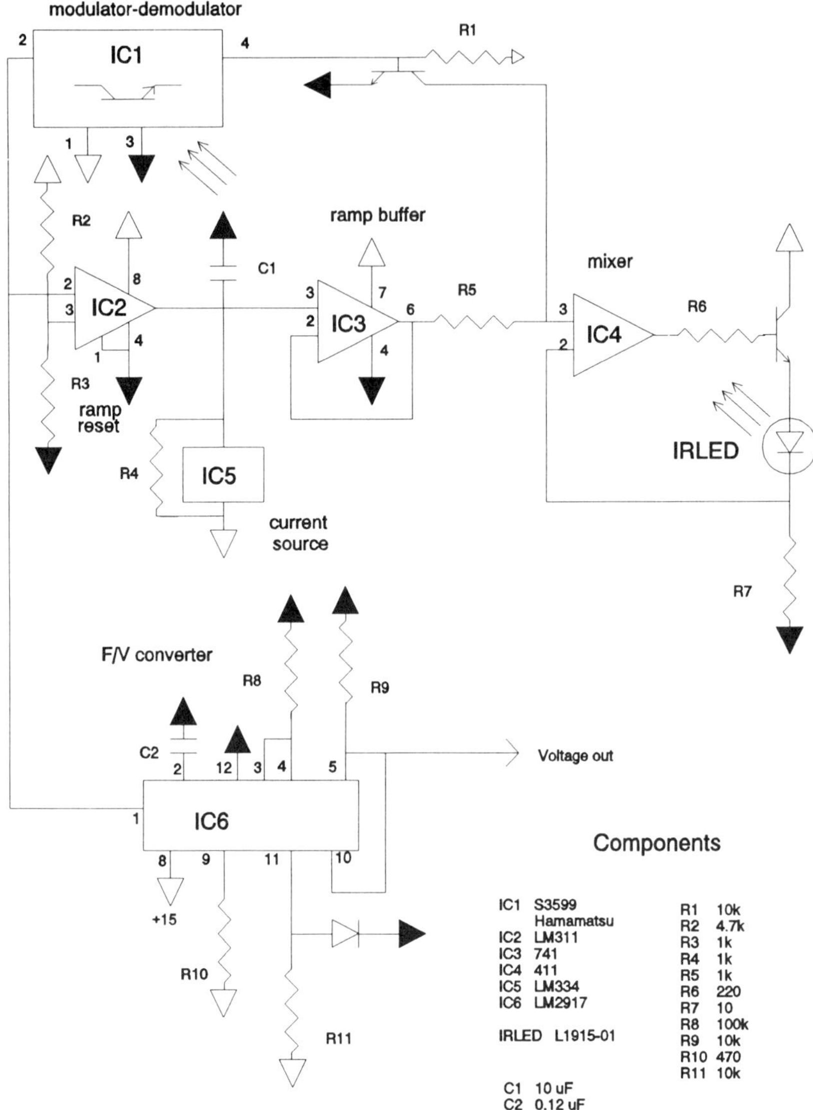

FIGURE 4. Light transmittance probe electronics.

B. Procedures

Shear stresses in the annular reactors and in the loop system were approximately matched at the start of an experiment. First, torque was measured for an annular reactor rotational speed of 175 rpm. Then shear stress was calculated at the inner cylinder surface using Equation 1:

$$\tau_{ar} = \frac{T}{R*A} \tag{1}$$

where τ_{ar} = shear stress at the cylinder wall ($ML^{-1}T^{-2}$)
A = surface area of the cylinder (L^2)
T = torque (ML^2T^{-2})
R = cylinder radius (L)

Shear stress in the pipe was then matched to shear stress in the annular reactor by adjusting the fluid velocity using the relationship in Equation 2,

$$\tau_{ar} = \tau_{pipe} = f\rho\frac{V^2}{8} \tag{2}$$

also using the Blasius equation:

$$f = \frac{0.316}{Re^{0.25}} \tag{3}$$

and the Reynolds number equation:

$$Re = \frac{Vd}{\nu} \tag{4}$$

If Equations 3 and 4 are substituted into Equation 2, then

$$\tau_{pipe} = \frac{0.316}{8\left(\frac{Vd}{\nu}\right)^{0.25}} \rho V^2 \tag{5}$$

or

$$\tau_{pipe} = 0.0395\ \rho^{1.75}\nu^{0.25}d^{-0.25} \tag{6}$$

where f = friction factor (dimensionless)
Re = Reynolds number (dimensionless)
V = velocity (LT^{-1})
d = tube diameter (L)
ν = kinematic viscosity (L^2T^{-1})
ρ = density of water (ML^{-3})
τ_{pipe} = shear stress at pipe wall ($ML^{-1}T^{-2}$)

Since both ρ and ν are temperature dependent, the appropriate values at 23.9°C ($\nu = 9.20E - 7$ m²/s and $\rho = 997.5$ kg/m³) are substituted into Equation 5.

Rearranging and solving for V resulted in the desired pipe flow velocity of 0.91 m/s. Due to an error in calculation, the actual velocity used was 1.18 m/s — 30% larger than the desired velocity. Residence times in each of the two subsystems were under 15 s. As a consequence, both systems were exposed to the same water quality throughout each experiment.

The majority of data was obtained over a period of several months. The system was run continuously over this period after inoculating with a single sample of an undefined mixed population. Between experiments the surfaces of interest — annular reactors, HTR section, DP section, mass accumulation section — were cleaned with a fiber brush until measured variables returned to baseline values. This procedure insured that rapid attachment and growth would occur.

III. RESULTS

Results are presented in two parts; the first part compares data measured at the end of each experiment, while the second part compares data collected continuously during each experiment. Each experiment lasted approximately 7 d.

A. End-of-Run Comparisons

A plot of annular reactor dry biofilm mass vs. tube dry biofilm mass on a per unit area basis is shown in Figure 5. The slope of the plot of annular reactor mass vs. tube mass was 0.69 and the r^2 is 0.94.

End-of-run HTR, DP, and torque (Figures 6 and 7), on the other hand, were not well correlated with mass accumulation, nor were HTR or DP well correlated with torque (Figure 8).

B. Continuous Annular Reactor and Loop System Comparisons

A graph of a typical run is seen in Figure 9. HTR rose rapidly, fell, and then started to rise again, often with considerable fluctuation. DP rose rapidly, but not as fast as HTR. It eventually leveled off or continued rising at a slower rate. Torque showed a constant gradual rise throughout the experiment.

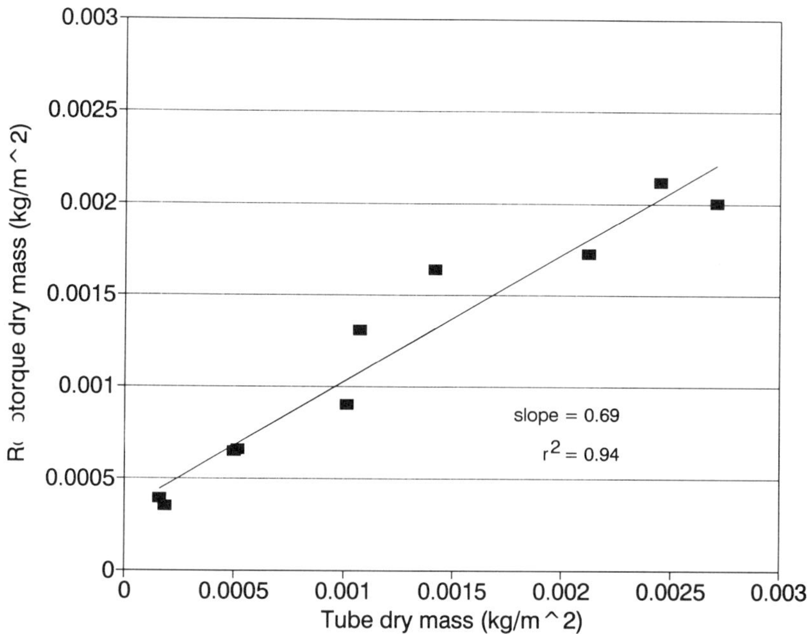

FIGURE 5. Annular reactor dry mass vs. tube dry mass.

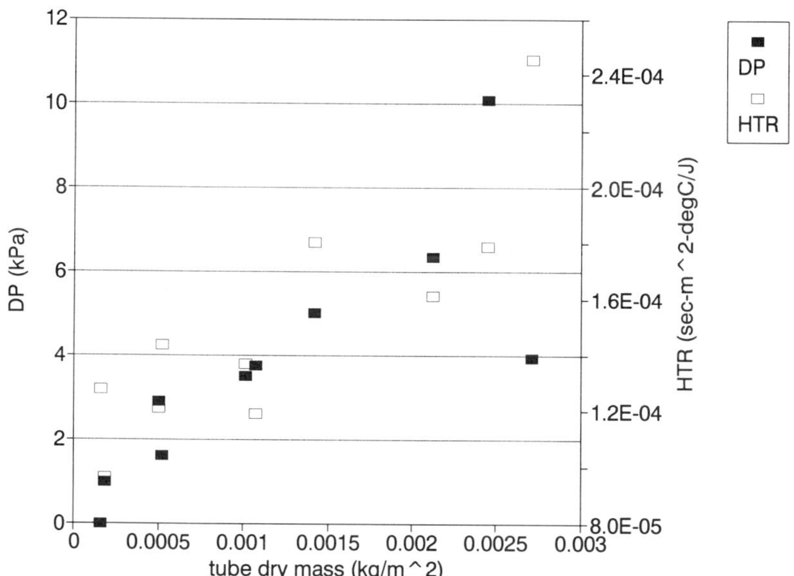

FIGURE 6. Differential pressure and heat transfer resistance vs. tube dry mass.

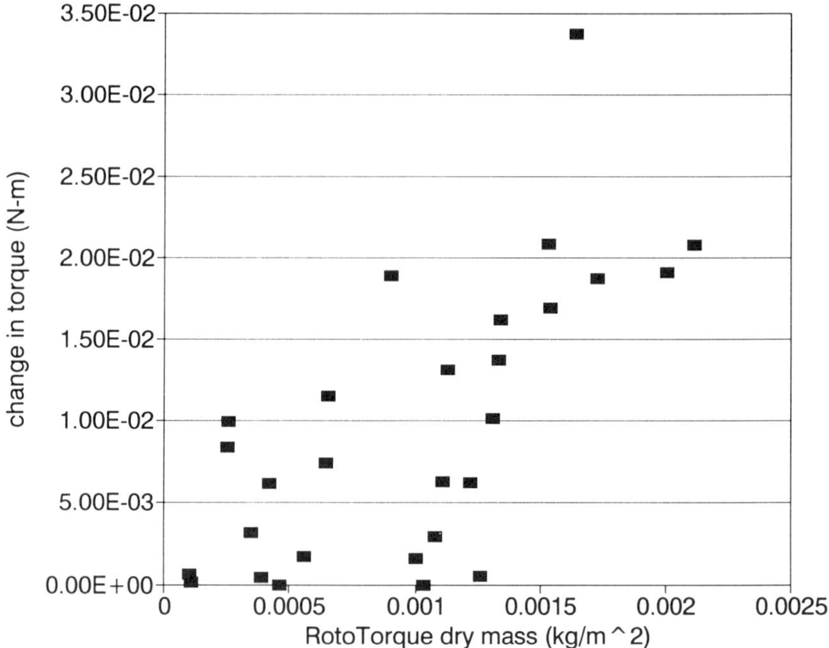

FIGURE 7. Torque vs. annular reactor dry mass.

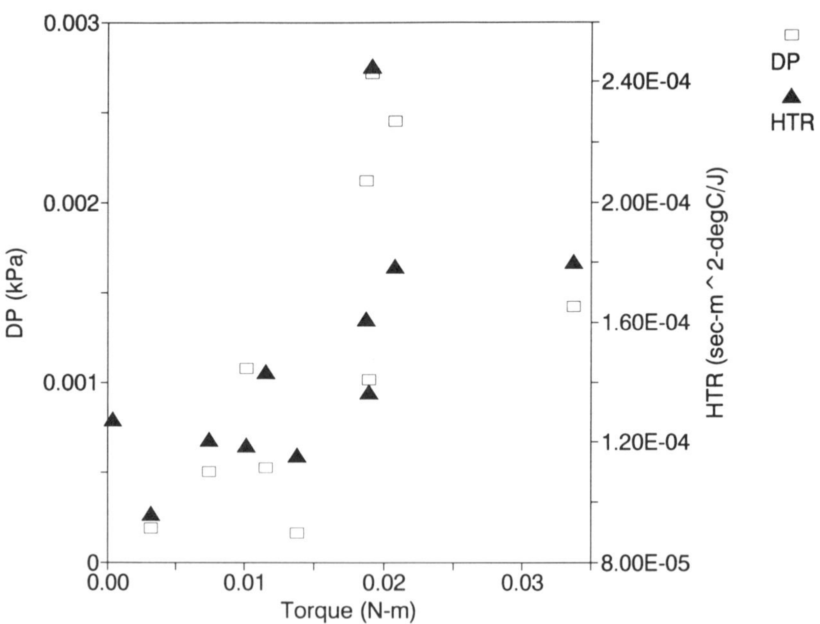

FIGURE 8. Differential pressure and heat transfer resistance vs. torque.

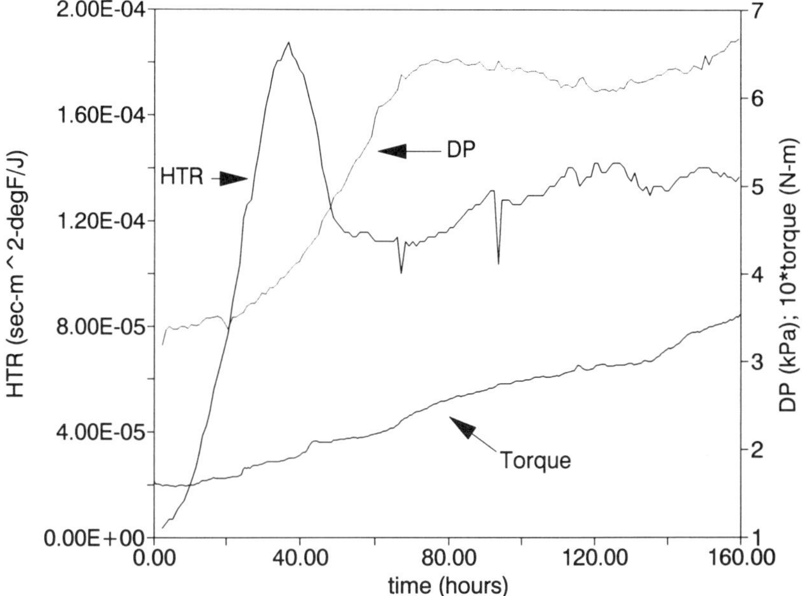

FIGURE 9. Typical experiment showing torque, heat transfer resistance, and differential pressure.

When two annular reactors were operated in parallel, their torque measurements tracked each other remarkably well (Figure 10). The computer-recorded data show very little noise.

Since the light-transmittance probe projected its beam against the rotating inner cylinder, it measured average biofilm density at the inner rotating cylinder wall. At the same time it also measured localized fouling at the outer wall, since it was mounted at a single point in that wall. Light-transmittance data points were a composite of these two measurements. Continually collected light-transmittance data, while showing little noise, did not correlate well with torque (Figure 11).

IV. DISCUSSION

Mass accumulation is a primary measure of fouling. It is important for developing models of fouling, and determining mass balances on fouling surfaces. The end-of-run areal mass comparisons in this investigation resulted in a slope of 0.69 (dry mass per unit area in the annular reactor/dry mass per unit area in the loop) with a correlation coefficient of 0.94. This relatively good correlation indicates that the annular reactors were successful at simulating biomass accumulation in the loop system and may be useful in predicting fouling in industrial tubular flow systems.

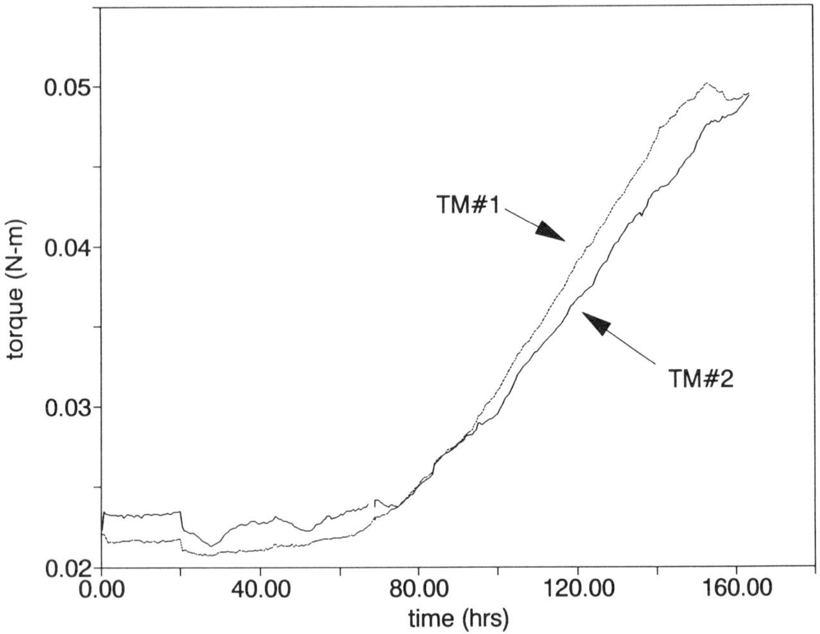

FIGURE 10. Two annular reactors operated in parallel.

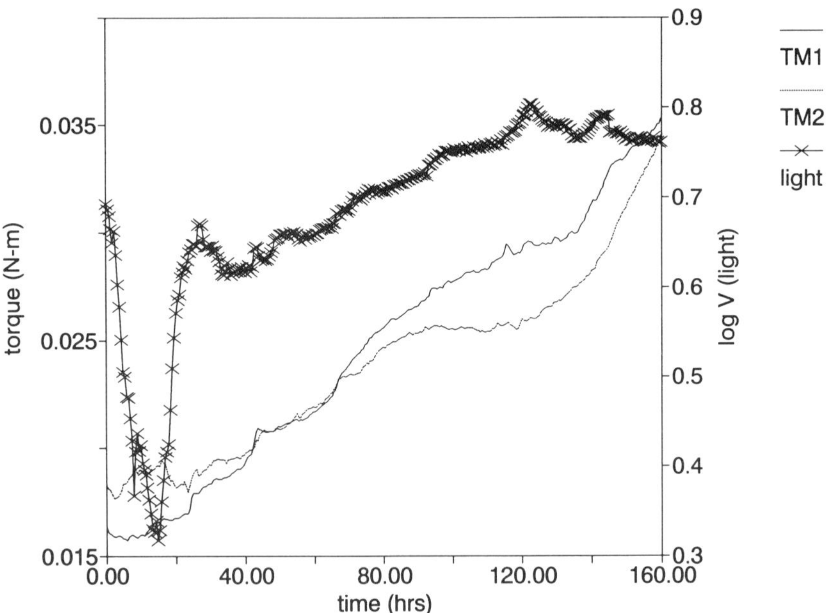

FIGURE 11. Torque and light transmittance during a run.

Ideally, a slope of 1.0 was expected, since chemical and hydrodynamic environments were matched. The low slope may have been a consequence of too high a flow rate in the loop system; a calculation error resulted in operating the loop at a flow rate 30% higher than intended. This high flow rate could have resulted in a thinner viscous sublayer and greater transport rates for nutrients and oxygen to the loop biofilm. The higher influx of nutrients and oxygen, and consequent higher growth rate, would have produced a higher mass accumulation in the loop system. Lewandowski and Walser[9] observed that biofilm thickness increased with increasing shear stress to a maximum, and then decreased with further increase in shear stress. Their maximum occurred at a shear stress of about 0.45 N/m^2. Since this was considerably below the 2.7 N/m^2 shear stress calculated for our system, we should have observed higher rates of detachment due to shear rather than increased mass accumulation. The difference could be due to individual properties of different biofilms.

Unfortunately, the successful correlation seen above was limited to mirroring mass accumulation. Plots of end-of-run HTR, DP, and torque as functions of the relevant mass concentrations resulted in significantly poorer correlations. Biofouling effects (BEs), such as HTR, DP, and torque, are functions not only of mass accumulation, but of the biofilm hydrodynamic environments as well. Density, thickness, and roughness are all biofilm characteristics which strongly influence BEs. Since biofilms and their hydrodynamic environments may modify each other in unpredictable ways, these BEs may not show strong correlations to mass. Consequently, it was not surprising to find that increase in torque (1 BE) was not strongly correlated with either HTR or DP.

A second goal, that of finding a continuously measurable annular reactor BE which correlates well with loop system BEs, was not successful. Biofilm light transmittance measured in the annular reactor was not successful at predicting HTR or DP in the loop system.

This study made it clear that one must be careful when interpreting biofouling monitor data. The reactor may successfully mirror growth or accumulation of biofilms in a tubular system, while individual biofouling probes in the same reactor may fail in predicting BEs. In this case the annular reactor was successful in simulating accumulation of mass in a pipe system, while the biofouling probes failed to correlate with PD or HTR.

An encouraging finding — that two torque monitors closely track each other when annular reactors are run in parallel — suggests that torque can be used in comparative studies, e.g., testing biocides or control strategies.

Torque might also be used as an empirical trigger, warning of onset of biofouling.

V. CONCLUSIONS

The annular reactor was successful in simulating biofouling in a circular conduit or tubular system. This was demonstrated by good correlation between mass accumulation in the two systems.

Torque, while showing good sensitivity to biofouling and close tracking in duplicate monitors, exhibited somewhat poorer correlation with mass accumulation and with secondary fouling parameters such as HTR and DP.

Light transmittance through biofilm in the annular reactor was also a sensitive measure of biofouling, but, being a localized measurement, it demonstrated considerably more variability than the other biofouling probes.

Annular reactors exhibited a potential for practical application in industry. They may be operated in parallel for comparative testing of biocides or biofouling control strategies. They may also be used to simulate mass accumulation in a tubular system.

REFERENCES

1. **Characklis, W. G., Trulear, M. G., Bryers, J. D., and Zelver, N.,** Dynamics of biofilm processes: methods, *Water Res.*, 16, 1207, 1982.
2. **Characklis, W. G. and Marshall, K. C.,** Biofilms: a basis for an interdisciplinary approach, in *Biofilms*, Characklis, W. G. and Marshall, K. C., Eds., John Wiley & Sons, New York, 1990, chap. 1.
3. **Warwood, B., Lee, W., Zelver, N. and Characklis, W. G.,** Evaluation of Instruments and Methods for Monitoring Fouling and Corrosion in Industrial Water Systems, Report prepared for Amoco Corporation, Betz Industrial, and Chevron Oil Field Research Company, by the Center for Interfacial Microbial Process Engineering, Montana State University, January 1991.
4. Biofouling Detection Monitoring Devices: Status Assessment, EPRI CS-3914, Project 2300–1, Final Report, March 1985.
5. **Characklis, W. G.,** Biofilms: a basis for an interdisciplinary approach, in *Biofilms*, Characklis, W. G. and Marshall, K. C., Eds., John Wiley & Sons, New York, 1990, chap. 3.
6. **Kornegay, B. H. and Andrews, J. F.,** Characteristics and Kinetics of Biological Film Reactors, FWPCA Final Report, Research Grant WP-01181, Department of Environmental Systems, Clemson University, Clemson, SC, 1967.
7. **Characklis, W. G.,** Effect of Hypochlorite on Microbial Slimes, Ph.D. thesis, Johns Hopkins University, 1970.
8. **Trulear, M. G.,** Dynamics of Biofilm Processes in an Annular Reactor, Ph.D. thesis, Rice University, Houston, 1980.
9. **Lewandowski, Z. L. and Walser, G.,** Influence of Hydrodynamics on Biofilm Accumulation, Environmental Engineering. Proceedings of the 1991 Specialty Conference. Krenkel, P. A., Ed., Sponsored by the Environmental Engineering Division of the American Society of Civil Engineering, Reno, NV, July 1991.

Section II

BIOFOULING

9

Analysis of Biofilm Disinfection by Monochloramine and Free Chlorine

Thomas Griebe, Ching-I Chen, Rohini Srinivasan, and Philip S. Stewart

I. INTRODUCTION

A biofilm consists of microbial cells embedded in an extracellular polymeric substance matrix attached to a substratum.[1] Prevention and control of biofilm accumulation is a challenging task to industry. The biofouling problem has usually been alleviated by application of oxidizing biocides, such as free chlorine (hypochlorous acid/hypochlorite), which have been used for decades. Although most of the microorganisms in industrial systems are associated with surfaces, biofilms have historically received less attention than have suspended (planktonic) microorganisms. Most disinfection studies have been carried out with suspended bacteria, thereby neglecting the influence of the substratum and cell aggregation.

It has been shown that various biocides are less effective against biofilm cells than dispersed cells of the same organism.[2–4] Chen et al.[5] demonstrated that *Pseudomonas aeruginosa* cells grown in a biofilm were less susceptible to biocide treatment, in part because the intact biofilm matrix constituted a diffusion barrier for the biocide. However, the mechanism of disinfection of biofilm organisms is not yet fully understood due to the lack of accurate methodology and to the misleading assumption that biocide reaction kinetics of planktonic and biofilm bacteria are similar.

0-87371-928-X/94/$0.00+$.50

Previous comparisons between free chlorine and monochloramine for the inactivation of biofilm bacteria in pipe systems[6–8] were inconclusive with respect to the superior disinfectant. The objective of this study was to compare the disinfection efficacy of the oxidative biocides, free chlorine and monochloramine, on intact biofilms and mechanically dispersed biofilms of *P. aeruginosa*. A further goal of this study was to develop an experimental methodology for the testing of biocide efficacy against biofilms using a bioreactor system.

II. MATERIALS AND METHODS

A. Experimental Biofilm System

A RotoTorque annular reactor was used for the biofilm experiments. As shown in Figure 1, the reactor consists of a stationary outer casing and a rotating inner cylinder, both made of polycarbonate. Twelve removable polished stainless steel slides (SS304, 1.7 × 19 cm) form an integral part of the inside wall of the outer cylinder and permit sampling of biofilms growing on them. The bulk liquid is completely mixed by virtue of the cylinder rotation and angled draft tubes bored through the solid inner cylinder. Prior to inoculation, the system was autoclaved at 121°C (17 psi) for 25 min. All influent solutions except dilution water were autoclaved at 121°C (17 psi) for 4 h. Dilution water was sterilized via filtration using two capsule filters (0.2 μm, Gelman Sciences) in series connected to the RotoTorque (Figure 1). Concentrated glucose solution was sterilized via filtration (0.2 μm) before it was added to the sterilized, concentrated mineral salt solution. Three reservoirs containing dilution water (distilled), substrate/mineral solution, and phosphate buffer solution (pH = 7.0), respectively, were connected to the RotoTorque through silicone tubing (Masterflex 6411–14 and -16) and peristaltic pumps (Masterflex 7553–30, Cole-Parmer). The flow rates of the dilution water, concentrated substrate-mineral solution, and buffer solution were 30, 1, and 1 ml/min, respectively. The composition of the resulting influent was the same as used by van der Wende[9] with glucose (20 mg/l) and potassium nitrate (13.6 mg/l) as the carbon and nitrogen sources, respectively. Temperature was maintained at 25 ± 0.5°C with a thermostated water bath. The rotation of the inner drum in the RotoTorque was set at 150 rpm (92 cm/s) throughout all experiments, corresponding to a shear stress of approximately 1.4 N/m^2.

B. RotoTorque Experiments

A frozen stock culture (1 ml) of *P. aeruginosa* (10^8/ml) was inoculated into a 250-ml Erlenmeyer flask containing 50 ml of medium with the same

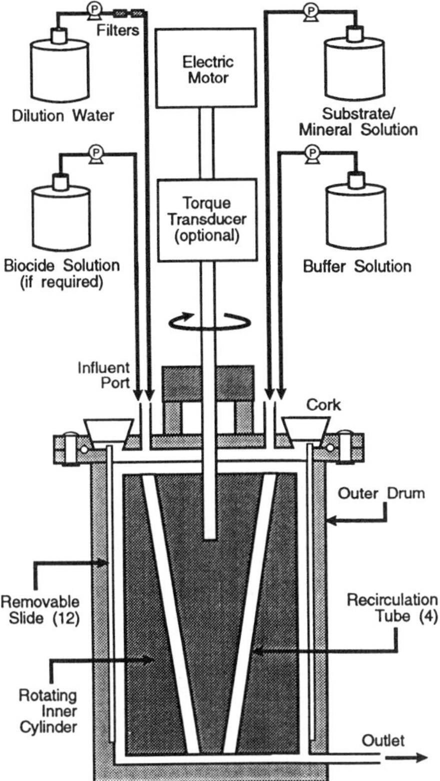

FIGURE 1. The RotoTorque system.

composition as the RotoTorque feed. The flask was incubated with shaking at 25 ± 0.5°C and 150 rpm (G24 Incubator Shaker, New Brunswick Scientific) for 24 h. The log phase culture (10 ml) was inoculated into the RotoTorque filled with medium (600 ml). After 24 h of batch cultivation at 25 ± 0.5°C and 150 rpm, all continuous influent flows were started. Since the dilution rate was very high ($3.2h^{-1}$), no planktonic cells were retained. Essentially all planktonic cells were detached biofilm cells.

Two RotoTorques were treated for 1 h with 4 mg/l monochloramine by a pulse injection of concentrated monochloramine, resulting in 4 mg/l monochloramine in the bulk fluid, followed immediately by a step feeding of 4 mg/l monochloramine (contained in the influent) after biofilms were accumulated to steady state (7 to 8 d after inoculation). One RotoTorque was used for *in situ* biofilm disinfection study and the other was used for biocide residual measurements.

C. Batch Experiments

Batch experiments were conducted to investigate disinfection of biofilm cells vs. suspended cells. Intact biofilms (slides) were removed from the RotoTorque, scraped into 150 ml of phosphate buffer (pH = 7.0, 5 m*M*), homogenized for 1 min with a Tekmar Tissumiser™, and gently stirred after addition of 1, 2, 3, or 4 mg/l monochloramine or 1 or 2 mg/l free chlorine (final concentration). At various times, samples were withdrawn, and the biocide was inactivated with 10 mg/l sodium thiosulfate and analyzed for viable cells. Aliquots were also removed to determine the residual monochloramine concentration.

D. Analytical Methods

Samples were taken daily from the effluent (representing the bulk fluid in the RotoTorque) and biofilm slides. The biofilm (7 to 8 d old) was scraped from the stainless steel slide into 150 ml of phosphate buffer (pH = 7.0, 5 m*M*) using a sterile cell scraper (Fisher Scientific) and homogenized via a Tekmar Tissumiser™ for 2 min with 100% power input. Effluent samples were also homogenized to disperse cell aggregates. Viable cells were counted by plating dilutions (in 5 m*M* phosphate buffer, pH = 7.0) of homogenized samples in triplicate (effluent and biofilm) on R2A™ agar (Difco). The results were expressed as number of colony forming units (cfu) per ml or m^2. Monochloramine-treated samples were diluted by phosphate buffer containing 10 mg/l sodium thiosulfate (biocide neutralizer) prior to plating. Total cells of homogenized biofilms and effluent samples were counted according to Griebe[10] by epifluorescent microscopy using a double staining procedure. Samples were filtered onto 25-mm-diameter (0.2-μm pore size), black polycarbonate membrane filters (Nuclepore) and stained successively with 4′,6-diamidino-2-phenylindole (DAPI, 10 μg/ml) for 10 min and acridine orange (0.1 μg/ml) for 3 min. The bacteria were counted randomly in at least 20 microscope fields using an Olympus BH2 epifluorescence microscope with DPlan Apo UV objectives and the U (DAPI) dichroic mirror cube.

E. Disinfectant Preparation and Detection

Concentrated monochloramine stock solution was prepared using a 3:1 molar ratio of ammonia (ammonium chloride, Fisher) to free chlorine (pH 9.0). Monochloramine concentration was measured by the DPD colorimetric method using a Hach test kit (Hach Co., Model CN-66) for total chlorine (detectability limit 0.05 mg/l). Since monochloramine was the only chlorine compound and there was no other strong oxidant in the system, monochloramine would not be expected to be transformed

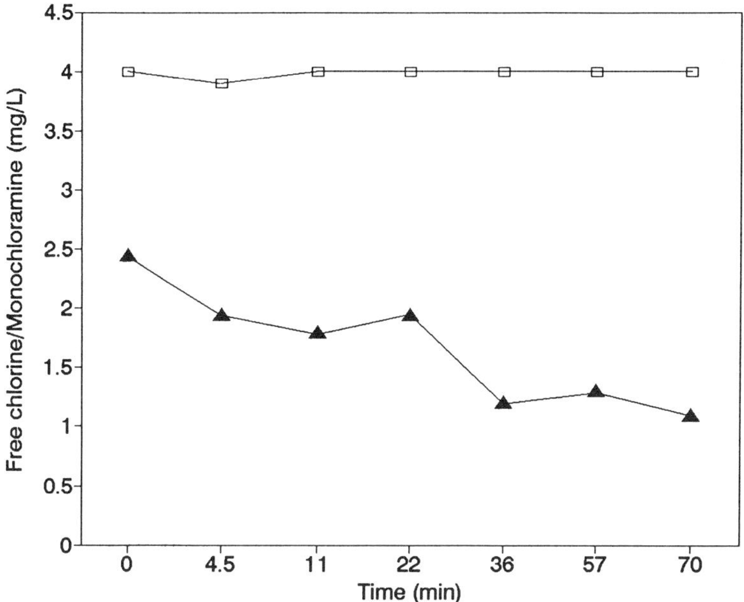

FIGURE 2. Biocide concentration in RotoTorque influent (stirred batch) after the addition of 4 mg/l monochloramine and 2.4 mg/l free chlorine (van der Wende[9]). (□) Monochloramine, (▲) free chlorine.

to dichloramine and trichloramine. Therefore, the total chlorine measurement was assumed to represent the concentration of monochloramine.

III. RESULTS AND DISCUSSION

A. Monochloramine and Free Chlorine Demands of the RotoTorque System

To determine the monochloramine demand of a sterile, clean RotoTorque system, experiments were conducted using exactly the same RotoTorque system, but with no microorganisms. The sterilized system was fed an influent containing 4 mg/l of monochloramine. The monochloramine concentration in the effluent remained at 4 mg/l, indicating that the RotoTorque system had no monochloramine demand under the stated operating conditions. The results were compared with those from van der Wende[9] treating the same system with free chlorine. Figure 2 shows the free chlorine demand of the RotoTorque system. A concentration of 2.5 mg/l free chlorine dropped about 50% within 60 min because free chlorine reacted with oxidative matter of the influent.

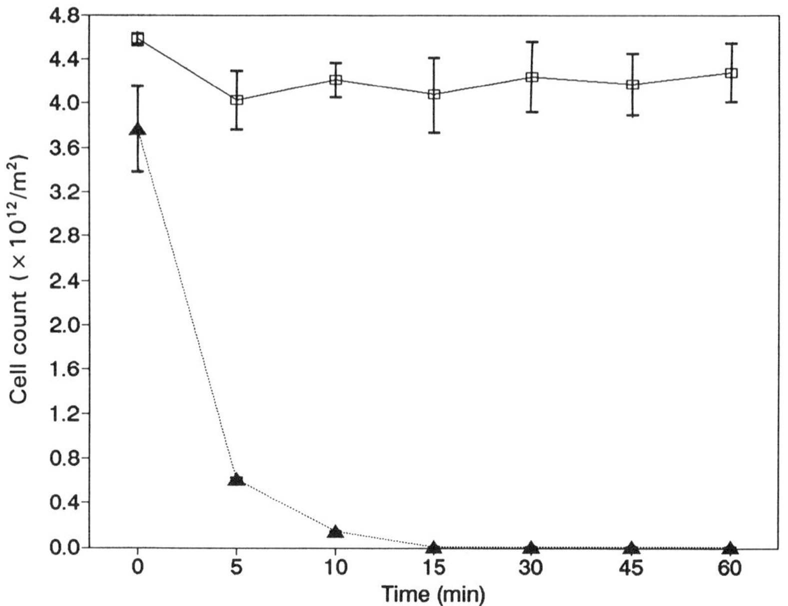

FIGURE 3. *In situ* biofilm dynamics during treatment with 4 mg/l monochloramine. (▲) Viable cell count, (□) total cell count. Time t = 0 corresponds to the beginning of biocide addition.

B. Disinfection of Biofilm Cells by Monochloramine and Free Chlorine

The results of *in situ* biofilm disinfection with 4 mg/l monochloramine are shown in Figure 3. The number of viable cells in the biofilm on the stainless steel surface decreased from 3.76 ($\pm$0.38) $\times$ 10^{12} cfu/m^2 to 1.68 ($\pm$0.13) $\times$ 10^{10} cfu/m^2 at 15 min, and to 2.52 ($\pm$0.24) $\times$ 10^{8} cfu/m^2 a decrease of about four orders of magnitude, at 60 min (Figure 3). Duplicate biofilm samples were taken at the first (t = 0 min) and last (t = 60 min) samplings to confirm the efficacy. There was no evidence of detachment or sloughing because the areal density of total cells (4.22 [$\pm$0.16] $\times$ 10^{12} cells/m^2) did not change during the biocide treatment. The residual monochloramine concentration in the bulk fluid was constant at 0.45 mg/l after the first sampling at 2.5 min through the end of the treatment.

The disinfection of biofilms by free chlorine is shown in Figure 4. A dose of 5.8 mg/l free chlorine reduced the areal density of viable cells in *P. aeruginosa* biofilms by one order of magnitude during the 1-h treatment. The residual free chlorine was 0.15 mg/l after the first sampling at 2.5 min. This indicates that the biofilm has higher free chlorine demand than monochloramine demand. Disinfection using a higher dose

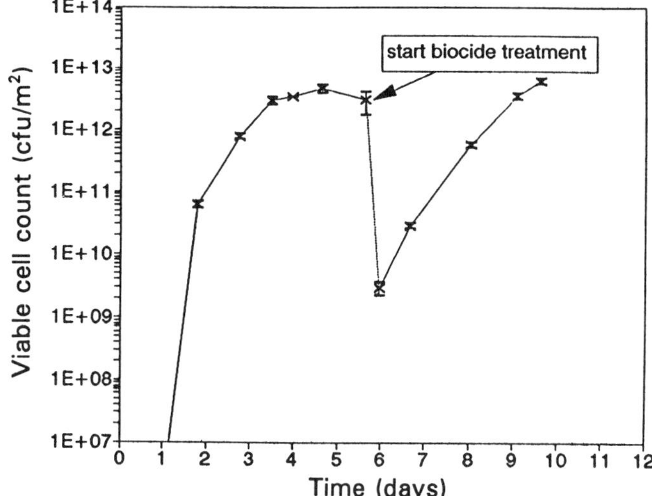

FIGURE 4. Viable cell numbers of *in situ* biofilm during treatment with a dose of 10.8 mg/l free chlorine (van der Wende).[9]

of free chlorine was also studied. A dose of 10.8 mg/l free chlorine inactivated viable cells by three orders of magnitude after the same interval. The residual free chlorine was at 0.5 mg/l after 2.5 min and slowly increased to 1.3 mg/l after 60 min. Therefore, a dose of 4 mg/l monochloramine was even more effective for biofilm inactivation than a dose of 10.8 mg/l free chlorine. van der Wende[9] also reported significant detachment and sloughing during the treatment with free chlorine, whereas no cell detachment was observed when applying monochloramine. The results of these studies indicate that monochloramine was more effective than free chlorine for inactivating biofilms of *P. aeruginosa*.

LeChevallier et al.[6] have demonstrated that biofilms in a model pipe macrosystem were successfully controlled using monochloramine levels ranging from 2 to 4 mg/l, but that free chlorine residuals from 3 to 4 mg/l were ineffective for biofilm disinfection. van der Wende et al.[11] hypothesized that the biofilm environment protects cells against the activity of chlorine by diffusional resistance and neutralization of chlorine through the reaction with biofilm and pipe wall materials. Planktonic cells find no such protection. LeChevallier et al.[3] also indicated that monochloramine and free chlorine might act differently at surfaces of biofilms of various bacteria.

C. Disinfection of Homogenized Biofilm Cells by Monochloramine and Free Chlorine

Intact biofilms on stainless steel from the RotoTorques were homogenized prior to disinfection. The disinfection of mechanically dispersed

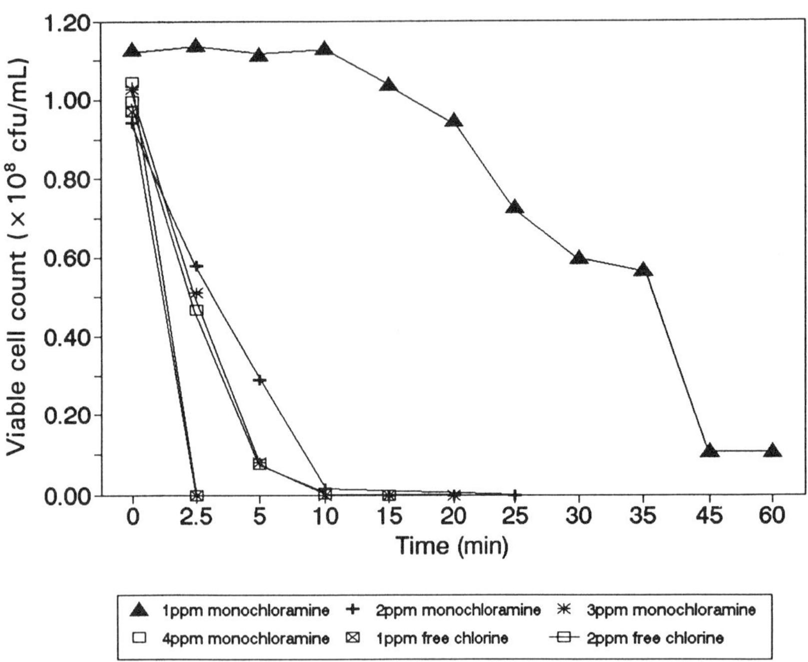

FIGURE 5. Disinfection of homogenized biofilms with monochloramine and free chlorine.

biofilm cells with either 3 or 4 mg/l monochloramine resulted in 99% killing of viable cells within the first 5 min of exposure (Figure 5). For all treatments (1, 2, 3, and 4 mg/l monochloramine), the biocide demand remained steady after 2.5 min at 0.45 mg/l. For the 2 mg/l monochloramine treatment, 99% of viable cells were killed at 15 min. Dispersed biofilm cells exposed to 1 mg/l monochloramine were not affected by the biocide within the first 15 min. However, the viability dropped moderately to 9.73% at 45 min and remained steady thereafter. These results indicate that there was a threshold level (between 1 and 2 mg/l) at which monochloramine was effective for a rapid inactivation of suspended biofilm cells.

Disinfection of suspended biofilm cells with 1 and 2 mg/l free chlorine (residuals 0.8 and 1.85 mg/l, respectively) resulted in 100% killing of viable cells within the first 2 min. Thus, free chlorine inactivated suspended biofilm cells more rapidly than monochloramine.

The commonly used CT coefficient (defined as biocide residual concentration [mg/l] multiplied by treatment time [min] to achieve 99% inactivation) enables comparison of various biocide activities. The disinfection data presented above were used to calculate CT values for monochloramine and free chlorine (Table 1). The mean monochloramine

TABLE 1. Comparative Efficiency of Monochloramine and Free Chlorine for 99% Inactivation of Intact Biofilms and Homogenized Biofilms

	CT value for monochloramine[a]	CT value for free chlorine
Intact biofilms	6.7 mg·min/l	69 mg·min/l[b]
Homogenized biofilms	28 mg·min/l	<1.6 mg·min/l

[a] CT is defined as biocide residual (mg/l) multiplied by treatment time (min) for achieving 99% inactivation of viable bacteria.
[b] Value was adopted from van der Wende.[9]

CT value for homogenized biofilm cells was 28 mg·min/l, whereas intact biofilms had a CT value of 6.75 mg·min/l. Nearly 100% inactivation of homogenized biofilm cells was achieved with a free chlorine residual of 0.8 or 1.85 mg/l after 2 min. The CT value for free chlorine was below 1.6 mg·min/l. Therefore, free chlorine is more effective than monochloramine for disinfecting homogenized (mechanically dispersed) biofilm cells. LeChevallier et al.[6] observed the same phenomena: monochloramine CT values for unattached bacteria were higher than for biofilms, and free chlorine was less effective for biofilm control.

D. Regrowth after Treatments with Monochloramine and Free Chlorine

The growth of biofilms reached steady state at 144 h, as revealed by steady viable cell and total cell counts in the bulk fluid (Figure 6). Viable and total cell counts were very consistent and indicated that about 95% of total cells were viable. After 1 h of treatment with 4 mg/l monochloramine, resulting in inactivation of 99.99% of viable cells, the regrowth of survived biofilm cells to pretreatment level required only 48 h (Figure 6, curve after 216 h). Based on the steady state biofilm-specific cell growth rate of 0.071 h^{-1}, calculated by Chen et al.,[5] and ignoring cell detachment during regrowth, predicted biofilm recovery would take about 141 h. If detachment occurs, this period would be longer. Thus, cell growth rates were considerably higher during biofilm recovery than at steady state.

The regrowth of biofilms after free chlorine treatment was similar. It took 48 h to reach the pretreatment level (Figure 4). Both regrowth studies showed that biofilm recovery to steady state was about three times faster than initial biofilm development. This was probably caused by structural change of biofilms after treatment, which facilitated nutrient transport through the film, resulting in a higher nutrient level within the biofilm and a higher specific growth rate for the survived cells. These results support the view that biofilms represent a protected

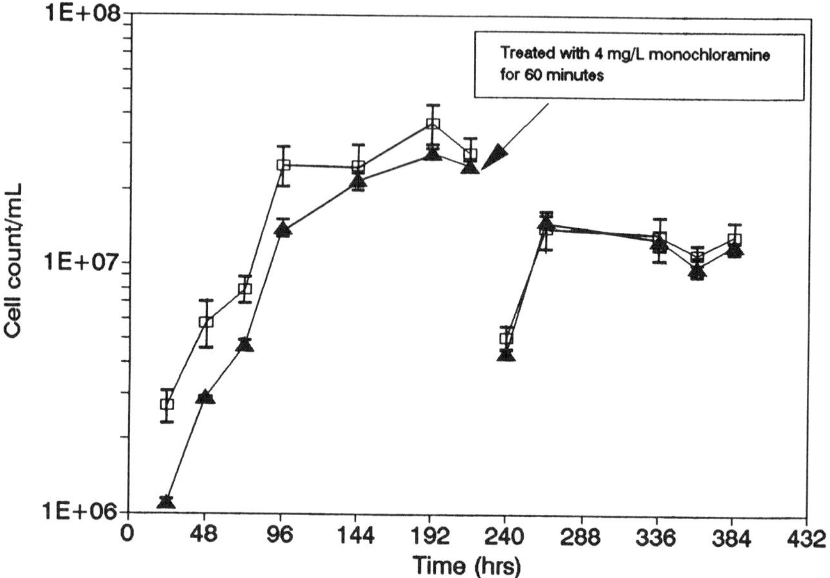

FIGURE 6. Changes of cell counts in the bulk fluid during growth of biofilms and regrowth after biocide treatment with 4 mg/l monochloramine, (▲) viable cell count, (□) total cell count.

niche where bacteria can survive stressful environmental changes. These results are consistent with field experiences which suggest that uncontrolled biofilm organisms can cause severe biofouling problems in industrial systems.

IV. CONCLUSION

1. Monochloramine was not reactive toward system components, whereas free chlorine showed a high reactivity with the medium.
2. Monochloramine was more effective than free chlorine for inactivation of *P. aeruginosa* biofilms. There was no significant biofilm detachment during monochloramine treatment, whereas free chlorine caused sloughing.
3. Mechanically dispersed biofilms were disinfected more rapidly by free chlorine than by monochloramine.

ACKNOWLEDGMENT

This research was funded by the National Science Foundation. Cooperative Agreement ECD-8907039 between the National Science Foundation and Montana State University is acknowledged. The authors would like to thank A. Camper and Dr. G. Geesey for providing several helpful suggestions. The authors are grateful to Dr. E. van der Wende for permitting the use of data from his dissertation.

REFERENCES

1. **Characklis, W. G. and Marshall, K. S.,** *Biofilms,* John Wiley & Sons, New York, 1990.
2. **LeChevallier, M. W., Cawthon, C. D., and Lee, R. G.,** Factors promoting survival of bacteria in chlorinated water supplies, *Appl. Environ. Microbiol.,* 54, 649, 1988.
3. **LeChevallier, M. W., Lowry, C. D., and Lee, R. G.,** Inactivation of bacterial biofilms, *Appl. Environ. Microbiol.,* 54, 2492, 1988.
4. **Sharma, A. P., Battersby, N. S., and Stewart, D. J.,** Techniques for the evaluation of biocide activity against sulphate-reducing bacteria, in *Preservatives in the Food, Pharmaceutical and Environmental Industries,* Board, R. G, Allwood, M. C., and Banks, J. G., Eds., Blackwell Scientific, Oxford, 1987.
5. **Chen, C.-I., Griebe, T., and Characklis, W. G.,** Biocide action of monochloramine on biofilm systems of *Pseudomonas aeruginosa, Biofouling,* 7, 1, 1993.
6. **LeChevallier, M. W., Lowry, C. D., and Lee, R. G.,** Disinfecting biofilms in a model distribution system, *J. Am. Water Works Assoc.,* 82, 87, 1990.
7. **Neden, D. G., Jones, R. J., Smith, J. R., Kirmeyer, G. J., and Foust, G. W.,** Comparing chlorination and chloramination for controlling bacterial regrowth, *J. Am. Water Works Assoc.,* 84, 80, 1992
8. **Wolfe, R. L., Stewart, M. H., Liang, S., and McGuire, M. J.,** Disinfection of model indicator organisms in a drinking water pilot plant by using peroxone, *Appl. Environ. Microbiol.,* 55, 2230, 1989.
9. **van der Wende, E.,** Biocide Action of Chlorine on *Pseudomonas aeruginosa* Biofilm, Ph.D. dissertation, Department of Civil Engineering, Montana State University, Bozeman, 1991.
10. **Griebe, T.,** Experimentelle Untersuchungen zur Aggregatbildung, *Diplomarbeit,* Institut für Hydrobiologie und Fischereiwissenschaft, Universität Hamburg, 1991.
11. **van der Wende, E., Characklis, W. G., and Smith, D. B.,** Biofilms and bacterial drinking water quality, *Water Res.,* 23, 1313, 1989.

10

Effects of Disinfection on Attachment Strength of Bacteria on PVC

Frederico Lage-Filho and Slawomir W. Hermanowicz

I. INTRODUCTION

Interactions between a disinfectant and attached microorganisms have been studied by several researchers.[1–5] A general conclusion reached in these works is that attached microorganisms exhibit an increased resistance to disinfection. However, very little is presently known about the effects of prior disinfection on subsequent attachment of bacteria to surfaces. In our recent work[6] it was found that disinfection with chlorine and monochloramine adversely affected attachment of a mixed bacterial population. The results reported in this chapter focus on the distribution of attached bacterial cells as a result of varying hydrodynamic forces close to the surface.

II. EXPERIMENTAL METHODS AND MATERIALS

Detailed description of experimental methods and materials can be found in Lage-Filho[7] and Hermanowicz and Lage-Filho.[6] A brief outline of the procedures is presented in the following paragraphs.

0-87371-928-X/94/$0.00+$.50

A. Bacterial Population

A mixed heterotrophic bacterial population was used in all experiments. The population was isolated from well water at the Richmond Field Station, Richmond, CA, and consisted of six major species: *Flavobacterium, Pseudomonas cepacia, P. florescence, P. aeruginosa, Agrobacterium,* and *Chromobacterium*. The isolated culture was stored at −70°C. Before each experiment, diluted R2A medium (dilution 1:20) was inoculated with the stored culture and allowed to grow for 30 to 40 d at 20°C. After growth the cells were harvested and used for the experiments.

B. Experimental Procedure

In each experiment, a suspension of bacterial cells in test water (carbon-free phosphate buffer, pH = 6.9, ionic strength I = 0.008 *M*) was exposed to a disinfectant:

Free chlorine
- concentrations, 0.2 and 0.6 mg/l
- contact time, 10 and 30 min
- number of experiments, 40

Monochloramine
- concentrations, 1 and 4 mg/l
- contact time, 10 and 40 min
- number of experiments, 46

Control (no disinfectant)
- number of experiments, 20

After the specified contact time, water was dechlorinated with sodium thiosulfate and transferred to the attachment test apparatus. The same procedure was followed with nondisinfected cells (control); number of experiments, 20).

C. Attachment Test

After disinfection, an attachment test was immediately performed on the disinfected and nondisinfected cells. For this purpose a rotating-disk apparatus was used. The apparatus consisted of a PVC test disk (5.8 cm in diameter) mounted on a shaft of a variable-speed motor and immersed in test water in a 4-l jar.

The rotating disk offers a number of advantages for attachment studies. Under laminar flow conditions, the hydrodynamics are well defined.[8] Theoretical and experimental investigations[9–11] indicated that the mass transfer rate is uniform on the whole surface of the disk. At the same time, hydrodynamic shear stress on the disk surface increases linearly with the distance from the disk center.

Before each experiment a new disk was thoroughly cleaned and autoclaved.[7] The disk was rotated at 305 rpm for a specified period of time (generally 30 min, but for some experiments from 2 min to 2 h), allowing the suspended cells to attach to its surface. Special care was taken to minimize possible attachment artifacts due to different concentration of cells at the air-water interface. After the attachment test the disk was removed and bacterial cells were enumerated on its surface.

D. Enumeration of Attached Bacterial Cells

Since the objective of the study was to evaluate the effects of disinfection on attachment of viable cells, a modification of the plate count method was used for cell counting. A disk with attached bacteria was incubated under a layer of R2A agar for 5 d at 20°C. The developed colonies were counted under 10× stereo microscope. The counting was conducted using a grid of 6×6-mm squares arranged into six concentric ring-like regions on the disk.

III. RESULTS AND DISCUSSION

A. Cell Distribution on Surface

Since the total number of attached cells varied with experimental conditions and disinfection parameters,[6] the observed surface concentration of attached cells in each of the concentric regions (denoted by n) was divided by the average attached cell surface concentration for the disk (denoted by n_{avg}). These normalized values n/n_{avg} are presented as functions of dimensionless position r/R (r = distance from disk center, R = disk radius, 2.9 cm) in Figures 1 through 3 for nondisinfected, chlorinated, and chloraminated cells, respectively. The dotted lines in the figures represent cell distributions in an individual test and the solid lines describe average dimensionless concentrations at each distance from the disk center. A general trend is clearly shown in all three cases. The ratio n/n_{avg} decreased from disk center towards the edge. Martin and Bouwer[12] and Martin[13] reported a similar tendency in their studies of attachment of bacterial cells and latex particles to a rotating disk.

The decline of cell concentration toward the disk edge could be attributed to increasing hydrodynamic shear stress in the same direction and limited ability of the cells to withstand the stress. However, some colonization occurred even close to the disk edge, and no clear limit for attachment could be established. Moreover, all three sets of data (nondisinfected, chlorinated, and chloraminated) exhibited an intermediary

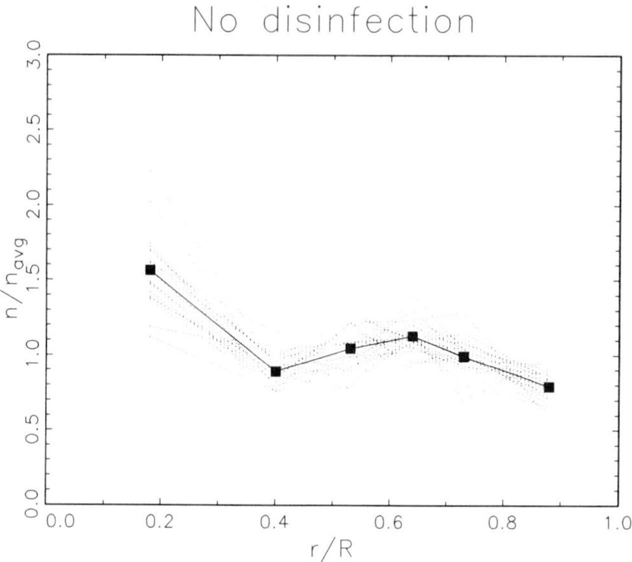

FIGURE 1. Normalized distribution of attached cells without disinfection.

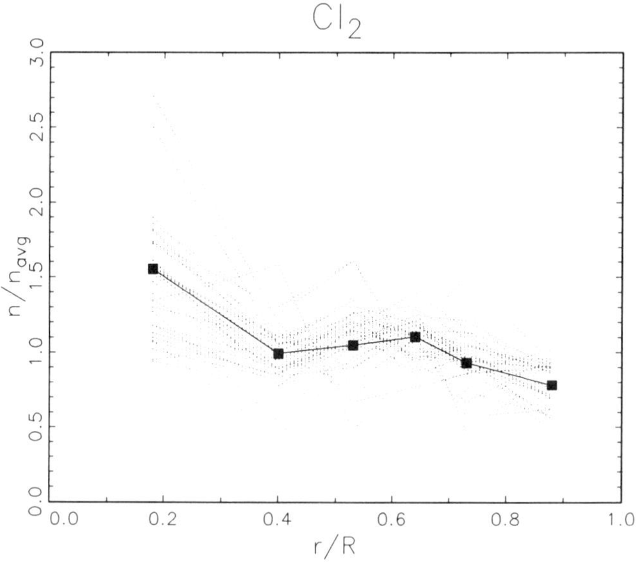

FIGURE 2. Normalized distribution of attached cells after chlorination.

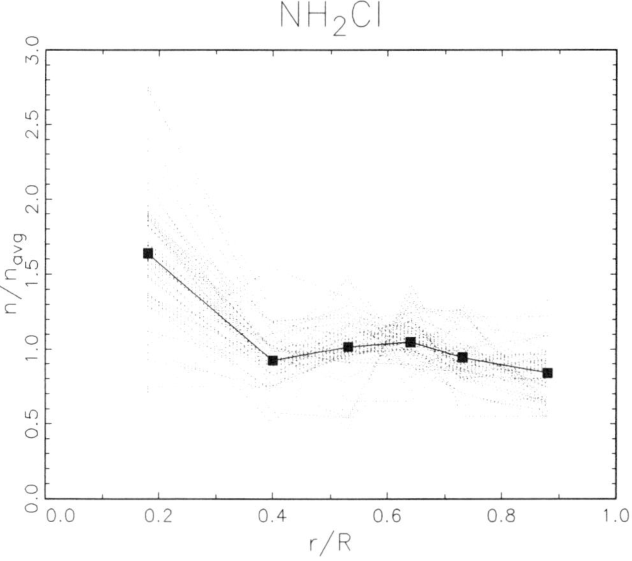

FIGURE 3. Normalized distribution of attached cells after chloramination.

maximum on n/n_{avg} at r/R of 0.62 or r = 1.8 cm. To test the existence of the intermediary maximum, the values of n/n_{avg} at r/R = 0.62 were compared with the n/n_{avg} values on both sides of this position, i.e., at r/R = 0.40 (r = 1.16 cm) and r/R = 0.72 (r = 2.09 cm) using paired-difference *t*-test. For each experiment the differences d_{ij} were calculated,

$$d_{ij} = (n/n_{avg})_i - (n/n_{avg})_j \tag{1}$$

where i and j refer to counting regions on the disk: i = 2 for r/R = 0.40, j = 4 for r/R = 0.62, and i = 5 for r/R = 0.72. The hypothesis that n/n_{avg} values were not statistically different was tested using the *t*-statistics:

$$t = \frac{d_{avg}\sqrt{N}}{S_d} \tag{2}$$

The results of the statistical analysis, presented in Table 1, indicate that the values of n/n_{avg} at r/R = 0.62 were significantly higher than either those at r/R = 0.40 or at r/R = 0.72. This finding, combined with a remarkably consistent shape of the distribution, justifies further analysis.

TABLE 1. Statistical Comparison of Bacterial Counts near Intermediary Maximum

Hypothesis	d_{avg}	S_d	N	t^a	P^b
No disinfection					
$d_{24} = 0$	−0.241	0.176	20	−6.12	$1.4 \cdot 10^{-6}$
$d_{45} = 0$	0.137	0.164	20	3.73	0.0003
Chlorination					
$d_{24} = 0$	−0.113	0.405	40	−1.76	0.02
$d_{45} = 0$	0.174	0.514	40	2.14	0.01
Chloramination					
$d_{24} = 0$	−0.127	0.280	46	−3.07	0.001
$d_{45} = 0$	0.105	0.275	46	2.59	0.003

[a] *t* is *t*-statistics (Equation 2).
[b] P is the significance level at which the tested hypothesis cannot be rejected.

B. Model of Initial Cell Attachment

In the rotating-disk system, the cells are transported to the surface by a combination of convective transport and diffusion. Once a cell approaches the surface, it is subjected to a range of attractive and repulsive forces which may result in cell adsorption. Initial adsorption is quite weak and cells may move about on the surface. Such movement, usually of a complex nature, has been observed by Lawrence and Caldwell.[14] Our analysis will focus on the initial stages of attachment. This analysis, presented in the following paragraphs, is speculative and overly simplistic. However, we feel it describes main features of the process (at least as we envision them).

It is assumed that after an initial approach to the surface, the cell can move parallel to the surface under the influence of hydrodynamic forces. In the same time, attachment forces increase and the cell stops when the attachment force component parallel to the surface counterbalances the parallel component of hydrodynamic forces. In our simple analysis, we assume that the cell initially moves with a velocity proportional to the velocity of the surrounding fluid. For the rotating disk, the radial component of liquid velocity u_1 is[8]

$$u_1 = \Omega \, F(z) \, r \tag{3}$$

where Ω denotes disk rotational velocity and F(z) is a generalized function of distance from disk surface z. Thus, radial velocity of an adsorbed cell u is

$$u = \alpha\, u_1 = \alpha\, \Omega\, F(z)\, r \tag{4}$$

where α is a proportionality coefficient between fluid velocity and cell velocity. It is important to note that for a constant distance from the surface, u increases linearly with radial distance r. Thus, the radial position of a cell which landed on the surface at a radial distance r_0 can be found by integration of Equation 4 and is given as the following function of time (of contact between the cell and the surface) t:

$$r = r_0 \exp(\alpha\, \Omega\, F\, t) \tag{5}$$

The contact time is measured from the first moment of interaction between the cell and the surface.

At this position the cell is subjected to shear stress with a radial component τ,

$$\tau = \rho\, F'(z)\, \Omega^{3/2}\, \nu^{1/2}\, r \tag{6}$$

where ρ is fluid density and ν its viscosity. F′(z) is the derivative of function F(z).[8] We consider shear stress as an indicator of hydrodynamic forces acting on the cell. These forces are countered by attachment/adsorption forces and the cell stops moving when the two are balanced.

We propose that the attachment forces can be also characterized by stress τ_{att}, which is a function of contact time between the cell and the surface. The contact time in our model depends very strongly on the parameter α, which at present cannot be characterized with any certainty. We further postulate that τ_{att} can be represented as a sum of two components with different time scales:

$$\tau_{att} = f_1(t) + f_2(t) \tag{7}$$

In our analysis we employ the following arbitrary forms of f_1 and f_2 expressed as functions of dimensionless time $\theta = t\alpha\Omega$

$$\tau_{att}(\theta) = \frac{k_{a1}\, \theta}{t_{a1} + \theta} + \frac{k_{a2}\, \theta^p}{t_{a2}^p + \theta^p} \tag{8}$$

with parameters k_{a1}, t_{as}, k_{a2}, t_{a2}, and p. Parameters k_{a1} and k_{a2} represent ultimate strength of attachment due to each component, while t_{a1} and t_{a2} characterize time scales of each component of Equation 8; p is an

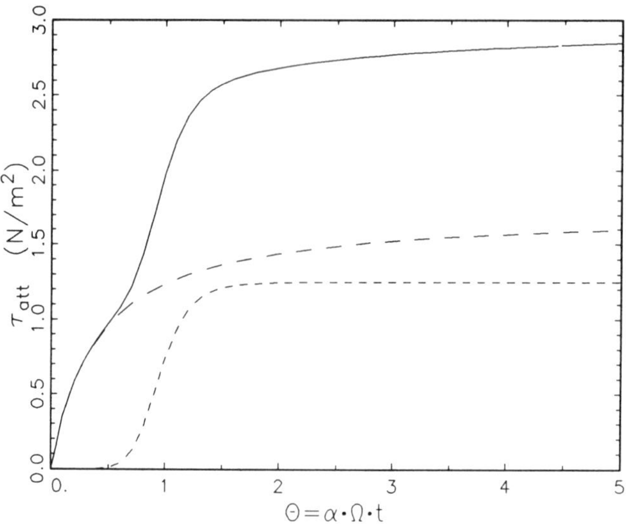

FIGURE 4. Example of attachment-strength function according to Equation 8.

adjustable, empirical parameter. For our analysis it is important that the time scales of the two components are sufficiently different. An example of τ_{att} according to Equation 8 is shown in Figure 4 for arbitrarily assumed values of $k_{al} = 1.72$ N/m^2, $t_{al} = 0.39$, $k_{a2} = 1.25$ N/m^2, $t_{a2} = 0.95$, $p = 7.5$. According to our model, the cell which moves in the radial direction according to Equation 5 stops at radial position r_1 when

$$\tau = \tau_{att} \tag{9}$$

Substituting Equations 8 and 6 into the above conditions, the final position of the cell r_1 can be calculated for each starting location r_0:

$$r_1 = g(r_0) \tag{10}$$

As a result of our postulated process, the cells which approached the surface will eventually attach at different locations, resulting in changes in attached-cell distribution. If the cells approach the surface in a uniform manner (as is the case for the rotating disk), then the distribution of cell concentration after the movement and attachment can be calculated from the function $g(r_0)$:

$$n/n_{avg} = dr_0/dr_1 = (dg/dr_0)^{-1} \tag{11}$$

Such distributions were calculated for a range of attachment parameters in Equation 8 and the following hydrodynamic parameters: $\Omega = 31.9$ s^{-1}

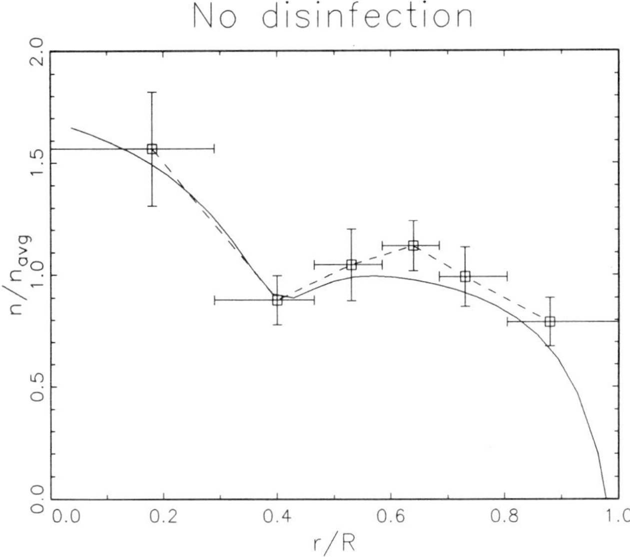

FIGURE 5. Calculated cell distribution compared with experimental results for nondisinfected cells.

(305 rpm), z = 1 μm, ν = 0.01 cm^2/s, and α = 1. Figures 5 through 7 show the calculated distributions compared with experimental distributions for nondisinfected, chlorinated, and chloraminated cells. The values of attachment parameters were close to those in Figure 4. It should be stressed that the calculated distributions were not rigorously fitted to the observed data, but only adjusted to give a reasonable visual resemblance.

It was not our goal to provide an estimate of attachment parameters (especially since the functions in Equation 8 were arbitrarily assumed), but to demonstrate the importance of mechanisms controlling initial cell adsorption and attachment and a possibility that cell movement can significantly affect the final distribution of attached cells. A simple model of cell movement combined with two-step adsorption/attachment can adequately explain observed results. The local maximum of cell concentration is a direct result of the inflection of the attachment-strength function in Figure 4.

C. Effects of Disinfection

Comparison of Figures 1 through 3 indicates that the distributions of attached cells were very similar regardless of disinfection. As the cell distribution on the disk is controlled by attachment phenomena, it appears that disinfected and nondisinfected culturable attached cells have

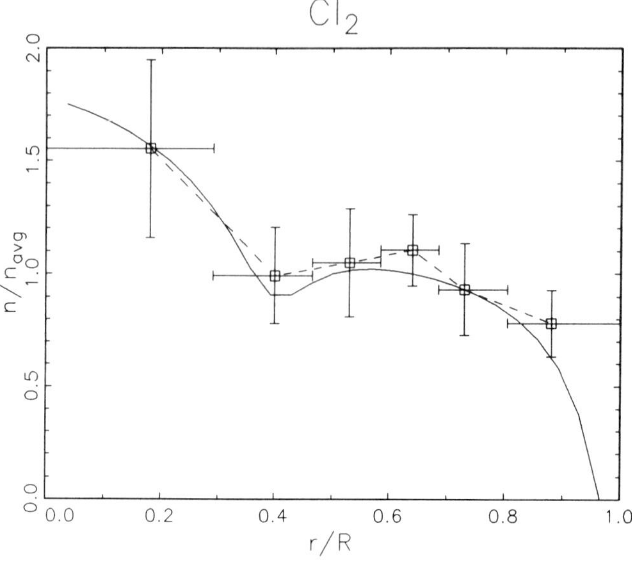

FIGURE 6. Calculated cell distribution compared with experimental results for chlorinated cells.

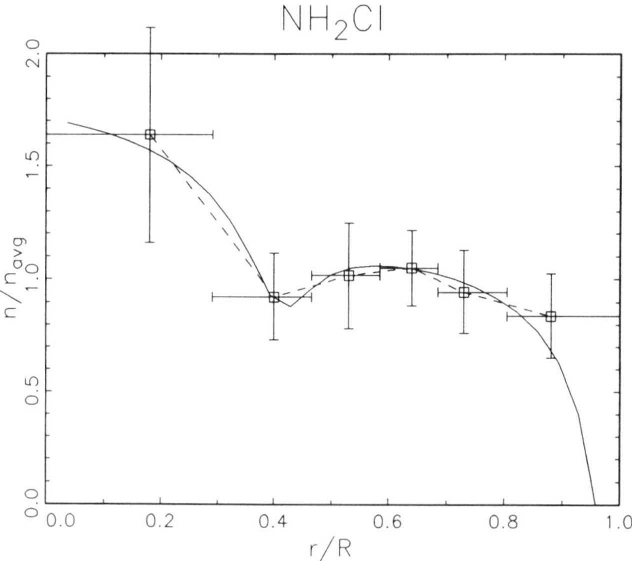

FIGURE 7. Calculated cell distribution compared with experimental results for chloraminated cells.

similar characteristics. However, the total number of culturable cells attached under identical conditions was much lower for disinfected cells (chlorinated and chloraminated).[6] In view of these results we postulate that the cells with better ability to attach are also more vulnerable to disinfection.

IV. SUMMARY

Distribution of culturable cells attached to the surface of a rotating disk exhibited an intermediary maximum at the distance from the center equal to approximately 60% of disk radius. It was postulated that such a distribution could be caused by displacement of bacterial cells on the surface after their initial weak attachment. A two-step attachment process combined with simple cell movement under the influence of hydrodynamic forces was able to model observed cell distributions. Disinfection did not have any significant effect on the shape of the distribution of bacterial cells on the surface, suggesting that adhesion characteristics of attached cells were not changed by disinfection.

ACKNOWLEDGMENTS

The research presented in this chapter was partially supported by the University of California, Water Resources Center, as part of Water Resources Center Project UCAL-WRC-W-747. The support of F. Lage-Filho by the Government of Brazil is gratefully acknowledged. S. W. Hermanowicz would like to thank Professor Peter Wilderer, Technical University of Munich, for his hospitality during final preparation of the paper. The authors appreciate the comments of two reviewers.

LIST OF SYMBOLS

d_{avg}	Average value of d_{ij}, dimensionless
d_{ij}	Paired difference of normalized surface concentrations of attached cells (Equation 1), dimensionless
$F(z)$	Generalized velocity function, dimensionless
k_{a1}	Ultimate attachment strength (Equation 8)
k_{a2}	
n	Observed surface concentration of attached cells, cm^{-2}
n_{avg}	Average surface concentration of attached cells, cm^{-2}
N	Number of paired differences, dimensionless
r	Distance from the disk center, cm

r_0 Distance of cell landing position from disk center, cm
R Disk radius, cm
s_d Standard deviation of d_{ij}, dimensionless
t Contact time between cell and disk surface, s
t_{a1} Time scales at attachment process (Equation 8), dimensionless
t_{a2}
u Velocity of bacterial cell on disk surface, cm/s
u_1 Radial liquid velocity, cm/s
z Distance from disk surface

Greek Letters

α Proportionality coefficient $u = \alpha \cdot u_1$, dimensionless
ν Water viscosity, cm^2/s
τ Hydrodynamic shear stress in radial direction, N/m^2
τ_{att} Attachment strength, N/m^2
Ω Rotational velocity of the disk, s^{-1}

REFERENCES

1. **Characklis, W. G.,** Bacterial Regrowth in Distribution Systems, Research Report, American Water Works Association Research Foundation, Denver, 1988.
2. **LeChevalier, M. W., Hassenauer, T. S., Camper, A. K., and McFeters, G. A.,** Disinfection of bacteria attached to granular activated carbon, *Appl. Environ. Microbiol.*, 48, 918, 1984.
3. **LeChevalier, M. W., Cawthon, C. D., and Lee, R. G.,** Inactivation of bacterial biofilms, *Appl. Environ. Microbiol.*, 54, 2492, 1988.
4. **LeChevalier, M. W., Cawthon, C. D., and Lee, R. G.,** Mechanisms of bacterial survival in chlorinated drinking water, Proc. Int. Conf. on Water and Wastewater Microbiology, Newport Beach, CA, February 8 to 11, 1988, 24.
5. **LeChevalier, M. W., Lowry, C. D., and Lee, R. G.,** Disinfecting biofilms in a model distribution system, *J. Am. Water Works Assoc.*, 82, 87, 1990.
6. **Hermanowicz, S. W. and Lage-Filho, F.,** Disinfection and attachment of bacterial cells, *Water Sci. Technol.*, 26, 655, 1992.
7. **Lage-Filho, F.,** Effects of Disinfection on Bacterial Attachment on Non-Biological Surfaces, Ph.D. thesis, University of California, Berkeley, 1992.
8. **Schlichting, H.,** *Boundary Layer Theory*, Prentice Hall, Englewood Cliffs, NJ, 1968.
9. **Levich, V.,** *Physicochemical Hydrodynamics*, Prentice Hall, Englewood Cliffs, NJ, 1960.
10. **Spielman, L. A. and FitzPatrick, J. A.,** Theory for particle collection under London and gravity forces, *J. Coll. Int. Sci.*, 42, 607, 1973.
11. **Dabros, T., Adamczyk, Z., and Czarnecki, J.,** Transport of particles to a rotating disk surface under an external force field, *J. Coll. Int. Sci.*, 62, 529, 1977.
12. **Martin, R. E. and Bouwer, E. J.,** The Deposition of Bacteria on Non-Biological Surfaces, Presented at Annual AIChE Meeting, Nov. 20, 1987, New York, NY.
13. **Martin, R. E.,** Quantitative Description of Bacterial Deposition and Initial Biofilm Development in Porous Media, Ph.D. thesis, Johns Hopkins University, Baltimore, 1990.
14. **Lawrence, J. R. and Caldwell, D. E.,** Behavior of bacterial stream population within the hydrodynamic boundary layer of surface microenvironment, *Microb. Ecol.*, 14, 15, 1987.

Section III
BIOCORROSION

11

Dissolved Oxygen Gradients Near Microbially Colonized Surfaces

Zbigniew Lewandowski

I. INTRODUCTION

Microorganisms frequently colonize inert water-immersed surfaces, forming biofilms that range in thickness from several to a few hundred micrometers. The microorganisms bind to the surfaces by means of exopolysaccharide polymers. From the initial growth stage, biofilms remain tightly bound to their host surfaces. The cells divide, forming sister cells bound within the polymeric matrix. The process leads eventually to formation of a continuous biofilm on the colonized surface. The biofilms alter the near-surface environment by affecting both the bulk media and the growth surface (substratum). The result is a complex network of reactions which may have many consequences, including biodeterioration of the surface.

The magnitude of chemical environment modification near the biofilms is directly related to the biofilm substrate uptake rate. Experimental techniques for determining biofilm reaction kinetics depend on two types of analyses: (1) chemical analysis of bulk water and (2) chemical measurements inside the biofilm using microsensors. Biofilm systems are diffusion limited, and chemical conditions near and inside biofilms can vary dramatically over a distance of only a few micrometers. Consequently, the information obtained from analysis of bulk water is limited and must be closely scrutinized before drawing any conclusions with regard to the biofilm.

0-87371-928-X/94/$0.00+$.50

Direct chemical measurements inside biofilms are severely restricted by the nature of the system. First, because biofilm thickness is usually quite small relative to the lateral surface dimensions, the space available for intrafilm instrumentation measurements is limited. Second, since biofilm respiration is diffusion limited, the substrate concentration varies across the film, forming concentration profiles. Each chemical constituent which is consumed or produced in the film forms a separate diffusivity-related profile. The heterogeneous and anisotropic nature of biofilm is enhanced by the fact that these films may be colonized at different locations by different microbial species. These spatial variations in surface coverage are particularly interesting when microorganisms colonize a metal substratum, which may result in the formation of local corrosion cells and contribute to a phenomenon called microbially influenced corrosion (MIC).

This chapter presents a technique of extracting kinetic parameters from substrate concentration profiles. The inherent heterogeneity of the biofilms causes the profiles taken at different locations to be different. Therefore, the kinetic parameters such as reaction rate, diffusivity, and half-saturation coefficient, evaluated at different locations, may contribute to describing this heterogeneity.

II. SUBSTRATE CONCENTRATION PROFILES USING MICROELECTRODES

Ion-selective and gas-sensing microelectrodes with tip diameters less than 10 μm are used for direct measurements of chemical constituents in biofilms. Advanced biofilm research depends upon sensor miniaturization. Microsensors, usually in the form of microelectrodes, are increasingly popular.[1–5] Microelectrodes are thus far the most accurate instruments for measuring the concentration profiles in biofilm systems. To make measurements effectively without disturbing the system to any significant extent, electrode tip diameter must be 10 μm or less. The size of the tip must be small because the electrode should penetrate the biofilm without physical damage to its structure. Larger electrodes would produce a dent in the biofilm, causing bulk water to penetrate and influence the measurement.

The dissolved oxygen (DO) profiles analyzed in this chapter were taken from a continuous-flow, open-channel reactor. A mixed-population biofilm was growing on a metal surface. Part of the deposits were abiotic corrosion products. The DO electrode was made of a 0.1-mm high-purity (99.99%) platinum wire etched electrochemically (with one end in KCN) to a tip diameter of about 2 μm. The wire was rinsed with concentrated HCl and ethanol and covered with soda-lime glass. The

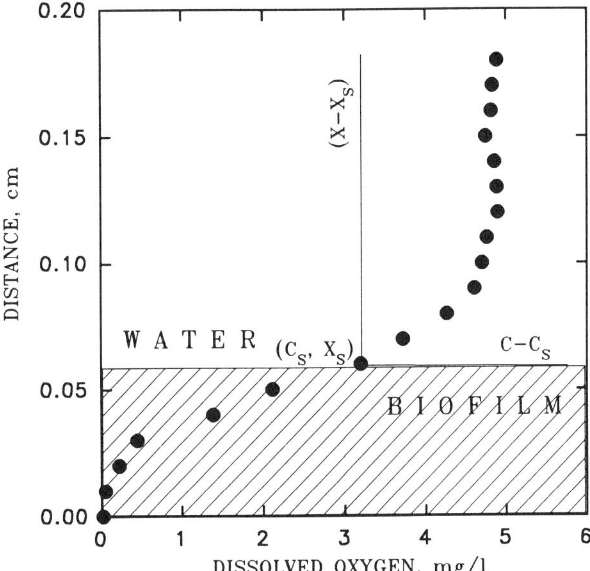

FIGURE 1. Profile of dissolved oxygen concentration across the biofilm system. The biofilm surface was positioned at the inflection point of the profile.

tip of the platinum wire was exposed by grinding on a rotating diamond wheel. The exposed platinum tip was subsequently etched in KCN to yield a recess of about 10 μm. Half of this recess was filled with gold by electrochemical plating. The operation was performed under a microscope with a mounted TV camera and observed on a video screen. The tip of the electrode was covered with a polymer (TePeX) to serve as the oxygen-permeable membrane. The measuring setup consisted of a picoammeter and a polarizing voltage source. The microelectrode was cathodically polarized to 800 mV against a silver-silver chloride reference electrode. The current in the circuit, in the range of picoamperes, is proportional to the concentration of DO. The electrode was calibrated in water by aeration and subsequent purging with pure nitrogen. Before the measurement the microelectrode was mounted on a micromanipulator and moved across the biofilm with predetermined increments. After each stop the current in the system was measured and compared with the calibration curve to calculate the DO concentration.

The result of measurement is a profile of DO vs. distance (Figure 1). The shape of the profile is determined by the following:

1. Microbial substrate uptake rate, which is a function of microorganism concentration and their affinities for the substrate

2. Substrate transport rate through the film, which depends on substrate diffusivity through the biofilm
3. Substrate transport rate to the biofilm, which is a function of microbial substrate uptake rate, substrate diffusivity through the water, and hydrodynamics near the biofilm surface

The shape of the concentration profile is simultaneously influenced by all these factors. Consequently, the interpretation of the profiles is complex. Some progress in extracting information from concentration profiles has been achieved. A procedure of kinetic-parameters calculation based on the substrate concentration profile was presented by Lewandowski et al.[6] This procedure, however, required that the water be stagnant. Stagnant water above the biofilm resulted in a linear substrate concentration profile across the diffusion boundary layer and simplified calculations. The procedure described in this chapter is not limited by this condition.

III. CHEMICAL CONDITIONS AT BIOFILM SURFACE

Fixing the position of the biofilm surface at the substrate concentration profile is essential to any calculations in biofilm systems. Perhaps the most important advantage of knowing this position is that the substrate profile can be divided into two parts, one in water and one in the biofilm. The profile above the film contains information about substrate transport to the biofilm. The profile below the biofilm surface contains the information about the microbial activity.

It was shown[7] that the diffusivity difference between the water and the biofilm along with the substrate utilization in the biofilm lead to a discontinuity in the substrate concentration profile at the biofilm surface. This discontinuity arises because the mass transfer of substrate in the water and the biofilm are governed by different equations, which are tied together by the requirement that the substrate flux at the fluid-biofilm interface be continuous at a steady state.

Lewandowski et al.[7] have developed a fiber optic microprobe which detects the position of the biofilm surface while measuring substrate concentration. The biofilm is penetrated with a microprobe, 15 μm in diameter, which simultaneously measures the substrate concentration and optical density. A single-cable, glass fiber optic was used to fabricate the optical density sensor. The tip was etched in hydrofluoric acid (HF) to obtain a sharp tip of 10 μm in diameter. The DO microelectrode was coupled with the optical fiber to form a dual-sensor microprobe. Care was exercised to place the sensor tips in adjacent positions so they would be at the same level in the biofilm. Biofilm sample (growing on a transparent polycarbonate) was placed in a petri dish. The probe was mounted in a motorized micromanipulator and positioned above the biofilm. A light-emitting diode (LED) was mounted below the petri dish in a hand-operated micromanipulator. The probe penetrated the biofilm at 5-μm increments and the DO concentration and light intensity were recorded at each stop. When the microprobe entered the biofilm the optical density of the environment changed, which was identified at the light intensity profile monitored by the fiber optic sensor. The change in optical density marked the position of the biofilm surface. At the same time the other part of the probe, the DO microelectrode, monitored changes in the DO concentration. Since the tips of both sensors were at the same level, the position of the biofilm surface evaluated from the optical density profile was superimposed on the DO profile. This technique, although accurate, was time consuming and work intensive.

Simplified procedures for biofilm surface positioning are used in routine measurements. A typical DO profile across a biofilm system is presented in Figure 1. The profile was measured in an open-channel biofilm reactor. A mixed-population biofilm was accumulated using a feeding solution based on glucose and mineral salts. The biofilm surface was positioned at the inflection point of the substrate concentration profile. This procedure is certainly less accurate than using a separate sensor, but is much more practical. How this experimental simplification influences the results of kinetic calculations remains to be tested.

IV. BIOFILM REACTION KINETICS

Assuming one-dimensional diffusion, the mathematical expression for the rate of change of substrate concentration in the biofilm is given as

$$\left[\frac{\partial C}{\partial t}\right]_f = D_f\left[\frac{\partial^2 C}{\partial x^2}\right]_f - \frac{V_{max}\,C}{K_s + C} \tag{1}$$

where D_f is the diffusion coefficient for the DO in the biofilm ($cm^2 \cdot s^{-1}$), C is the DO concentration at a point x ($mg \cdot l^{-1}$), V_{max} ($mg \cdot l^{-1} s^{-1}$), and K_s ($mg \cdot l^{-1}$) have the usual meaning in the Michaelis-Menten equation.

The steady-state concentration within a biofilm $(dC/dt = 0)_f$ is achieved when the consumption rate is equal to the rate of transport due to diffusion. Disregarding time dependence allows the removal of the partial derivatives, giving

$$0 = D_f\left[\frac{d^2 C}{dx^2}\right]_f - \frac{V_{max}\,C}{K_s + C} \tag{2}$$

This nonlinear equation cannot be solved exactly. However, as presented by Frank-Kamenetski,[8] it is often beneficial to integrate it once. The procedure follows:

$$p = \left[\frac{dC}{dx}\right] \tag{3}$$

$$\frac{d^2C}{dx^2} = \frac{dp}{dx} = \frac{dC}{dx}\frac{dp}{dC} = p\frac{dp}{dC} \tag{4}$$

since

$$\frac{1}{2}p^2\frac{d}{dC} = \frac{1}{2}\frac{d}{dp}p^2\frac{dp}{dC} = p\frac{dp}{dC} \tag{5}$$

then

$$\frac{d^2C}{dx^2} = \frac{1}{2}p^2\frac{d}{dC} \tag{6}$$

Substituting from Equation 2:

$$\frac{1}{2}p^2\frac{d}{dC} = \frac{V_{max}}{D_f}\frac{C}{K_s + C} \tag{7}$$

which combined with (3) yields:

$$\left[\frac{dC}{dx}\right]_f = \sqrt{2\frac{V_{max}}{D_f}\int\frac{C}{K_s + C}dC} \tag{8}$$

Evaluation of the integral in Equation 8 gives

$$\int\frac{C}{K_s + C}\,dC = C - K_s\ln(K_s + C) + \text{const} \tag{9}$$

The integration constant in Equation 9 can be determined from the boundary conditions. If the film is not totally penetrated by the constituent, then $(dC/dx)_f = 0$ for $C = 0$ (i.e., there is no mass transport beyond this depth). If the biofilm is totally penetrated, and the concentration at the bottom equals C_0, then $(Dc/dx)_f = 0$ for $C = C_0$ and there is no penetration at the substratum surface. For partially penetrated biofilms, the constant is

$$\text{const} = K_s \ln K_s \tag{10}$$

which, when substituted into Equation 9 and combined with Equation 8, gives

$$\left[\frac{dC}{dx}\right]_f = \sqrt{2\frac{V_{max}}{D_f}\left[C - K_s \ln\frac{K_s + C}{K_s}\right]} \tag{11}$$

Likewise, for totally penetrated films, the constant is

$$\text{const} = K_s \ln(K_s + C_0) - C_0 \tag{12}$$

which, through the same substitutions, gives

$$\left[\frac{dC}{dx}\right]_f = \sqrt{2\frac{V_{max}}{D_f}\left[C - C_0 - K_s \ln\frac{K_s + C}{K_s + C_0}\right]} \tag{13}$$

The reaction rate (R) can be directly calculated from the above equations by recognizing that it is equal to the rate of mass transfer across the biofilm surface,

$$R = F \cdot D_f \left[\frac{dC}{dx}\right]_f \tag{14}$$

where F is the biofilm surface area. Equation 13 can then be considered a general form of Equation 11. This means that the partially penetrated biofilms should be considered a specific case of totally penetrated biofilms. The reaction rate at the biofilm surface is given as

$$R = F\sqrt{2V_{max}D_f\left[C_s - C_0 - K_s \ln\frac{K_s + C_s}{K_s + C_0}\right]} \tag{15}$$

where C_s is the substrate concentration at the biofilm surface, F is the biofilm surface area. When the biofilm is totally penetrated by the substrate ($C_0 = 0$), Equation 15 is reduced to the case described by Equation 11.

For further calculations, the first and the second derivatives should be known. Therefore, the part of the substrate concentration profile below the biofilm surface is extracted from Figure 1 and approximated using a third-order polynomial.

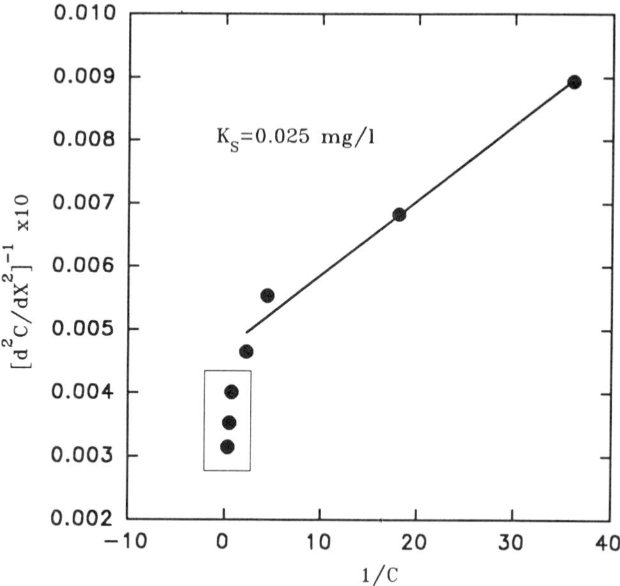

FIGURE 2. Evaluation of the half-saturation coefficient.

V. HALF SATURATION COEFFICIENT, K_s

Inversion of Equation 2 permits estimation of the K_s value:

$$\left[\frac{d^2C}{dx^2}\right]_f^{-1} = \frac{D_f * K_s}{V_{max}} \frac{1}{C} + \frac{D_f}{V_{max}} \tag{16}$$

The slope divided by the intercept in the plot (Figure 2) of the inverse of the second derivative against the inverse of the concentration gives K_s. The second derivative is calculated from the third-order polynomial representation of the DO profile in the biofilm. The data in Figure 2 indicate that the procedure of linearizing is efficient when using the DO concentrations near the K_s value. When the concentration of DO becomes much higher than the K_s (small 1/C value), the profile in Figure 2 becomes nonlinear. Although this explanation is by no means conclusive it is possible that the problem is similar to that encountered in planktonic cultures. It is well known from evaluating the K_s in planktonic cultures of microorganisms that the substrate concentration should not be much different than the K_s itself, otherwise the result may be meaningless.[9] Using the concentration values near the bottom of the biofilm yielded a "reasonable" value of $K_s = 0.025$ mg·l^{-1}. This value is similar to what may be expected from planktonic cultures.

VI. SUBSTRATE PROFILES ACROSS THE DIFFUSION BOUNDARY LAYER

The substrate molecules, before reaching the biofilm surface, must travel across the diffusion boundary layer. In an idealized, perfectly stagnant water the process would be entirely dependent on molecular diffusion. Consequently, the profile of substrate concentration would be linear. In natural systems, convection occurs. The magnitude of convection depends on turbulence, which in turn depends on flow velocity. The flow velocity is maximum away from the biofilm. It decreases near the biofilm and probably reaches zero at the surface or just below the biofilm surface. The velocity profile near the biofilm surface is nonlinear, and so is the convective transport rate. As a result, the concentration profile above the biofilm surface is also nonlinear.

The biofilm surface was located at the inflection point of the substrate concentration profile (Figure 1). For further considerations this point constitutes an origin of a new system of coordinates, (x_s, C_s). The new system of coordinates divides the profile into two parts. The part in the biofilm, below the interface, contains information about biofilm reaction rate and substrate diffusivity through the film. The part of the profile above the interface is analyzed to reveal information on substrate diffusivity through water and hydrodynamics. To calculate the diffusivity of DO in the biofilm, the profile just above the biofilm surface (bulk water side) is compared to the profile just below the biofilm surface (biofilm side). Since the flux of DO across this interface must be preserved, the first derivatives of concentration along the distance multiplied by the relevant diffusivities must be equal on both sides of the water-biofilm interface. The first derivative of concentration along the distance from the biofilm side can be calculated from Equation 11 or 13. The first derivative of DO concentration along the distance from the bulk water side can be calculated from the following analysis.

It was found that the DO profile above the biofilm surface can be adequately described by an empirical exponential function,

$$\frac{C - C_s}{C_b - C_s} = 1 - \exp[-B(x - x_s)] \tag{17}$$

where C is the local substrate concentration, C_b is the bulk substrate concentration, B is an experimental coefficient, and x is the distance in the new system of coordinates. This equation can be linearized to conveniently find the coefficient B from experimental data:

$$\ln\left[1 - \frac{C - C_s}{C_b - C_s}\right] = -B(x - x_s) \tag{18}$$

Coefficient B, calculated as the slope of the line when presenting the data in coordinates $(x-x_s)$ vs. $\ln[1-(C-C_s)/(C_b-C_s)]$, equals 54 cm^{-1} (Figure 3a). The model adequately reflects the distribution of the experimental data (Figure 3b).

The first derivative of concentration along the distance at the biofilm surface from the bulk water side can be calculated as follows:

$$\left[\frac{dC}{dx_{(x=x_s)}}\right]_w = B(C_b - C_s) \tag{19}$$

Substituting the numerical values of $C_b = 4.9$ mg/l and $C_s = 3.2$ mg/l yields $(dC/dx)_w = 92$ $mg \cdot l^{-1} \cdot cm^{-1}$.

VII. DIFFUSION COEFFICIENT

The biofilm is partially penetrated with oxygen (Figure 2). Thus the first derivative dC/dx should be linearly related to the $\{C-K_s \ln[(C + K_s)/K_s]\}^{1/2}$ (Figure 4). The regression line is as follows:

$$\left[\frac{dC}{dx}\right]_f = 75.2\sqrt{C - K_s * \ln\frac{K_s + C}{K_s}} \tag{20}$$

$(dC/dx)_f$ at the biofilm-water interface can be calculated from Equation 11. For the DO concentration at the biofilm-water interface, $C = C_s = 3.2$ mgl^{-1} and $K_s = 0.025$ mgl^{-1}, $(dC/dx)_f = 132$ $mgl^{-1}cm^{-1}$.

The flux of oxygen through the biofilm-water interface is

$$J_f = D_f\left[\frac{dC}{dx}\right]_f \tag{21}$$

where subscript "f" stands for "film". The flux of oxygen through the diffusion layer is

$$J_w = D_w\left[\frac{dc}{dx}\right]_w \tag{22}$$

where subscript "w" stands for "water". The flux continuity must be preserved at the biofilm-water interface ($J_f = J_w$):

$$D_f\left[\frac{dC}{dx}\right]_f = D_w\left[\frac{dC}{dx}\right]_w \tag{23}$$

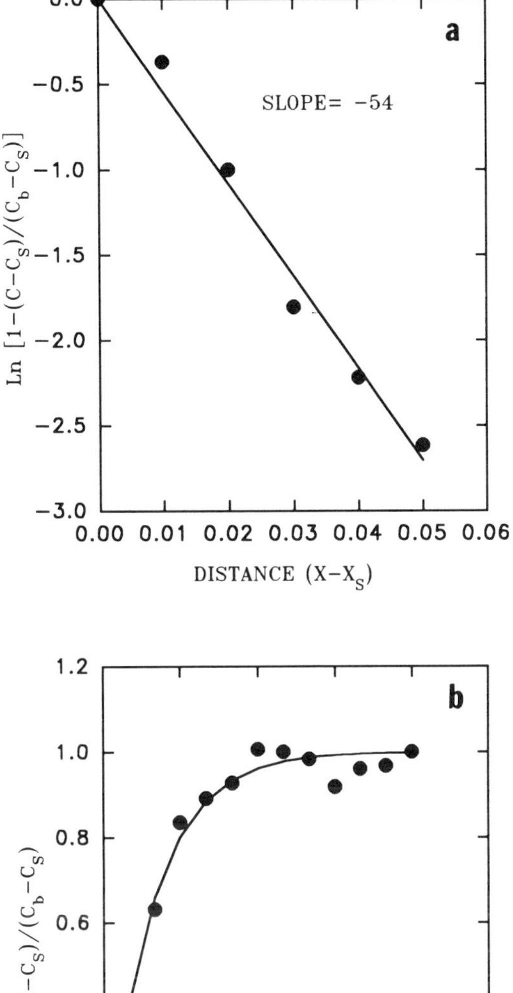

FIGURE 3. (a) Linearized dissolved oxygen profile (above the biofilm surface) together with (b) model and experimental data.

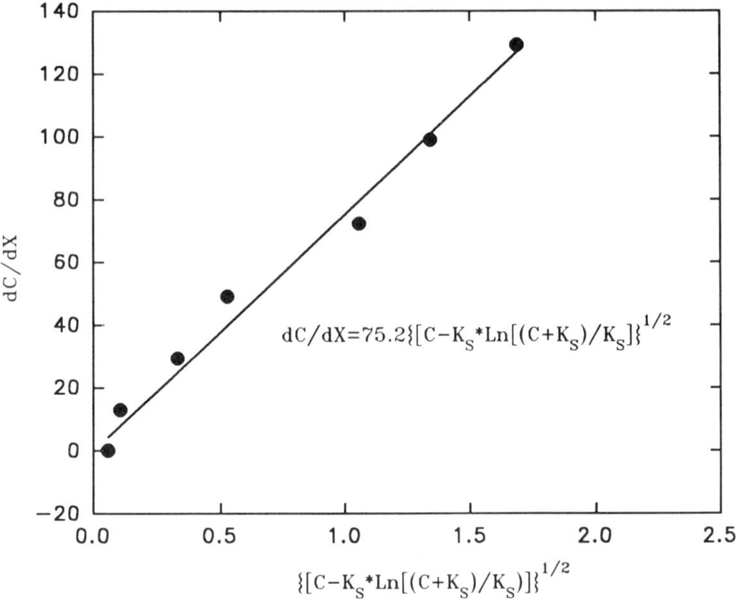

FIGURE 4. Linearized dissolved oxygen profile (below the biofilm surface).

The $(dC/dx)_w$ calculated from Equation 19 equals 91.8 $mgl^{-1}cm^{-1}$.

The biofilm diffusion coefficient can be calculated as follows:

$$D_f = D_w \frac{\left[\frac{dC}{dx}\right]_w}{\left[\frac{dC}{dx}\right]_f} \tag{24}$$

Substituting the calculated values, D_f is estimated as $0.70D_w$.

Using the DO diffusion coefficient in water at 21°C[10], $D_w = 2.0 \times 10^{-5}$ cm^2s^{-1}, the DO diffusion coefficient in the biofilm is $D_f = 1.31 \times 10^{-5}$ cm^2s^{-1}.

VIII. OTHER KINETIC PARAMETERS

By substituting the known values of D_f, $(dC/dx)_w$, $(dC/dx)_f$, and K_s, other important parameters can be calculated from the presented equations. For example, flux of oxygen equals $D_f \cdot (dC/dx)_f = 1.73 \times 10^{-6}$ $mg \cdot cm^{-2} \cdot s^{-1}$.

IX. DISCUSSION AND CONCLUSIONS

Kinetic parameters of biofilm reaction can be evaluated from the substrate concentration profile for flowing water systems. The result is site specific, which means that the evaluation is valid only for the site penetrated by the microsensor.

The measurements presented in this chapter stress the importance of the conditions at and near biofilm surfaces. All calculations are related to the position of the biofilm surface. Such treatment requires that the biofilm surface is precisely positioned. Biofilms, however, lack precisely positioned surfaces. In fact, the biofilm-water interface is a layer of variable thickness rather than a defined plane, something existing models do not tolerate. In extreme cases when the biofilm surface is colonized with filamentous microorganisms the position of the surface can be a function of flow velocity. The higher the flow velocity, the more the filamentous microorganisms will yield to the flow and smooth the surface, resulting in a simultaneous change in the outer biofilm boundary. The biofilm-water interface is not a clearly defined concept, and problems arise when this fact is imposed on existing models of diffusion and hydrodynamics. Because of all these limitations, the analytical solutions obtained can only have an approximate character. The precision of the solutions obtained is directly related to the precision with which the biofilm-water interface can be located. New approaches to the concept of the biofilm-water interface are needed before further progress can be obtained.

The calculated DO diffusivity in the biofilm considered in this study is $D_f = 1.31 \times 10^{-5}$ cm^{2}s^{-1}. This result is very close to the diffusivity of DO in marine sediments reported by Revsbech et al.[10] In sediment cores, incubated under 0.1 *M* solution of $HgCl_2$ for 3 weeks to eliminate microbial activity, Revsbech measured $D_f = 1.40 \times 10^{-5}$ cm^{2}s^{-1}. The results are very close, but the technique described here does not require any treatment to exclude the microbial activity and the measurement can be conducted *in situ*.

The locally measured values of substrate concentration are subject to experimental error which, considering the complexity of the equipment, can also be a function of many variables. Because of high impedance of microelectrodes, such measurements are always subject to electromagnetic noise. Noise elimination without signal distortion is an activity which is continuously exercised while using microelectrodes. How precisely the local conditions can be described depends on many factors and cannot be generalized.

The procedures presented here yield the kinetic parameters of the biofilm system as diffusivity, maximum reaction rate, half-saturation coefficient, and DO flux. The results are obtained from *in situ* microelectrode measurements. The experimental and theoretical limitations, however, are imposed on their validity.

AKNOWLEDGMENTS

The author acknowledges the support from the Center for Biofilm Engineering at Montana State University, a National Science Foundation-sponsored Engineering Research Center, and the Center's Industrial Associates.

REFERENCES

1. **Whalen, W. J., Bungay, H. R., III, and Sanders, W. M., III,** Microelectrode determination of oxygen profiles in microbial slime systems, *Environ. Sci. Technol.*, 3, 2297, 1969.
2. **Bungay, H. R., III, Whalen, W. J., and Sanders, W. M.,** Microprobe techniques for determining diffusivities and respiration rates in microbial slime systems, *Biotechnol. Bioeng.*, 22, 765, 1969.
3. **Revsbech, N. P. and Jorgensen, B. B.,** Microelectrodes: their use in microbial ecology, in *Advances in Microbial Ecology*, Marshall, K. C., Ed., Plenum Press, New York, 1986.
4. **Reithues, M., Buchholtz, R., Onken, O., Baumgartl, H., and Lubbers, D. W.,** Determination of oxygen transfer from single air bubbles to liquids by oxygen microelectrodes, *Chem. Eng. Process*, 20, 332, 1986.
5. **Baumgartl, H.,** Systemic investigations of needle electrode properties in polarographic measurements of local tissue PO2, in *Clinical Oxygen Pressure Measurement*, Ehrly, A. M., Hauss, J., and Huch, R., Eds., Springer-Verlag, 1987, 17.
6. **Lewandowski, Z., Walser, G., and Characklis, W. G.,** Reaction kinetics in biofilms, *Biotechnol. Bioeng.*, 38, 877, 1991.
7. **Lewandowski, Z., Walser, G., Larsen, R., Peyton, B., and Characklis, W. G.,** Biofilm Surface Positioning, Environmental Engineering Proceedings 1990, EE Div/ASCE, Arlington, VA, July, 1990, 17.
8. **Frank-Kamenetski, D. A.,** *Diffusion and Heat Transfer in Chemical Kinetics*, Plenum Press, New York, 1969.
9. **Andrews, G. F.,** Parameter estimation from batch culture data, *Biotechnol. Bioeng.*, 26, 824, 1984.
10. **Revsbech, N. P., Madsen, B., and Jorgensen, B. B.,** Oxygen production and consumption in sediments determined at high spatial resolution by computer simulation of oxygen microelectrode data, *Limnol. Oceanogr.*, 31(2), 293, 1986.

Section III
BIOCORROSION

12

Anaerobic SRB Biofilms in Industrial Water Systems: A Process Analysis

S. Okabe, W. L. Jones, W. Lee, and W. G. Characklis

I. INTRODUCTION

Biofilm development in natural and industrial water systems depends on the types and concentrations of electron acceptors (e.g., oxygen, nitrate, and sulfate) and electron donors (organic substrates), as well as on environmental factors including temperature, pH, salinity, and fluid dynamics. Development of sulfate-reducing bacterial (SRB) biofilms can be expected whenever environmental conditions such as redox potential or oxygen tension and nutrients are suitable for SRB growth. For example, SRB biofilms may develop in anaerobic bulk water systems, such as in petroleum-producing facilities, but are also found in aerobic bulk water systems such as cooling water systems. In aerobic bulk water systems, anaerobic microniches and/or anaerobic layers exist in biofilms due to depletion of oxygen by aerobic bacterial activity.[1,2] In all natural and industrial aquatic environments, SRB show a pronounced tendency to adhere to available surfaces and to proliferate to form biofilm.[3–5] Because of this sessile mode of growth, bacteria within these biofilms are often undetected by conventional sampling techniques, which analyze bulk fluid conditions. Nevertheless, it is these biofilm (sessile) SRB that are responsible for much of the anaerobic activity in natural and industrial water systems.

At present, quantitative prediction of SRB activity and growth in industrial water systems is essentially impossible because data on rate

0-87371-928-X/94/$0.00+$.50

and extent of SRB growth under relevant environmental conditions are not available. Therefore, it is necessary to determine effects of environmental factors on the activity and growth of SRB to develop a comprehensive model and to use this model to predict the SRB behavior in given environments. By comparing data from a variety of environments, a conceptual model is first developed which includes factors such as growth substrate limitation, sulfide inhibition, and the effects of attachment. Translation of the conceptual model into a mathematical form requires much more controlled experimentation to determine kinetic and stoichiometric coefficients (e.g., growth rate, yield) and the specific effects of external factors such as temperature on these coefficients.

As will be shown below, the initial conceptual model must include the following:

- Nutrient availability (including electron donor and acceptor, N, P)
- Effect of temperature
- Sulfide inhibition
- Attachment to surfaces (i.e., biofilm vs. planktonic growth)

Nutrient availability can affect both the growth of the organisms (through energy limitation or through limitations in biosynthetic precursors) as well as the amounts and types of products (e.g., cell material vs. extracellular products). Temperature can affect both kinetics and stoichiometry via phenomena ranging from thermodynamic activity changes through physical enzyme conformation changes. Product inhibition (sulfide) reduces biochemical activity through numerous mechanisms. Finally, attachment to surfaces has been shown to affect bacterial metabolism in a variety of ways, although it is difficult to specify whether these changes are due to a physiological response to attachment or to an altered extracellular environment resulting from diffusion limitations.[6]

A. Effects of SRB on Industry

SRB are very important microorganisms from an environmental and industrial standpoint. The anaerobic corrosion of metals is enhanced by the activities of SRB and is a universal industrial problem where aqueous process fluids contact equipment. The cost related to corrosion is estimated to be $200 billion per year in the U.S.[7,8] The costs to the U.S. Navy with regard to corrosion has been estimated to be $5 billion per year.[8] Extensive sulfide corrosion problems with concrete sewer pipes and wastewater treatment have also been reported.[9]

In the petroleum industry, SRB cause serious problems including corrosion of equipment, plugging of the petroleum formation, and reservoir souring (contamination of petroleum with H_2S).[4,10–13] Sulfide production by SRB increases the sulfur content of the crude oil, which decreases its value and increases refining costs. Costs for downtime, resulting in loss of production, to clean and replace fouled or corroded equipment easily extend to $10 million per day.[14] Hydrogen sulfide production by SRB leads to the corrosion of down-hole drill strings and casings as well as production facilities.[15] SRB growth in seawater injection systems can lead to corrosion as well as contamination of oil and gas with H_2S and viable SRB. Cord-Ruwisch et al.[16] reported that an increase in H_2S was observed during several years of operation at an oil field in northern Germany, and that H_2S formation resulted in plugging of the injection well by FeS flocs. Comprehensive lists of references regarding SRB causing problems in petroleum industries were reported by Postgate,[17,18] Sanders and Hamilton,[15] and Hamilton.[19] Biofilm accumulation also increases capital costs for equipment in power plants. For example, a nuclear power plant had to replace a condenser after approximately 6 years operation because of severe corrosion attributed partially to microbial activity.[14]

B. Control Strategies

1. *Biocides*

Extensive research has been conducted to develop effective biocides with the goal of inhibiting SRB growth and hence sulfide production. For example, in the secondary production of petroleum, injection water used in flooding operations is treated routinely with a biocide (typically glutaraldehyde) to control SRB growth in the injection well, reservoir, and piping.[20] Eagar et al.[21] reported that glutaraldehyde was an effective agent for controlling biofilm growth and activity in test waterflood systems. Also, the results of field study indicated that glutaraldehyde was sufficiently persistent in the distribution system to remain at an efficacious level and to be capable of reducing corrosion to an acceptable rate. Gaylarde and Johnston[22] strongly recommended that biocide test methods for SRB activity should employ mixed sessile SRB in the presence of metal coupons, because sessile SRB on the metal coupon surfaces survived at twice the recommended dose for both biguanide and nitropropanediol.

Biocide addition is often of limited effectiveness, since SRB are associated with other aerobic and anaerobic bacteria in biofilms which coat the surfaces of pipes and other materials. Within these biofilms, SRB are somewhat protected because biocides do not effectively penetrate through the biofilm. All of the reported data have shown that bacteria within biofilms are much more difficult to control with biocides than their planktonic counterparts in these systems.[4,22,23] Thus, biocide treatment may not be an ultimate means to control SRB activity because of rapid microbial regrowth, poor cost effectiveness, and environmental concerns.

The use of ionizing radiation to control SRB activity and growth has recently attracted attention. Ultraviolet radiation was used to kill SRB in injection waters by Ege et al.[23] Gamma radiation was also applied to control SRB at the bottom of the well bores as the water entered the oil reservoir.[24]

2. *Nutrient Removal*

The reduction of the concentration of an essential nutrient (e.g., phosphorous, nitrogen, and/or sulfate) to below the limiting concentration is a potential means of controlling SRB activity, because the essential nutrients control activity and growth of SRB when they become limiting. Maree and Strydom[25] reported the feasibility of microbial sulfate removal from industrial effluent using an upflow packed bed reactor with photosynthetic sulfur oxidation to prevent the emission of sulfide and confirmed the successful performance of the reactor. There is no information in the literature which addresses control of SRB activity and growth by removing required nutrients. Nutrient removal may be a possible means of controlling SRB activity and growth. This would be of benefit both in environmental and economic terms.

3. *Microbial Competition*

Microbial control of sulfide production by SRB using *Thiobacillus denitrificans* has attracted considerable attention lately.[26–28] *T. denitrificans* is an autotroph and a facultative anaerobe which oxidizes sulfide to sulfate using oxygen or nitrate as the electron acceptor. The introduction of viable cells of *T. denitrificans* into environments with SRB has the potential of controlling sulfide production so long as nitrate concentration remains high. The application of this method is an attempt to control sulfide production at or near the water injection well in an oil reservoir. A mutant of *T. denitrificans* (strain F) resistant to glutaraldehyde and sulfide was obtained by McInerney et al.[27] This mutant strain would allow a combined microbial and biocidal (glutaraldehyde) treatment of SRB-contaminated industrial systems. Sublette and Sylvester[29–31] and Sublette[32] have demonstrated that *T. denitrificans* may be readily cultured aerobically and anaerobically in batch and continuous reactors on gaseous H_2S under sulfide-limiting conditions. A microbial process for the removal of H_2S from gases has been proposed based on mixing the gas with a culture of *T. denitrificans*.[29] A practical difficulty is efficiently inoculating *T. denitrificans* into the well-bore area. Further, their activity reproduces sulfate which can then be reduced downgradient. In this case, there would be no net benefit from *T. denitrificans* activity.

The competition for the available electron donors between SRB and methane-producing bacteria (MPB) has also received considerable attention.[33–37] The SRB apparently have a higher affinity (low K_m) for hydrogen and acetate relative to the MPB. Thus, SRB normally dominate both in natural ecosystems, such as freshwater and marine sediments, where methanogenesis was found to be inhibited by the presence of sulfate. Yoda et al.[37] reported that in an anaerobic fluidized bed the methane production rate and MPB biomass decreased after several months of operation at low acetate concentration, whereas sulfate reduction rate increased. On the other hand, MPB were able to form a biofilm faster than SRB at high acetate concentrations, presumably due to the greater ability of MPB to adhere to carrier surfaces than SRB. Hilton and Oleszkiewicz[38] reported that SRB are more sensitive than MPB to the elevated total sulfide concentrations, while both are sensitive to elevated molecular H_2S concentrations. Thus, at high total sulfide concentrations and high pH the MPB should be able to outcompete the SRB for substrate.

4. Aeration

Oxygen is the cheapest and most effective inhibitor of SRB activity. If any system can be maintained in an aerated condition, even though the dissolved oxygen concentration is vanishingly small, SRB remain dormant. They are not killed, however, and may become active once anaerobic conditions are reestablished.[39,40] In practice, aeration is of limited effectiveness because SRB are generally associated with other bacteria in biofilms. Oxygen does not effectively penetrate through these biofilms as it is consumed by aerobes.[1] As clearly demonstrated by Lee et al.,[2] SRB activity at the substratum beneath a biofilm can be extensive, even at high dissolved oxygen in the bulk water. Furthermore, introducing oxygen into some industrial water systems increases the extent of pitting corrosion of facilities due to the formation of corrosive compounds such as elemental sulfur and/or pyrite.[2]

II. PROCESS ANALYSIS AND MODELING

SRB biofilm accumulation is a complex phenomenon resulting from several processes occurring in parallel and in series. The rate and extent of these processes, in turn, are influenced by numerous physical, chemical, and biological factors. Thus, a process analysis must be applied to solve biofilm-related problems. The process analysis generally requires (1) development of a conceptual model, (2) development of a mathematical model, and (3) experimental testing, calibration, and validation of the model.

A. Process Analysis

From the viewpoint of a process analysis of a reaction system, the most important components are expressions that quantitatively describe the rate (kinetics) and extent (stoichiometry) of the fundamental processes contributing to biofilm accumulation. Stoichiometry indicates the relationship between the extent of microbial growth and the uptake and production of the chemical species involved. Rate describes how fast the reactions will occur. Both stoichiometry and rate must be known to effectively design and control technical scale processes. The stoichiometric relationships are important since they permit estimation of the rate and extent of biomass and product formation (e.g., hydrogen sulfide) by measuring change in substrate (e.g., sulfate) concentration with time.

A conceptual model describing biofilm accumulation processes would be beneficial in interpreting available historical data and be invaluable in designing future experiments. If the conceptual model could be stated in mathematical terms, a mathematical simulation of biofilm accumulation could be performed on the computer at considerably less expense than laboratory experiments. Furthermore, the expected influence of process variables such as temperature and substrate concentrations could be determined on the computer *prior to* conducting laboratory experiments. The mathematical description of the individual processes could be combined to develop models to extrapolate and generalize experimental results. Many of these fundamental processes have been described mathematically by Characklis and Marshall.[41]

B. Experimental Approach

It is important to proceed in stages, beginning with pure culture work where precisely defined growth conditions and conclusions relevant to those conditions can be made. Understanding of the behavior of single species can then lead to a more rational image of the behavior of a mixed population.

Rate and stoichiometry are often determined in chemostat experiments due to the specified relationships that develop between substrate depletion rate, product formation rate, growth rate, and dilution at steady state. Although batch cultures are sometimes useful for stoichiometric estimates, rate coefficients of microbial sulfate reduction are difficult to measure in a batch culture because pH, sulfide concentration, and limiting substrate are not maintained at the same levels over many generations. On the other hand, chemostat analyses of mixed cultures are made difficult by the tendency of slower-growing organisms to be washed out. Analysis of rate and stoichiometry of processes within a biofilm are frequently complicated by significant mass transfer resistances in the liquid or diffusional resistances within the biofilm.

Nevertheless, biofilm experiments may be necessary in order to understand the changes in stoichiometry and rate which may occur due to physiological changes in the organisms following attachment. Rate and stoichiometry data determined from the pure-culture chemostat experiments (planktonic cells) can be invaluable in designing trials with biofilms (sessile cells) to establish whether or not rate and stoichiometry of the planktonic cells can be used to predict bacterial behavior within the biofilm. Once factors affecting growth and activity of planktonic cells are determined in the chemostat, their quantitative effect on accumulation and activity of biofilms must be determined. Interactions between populations in mixed culture are probably best evaluated at this level as well. Finally, all these data can be incorporated into a model which will permit prediction of SRB behavior in various environments. Sensitivity analyses using the model will identify critical parameters and lead to the development of means to control SRB growth and activity.

III. FACTORS AFFECTING SRB ACTIVITY AND GROWTH

In this section, major environmental factors affecting SRB activity and growth are discussed, with emphasis on rate and stoichiometry.

A. Temperature Effects

Temperature is a major process variable in microbial reactions. Temperature affects the rate of growth and substrate utilization in two ways: (1) the rates of biochemical reaction and (2) the transport rate of substrate into microorganisms. In chemostat studies, the maximum specific growth rate (μ_{max}) of *Desulfovibrio desulfuricans* was relatively constant between 25 and 43°C and dramatically decreased outside this temperature range, with the optimum temperature range between 35 and 43°C.[42] The activation energy for μ_{max} for planktonic cells was 104 kJ mol^{-1} in the range 12 to 25°C. By comparison, Nielsen[43] determined the activation energy for sulfate reduction rate to be 85 kJ mol^{-1} in a mixed-population SRB biofilm. These data suggest that the effect of temperature on microbial sulfate reduction in suspended pure-culture systems is not significantly different from the mixed-population biofilm. At higher temperatures, thermophilic SRB with temperature optima in the 55 to 70°C range would be expected to proliferate. Over a broad range of temperatures, significant population shifts might be expected.

The half-saturation coefficient (K_{Lac}) for lactate and the cell yield ($Y_{c/Lac}$) are dependent on temperature.[42] Because cell yields are low, however (see below), the stoichiometry between lactate uptake and sulfate uptake almost entirely reflects the catabolic activity of the cells. In *D. desulfuricans* suspended culture, this stoichiometry was relatively unaffected by changes in temperature.[42] This latter observation is not surprising in light of the redox stoichiometry (electrons accepted by sulfate, donated by lactate) dictated by catabolism. Thus, with the exception of cellular synthesis, temperature primarily affects the rates of SRB-mediated reactions, but not their stoichiometry.

B. Nutrient Requirements

By operating a chemostat under nitrogen- and phosphorous-limiting conditions, stoichiometric limiting C:P and C:N ratios (w/w) for *D. desulfuricans* were determined to be 400:1 to 800:1 and 45:1 to 120:1, respectively.[42,44] These stoichiometric limiting ratios were much higher than typical aerobic population values, because oxidation of lactate to acetate with sulfate as an electron acceptor yields about only 6% of the theoretical free energy of complete oxidation with oxygen as an electron acceptor, as described below:

$$CH_3CHOCOO^- + 0.5\ SO_4^{2-} \rightarrow CH_3COO^- + HCO_3^- + 0.5\ HS^- + 0.5\ H^+ \qquad \Delta G° = -14\ kJ$$

$$CH_3CHOCOO^- + 3\ O_2 \rightarrow 3\ HCO_3^- + 2\ H^+ \qquad \Delta G° = -247 kJ$$

Since the biomass yield from lactate with sulfate ($Y_{c/Lac}$ = 0.03 [g cell] [g lactate]$^{-1}$)[42] is approximately 6% of an aerobic system ($Y_{x/s}$ = 0.50 [g cell][g lactate]$^{-1}$),[41] phosphorous and nitrogen requirements are expected to be concomitantly reduced. Postgate[18] reported that SRB have the same cell elemental composition as most other bacteria. Typical analytical data for *D. vulgaris* from continuous culture are C, 46.5%; H, 7.2%; N, 12.5%; S, 1.3%; and P, 0.23%. Using these data and the cell yield of 0.03 (g cell)(g lactate)$^{-1}$, limiting C:P and C:N ratios would be 15,000:1 and 270:1, respectively, based on total amounts required for balanced growth. In comparison with the measured ratios above, observed growth limitation occurs at C:N and C:P ratios well below those required merely for balanced growth. It should be noted, however, that

TABLE 1. Stoichiometries of Microbial Sulfate Reduction by a Mixed-Population Biofilm Containing SRB and Planktonic *D. desulfuricans* in Pure Culture

Mixed-population biofilm[7]
$CH_3CHOHCOO^- + 0.48\ SO_4^{2-} \rightarrow 0.24\ CH_{1.4}N_{0.2}O_{0.4} + 0.33\ CH_3COO^- + 0.70^a\ S^{-2} +$ 2.1 Carbon products

Planktonic population of *D. desulfuricans*[42]
$CH_3CHOHCOO^- + 0.46\ SO_4^{2-} \rightarrow 0.098\ CH_{1.4}N_{0.2}O_{0.4} + 0.98\ CH_3COO^- + 0.50\ S^{-2} + 0.99\ CO_2$

[a] Sulfide measurements compromised due to FeS accumulation in biofilm.

energy requirements for uptake would be expected to increase when concentrations of these nutrients become extremely low.

The production of extracellular polymeric substances (EPS) by *D. desulfuricans* increased when phosphorous and nitrogen concentrations became limiting.[42,44] As a result of increased EPS production, cell production decreased by about 50% when both N and P became limiting. In biofilm systems, increased EPS production may offset any advantage gained by reducing the biomass production rate. Increased EPS production may influence injection well plugging of oil reservoirs as described by Lappan and Fogler.[45] An increase in EPS production also influences biocide performance in industrial water systems because the effectiveness of the biocide treatment is significantly reduced by the protective and highly adsorptive nature of the EPS.[4,46]

The specific growth rate of planktonic *D. desulfuricans* cells was shown to be a function of both sulfate and lactate concentrations in the bulk water.[42,44] The specific growth rate was influenced by the sulfate and lactate concentrations below about 10 and 6 mg l^{-1}, respectively. When sulfate became limiting, the observed specific growth rate and cell yield decreased due to an increase in the maintenance energy requirement. The increase in maintenance energy requirement was attributed to the fact that *D. desulfuricans* might ferment lactate at low levels of sulfate.

When SRB are part of mixed-population biofilms, their activity may become limited by the transport of substrate; SRB rate and stoichiometry may not be accurately reflected by measurements of bulk solution concentrations.[43] Lee[7] reported the stoichiometry of microbial sulfate reduction in a mixed-population anaerobic biofilm containing SRB (Table 1). The stoichiometric coefficient for bacterial cells (0.240) was relatively high compared to that obtained from planktonic monopopulation SRB (0.098). Further, the mixed population system produced only a third of the acetate produced by the monopopulation system. The reduction of net acetate production as well as the increase in biomass production may be attributed to acetate uptake by general anaerobic bacteria. These results are consistent in that a mixed culture would be expected to make better use of available substrate than a pure culture. What is surprising is that the stoichiometric ratio of lactate: SO_4^{2-} was 2:1 in both systems. The stoichiometric ratio between lactate and sulfate may thus be applicable to a variety of mixed-population SRB biofilm systems. Because community interaction influences the fate of carbon sources, however, stoichiometric coefficients based solely on cell yield per unit of organic substrate are not expected to remain constant between systems.

C. Sulfide Inhibition

Sulfate reduction generates hydrogen sulfide that may be inhibitory to the organisms that produce it. Okabe et al.[44] reported that 50% inhibition of lactate utilization by *D. desulfuricans* occurred at approximately 500 mg total sulfide per liter, but that the effect was largely reversible. Reis et al.[47] reported that a batch culture of the genus *Desulfovibrio* was directly inhibited by the H_2S produced and that complete inhibition occurred at approximately 550 mg l^{-1} H_2S concentration. Shimada[48] reported that 100 mg l^{-1} of H_2S inhibited the growth of a mixed SRB population in batch cultures and that no SRB growth was observed at 500 mg l^{-1} of H_2S. The overall effect of sulfide on the microbial sulfate reduction is qualitatively described by some authors.[18,38,49–51] Sulfide toxicity is strongly dependent on pH because molecular H_2S can pass through the bacterial cell membrane.[52] Reis et al.[47] reported an inhibitory effect of sulfide on the specific growth rate of *Desulfovibrio* and found that H_2S is the most inhibitory form of sulfide. Collectively, these results suggest that analysis and prediction of SRB growth and activity should account quantitatively for inhibition and toxicity effects of sulfide, particularly in the nonionized form of H_2S.

D. Effects of Attachment on Activity and Growth

Attachment of SRB to surfaces may result in observed changes in activity and/or growth. It is often unclear whether this change is due to changes in the local environment of the cell or a more fundamental change in cell metabolism. There is some experimental evidence of altered metabolism of aerobic cells in biofilms,[53] but little evidence is available for SRB. For instance, Fukui and Takai[54] reported that free-living *D. desulfuricans* cells produced colonies more rapidly than particle-associated ones, suggesting an attachment-mediated physiological change. For more mature biofilms, the distinction becomes much less clear, as H_2S may accumulate due to mass transfer limitations.[43] The resulting inhibition may produce effects which are indistinguishable from physiological response to attachment, such as development of maintenance requirements (increased activity) and retardation of growth. Thus, an accurate description of the effects of growth on a surface requires a distinction between growth and activity responses. Further study is needed to distinguish between *a priori* physiological response to attachment and responses to microenvironmental changes.

IV. SRB BIOFILMS

A. SRB Biofilm Accumulation Models

There is very little quantitative information available related to the rate and extent of SRB biofilm accumulation. Nielsen[43] studied a mixed-population biofilm containing SRB in an annular biofilm reactor. He reported that the biofilm thicknesses reached 300 to 400 μm, but that biofilm was no longer fully penetrated by sulfate at concentrations less than 100 mg l^{-1}. Typical zero-order volumetric rate constants for sulfate reduction in a biofilm without sulfate limitation were determined to be 5.4 to 8.9 mg SO_4^{2-} cm^{-3} h^{-1} at 20°C. The sulfide production from biofilms grown on domestic wastewater was modeled using biofilm kinetics and found to agree with experimental results.[55] Sulfate limitation in a typical sewer biofilm, with a thickness of 200 to 300 μm, was shown to occur around sulfate concentrations of 3 to 5 mg SO_4–S l^{-1}, but zero-order rate constants for sulfate reduction were also much lower: 0.12 to 0.17 mg SO_4^{2-} cm^{-3} h^{-1} at 20°C. These results are consistent in that diffusion limitations become less important as the reaction rate decreases.

Lee[7] reported that accumulation of a mixed-population biofilm containing SRB was strongly dependent on substrate loading rate. The biofilm thickness easily reached about 1000 μm at high substrate loading (100 mg C m^{-2} h^{-1}), whereas the thickness was about 5 μm at low substrate loading rate (5.4 mg C m^{-2} h^{-1}). Under both loadings, the biofilm was uniformly distributed and very rigid.

Empirical models for the prediction of sulfide production from sewer systems have been published.[55–57] However, the biofilm kinetics and effects of nutritional and physical factors are not taken into account in these models. More quantitative and comprehensive prediction models for sulfide production are necessary for more accurate prediction. To design new wastewater treatment systems to minimize sulfide production or to efficiently control sulfide production in industrial water systems, a reliable method that predicts the sulfide production rate is needed.

V. CONTROL STRATEGIES BASED ON NUTRIENT REQUIREMENT

As all SRB require sources of N, P, and S to support and maintain growth, reduction of one or all of these nutrients to below limiting levels may provide a means to control SRB activity. Laboratory experimental results indicate that the removal of sulfate from industrial water has the greatest potential to control SRB activity. Conversely, phosphorous and nitrogen requirements for SRB growth were so small (below a few mg l^{-1}) that effective removal by existing biological and/or chemical treatments becomes prohibitively costly. In petroleum reservoirs, for example, formation waters contain sufficient quantities of measurable ammonium ions to support balanced growth on the available carbon. Ammonium ion levels as high as 250 mg l^{-1} have been measured.[58] Ammonium levels up to 87 mg l^{-1} were measured in formation waters in the North Sea oil fields operated by Shell Expro.[58] The major source of nutrients in the oil reservoir is likely to be the formation water itself. Thus, removal of nitrogen from the injection water may not be a feasible means to control SRB activity in the oil field. Ironically, scale- and corrosion-inhibitor chemicals typically added to the injection water can enrich the system in C, N, and P.

Sulfate is sometimes absent in the oil reservoir *prior to* introduction of seawater in secondary recovery operations. Removing sulfate from the injection water may therefore be a feasible way to control SRB activity in oil reservoirs. However, biological processing for sulfate removal from seawater or produced water would be hindered by the same operating difficulties that recovery systems currently experience: high sulfide concentrations, a need for a treatment method for the produced sulfide, and high corrosion rates. Further technology development is needed in order to make such a system economically feasible.

VI. SUMMARY

Most of the kinetic and stoichiometric data reported above were obtained from pure cultures under well-controlled laboratory conditions. Real environments are much more complex and dynamic; many interactions between SRB and other components (the substratum, other bacterial species, and metabolic products) occur. Future investigations must be conducted with mixed populations containing SRB and in environments which are representative of industrial and natural systems. Comparison with pure culture data will provide information critical to further model development and calibration.

When the nutritional requirements of SRB and the effects of the physicochemical factors on SRB growth and activity are understood and quantified, a more rational image of SRB behavior in environments such as oil reservoirs during waterflooding can be obtained. This approach will eventually lead to the development of a comprehensive model that will permit more accurate prediction of SRB behavior. It will also permit means of controlling SRB growth and activity in industrial water systems, where SRB cause many problems.

VII. ACKNOWLEDGMENT

The authors gratefully acknowledge support from the cooperative agreement ECD-8907039 between Montana State University and the National Science Foundation, from the Center Industrial Associates Pooled Funds, and from the Montana Science and Technology Alliance.

REFERENCES

1. **Jorgensen, B. B.,** Bacterial sulfate reduction within reduced microniches of oxidized marine sediments, *Mar. Biol.*, 41, 7, 1977.
2. **Lee, W., Lewandowski, Z., Okabe, S., Characklis, W. G., and Avci, R.,** Corrosion of mild steel underneath aerobic biofilms containing sulfate-reducing bacteria, *Proc. Corrosion '92*, Paper No. 190, The NACE Annual Conference, Nashville, TN, 1992.
3. **Costerton, J. W. and Geesey, G. G.,** Microbial contamination of surfaces, in *Surface Contamination*, Vol. 1, Mittal, K. L., Ed., Plenum Press, New York, 1979, 57.
4. **Dewar, E. J.,** Control of microbiologically induced corrosion and accumulation of solids in a seawater flood system, *Mater. Perform.*, 25, 39, 1986.
5. **Rosnes, J. T., Graue, A., and Lien, T.,** Activity of sulfate-reducing bacteria under simulated reservoir conditions, Publication SPE 19429, presented at the SPE Formation Damage Control Symposium, Lafayette, LA, SPE, Houston, 1990, 231.
6. **Van Loosdrecht, M. C. M., Lyklema, J., Norde, W., and Zehnder, A. J. B.,** Influence of interfaces on microbial activity, *Microbiol. Rev.*, 54, 75, 1990.
7. **Lee, W.,** Corrosion of Mild Steel Under an Anaerobic Biofilm, Ph.D. dissertation, Montana State University, Bozeman, 1990.
8. **Walch, M.,** Introduction: microbiological influences on marine corrosion, *Mar. Technol. Soc. J.*, 24(3), 3, 1990.
9. **Witzgall, R. A., Horner, I. S., and Schafer, P. L.,** Case histories of sulfate corrosion: problems and treatments, *Water Environ. Technol.*, July, 42, 1990.
10. **Burger, E. D., Addington, D. V., and Crews, A. B.,** Reservoir souring: bacterial growth and transport, Proc. Fourth International IGT Symposium on Gas, Oil, and Environmental Biotechnology, Colorado Springs, CO, December 1991, IGT, Chicago, 1992.
11. **Cochrane, W. J., Jones, P. S., Sanders, P. F., Holt, D. M., and Mosley, M. J.,** Studies on the thermophilic sulfate-reducing bacteria from a souring north sea oil field, Publication SPE 18368, presented at SPE European Petroleum Conference, 1988, London, SPE, Houston, 1988, 301.

12. **Frazer, L. C. and Bolling, J. D.,** Hydrogen sulfide forecasting techniques for the Kuparuk river field, Publication 22105, Proc. International Arctic Technology Conference, Anchorage, AK, May 1991, SPE, Houston, 1991, 399.
13. **Herbert, B. N.,** Reservoir souring, in *Proc. Microbial Problems in the Offshore Oil Industry, Hill, E. C., Shennan, J. L., and Watkinson, R. J., Eds.,* John Wiley & Sons, New York, 1986, 63.
14. **Characklis, W. G.,** Biofouling: effects and control, in *Biofouling and Biocorrosion in Industrial Water Systems,* Flemming, H.-C. and Geesey, G. G., Eds., Springer-Verlag, Berlin, 1991, 7.
15. **Sanders, P. F. and Hamilton, W. A.,** Biological and corrosion activities of sulphate-reducing bacteria in industrial process plant, in Conference Proceedings, *Biologically Induced Corrosion,* Dexter, S., Ed., NACE, Washington, D.C., 1983, 47.
16. **Cord-Ruwisch, R., Kleinitz, W., and Widdel, F.,** Sulfate-reducing bacteria and their activities on oil production, *J. Petrol. Technol.,* January, 97, 1987.
17. **Postgate, J. R.,** Economic importance of sulphur bacteria, *Phil. Trans. R. Soc. London B.,* 298, 583, 1982.
18. **Postgate, J. R.,** *The Sulfate-Reducing Bacteria,* 2nd ed., Cambridge University Press, Cambridge, 1984.
19. **Hamilton, W. A.,** Sulfate-reducing bacteria and anaerobic corrosion, *Annu. Rev. Microbiol.,* 39, 195, 1985.
20. **Brunt, K. D.,** Biocides for the oil industry, in *Proc. Microbial Problems in the Offshore Oil Industry,* Hill, E. C., Shennan, J. L., and Watkinson, R. J., Eds., John Wiley & Sons, New York, 1986, 201.
21. **Eagar, R. G., Leder, J., Stanley, J. P., and Theis, A. B.,** The use of glutaraldehyde for microbiological control in waterflood systems, *Mater. Perform.,* 27(8), 40, 1986.
22. **Gaylarde, C. C. and Johnston, J. M.,** Some recommendations for sulfate-reducing bacteria biocide tests, *J. Oil Colour Chem. Assoc.,* 12, 305, 1983.
23. **Ege, S. L., Houghton, C. J., and Tucker, B. M.,** UV-control of SRB in injection water, *Conference Proceedings of U. K. Corrosion '85,* Harrogate, November 1985, 243.
24. **Agaev, N. M., Smorodin, A. E., and Guseinov, M. M.,** Influence of gamma radiation on the metabolic activity of sulfate-reducing bacteria, *Prot. Met.,* 21(1), 111, 1985.
25. **Maree, J. P. and Strydom, W. F.,** Biological sulfate removal from industrial effluent in an upflow pack bed reactor, *Water Res.,* 21, 141, 1987.
26. **Buisman, C. J., Wit, B., and Lettinda, G.,** Biotechnological sulfide removal in three polyurethane carrier reactors: stirred reactor, biorotor, and upflow reactor, *Water Res.,* 24, 245, 1990.
27. **McInerney, M. J., Sublette, K. L., and Montgomery, A. D.,** in *Proc. Microbial Problems in the Offshore Oil Industry,* Hill, E. C., Shennan, J. L., and Watkinson, R. J., Eds., John Wiley & Sons, New York, 1986, 441.
28. **Montgomery, A. D., McInerney, M. J., and Sublette, K. L.,** Microbial control of the production of hydrogen sulfide by sulfate-reducing bacteria, *Biotechnol. Bioeng.,* 35, 533, 1990.
29. **Sublette, K. L. and Sylvester, N. D.,** Oxidation of hydrogen sulfide by *Thiobacillus denitrificans*: desulfurization of natural gas, *Biotechnol. Bioeng.,* 29, 249, 1987.
30. **Sublette, K. L. and Sylvester, N. D.,** Oxidation of hydrogen sulfide by continuous culture of *Thiobacillus denitrificans, Biotechnol. Bioeng.,* 29, 753, 1987.
31. **Sublette, K. L. and Sylvester, N. D.,** Oxidation of hydrogen sulfide by mixed cultures of *Thiobacillus denitrificans* and heterotrophs, *Biotechnol. Bioeng.,* 29, 759, 1987.
32. **Sublette, K. L.,** Aerobic oxidation of hydrogen sulfide by *Thiobacillus denitrificans, Biotechnol. Bioeng.,* 29, 690, 1987.
33. **Abram, J. W. and Nedwell, D. B.,** Inhibition of methanogenesis by sulfate reducing bacteria competing for transferred hydrogen, *Arch. Microbiol.,* 117, 89, 1978.

34. **Isa, Z., Grusenmeyer, S., and Verstraete, W.,** Sulfate reduction relative to methane production in high-rate anaerobic digestion: microbiological aspects, *Appl. Environ. Microbiol.*, 51, 580, 1986.
35. **Krisjansson, J. K., Schönheit, P., and Thauer, R. K.,** Different K_s values for hydrogen of methanogenic bacteria and sulfate reducing bacteria: an explanation for the apparent inhibition of methanogenesis by sulfate, *Arch. Microbiol.*, 131, 278, 1982.
36. **Schönheit, P., Kristjansson, J. K., and Thauer, R. K.,** Kinetic mechanism for the ability of sulfate reducers to out-compete methanogens for acetate, *Arch. Microbiol.*, 132, 285, 1982.
37. **Yoda, M., Kitagawa, M., and Miyaji, Y.,** Long term competition between sulfate-reducing and ethane-producing bacteria for acetate in anaerobic biofilm, *Water Res.*, 21, 1547, 1987.
38. **Hilton, B. L. and Oleszkiewicz, J. A.,** Sulfide-induced inhibition of anaerobic digestion, *J. Environ. Eng.*, 114, 1377, 1989.
39. **Cypionka, H., Widdel, F., and Pfenning, N.,** Survival of sulfate-reducing bacteria after oxygen stress, and growth in sulfate-free oxygen-sulfide gradients, *FEMS Microbiol. Ecol.*, 31, 39, 1985.
40. **Hardy, J. A. and Hamilton, W. A.,** The oxygen tolerance of sulfate-reducing bacteria isolated from North Sea waters, *Curr. Microbiol.*, 6, 259, 1981.
41. **Characklis, W. G. and Marshall, K. C.,** *Biofilms*, Wiley-Interscience, New York, 1990.
42. **Okabe, S. and Characklis, W. G.,** Effects of temperature and phosphorous concentration on microbial sulfate reduction by *Desulfovibrio desulfuricans*, *Biotechnol. Bioeng.*, 39, 1031, 1992.
43. **Nielsen, P. H.,** Biofilm dynamics and kinetics during high-rate sulfate reduction under anaerobic conditions, *Appl. Environ. Microbiol.*, 53, 27, 1987.
44. **Okabe, S., Nielsen, P. H., and Characklis, W. G.,** Factors affecting microbial sulfate reduction by *Desulfovibrio desulfuricans* in continuous culture: limiting nutrients and sulfide concentration, *Biotechnol. Bioeng.*, 40, 725, 1992.
45. **Lappan, R. E. and Fogler, H. S.,** The effects of bacterial polysaccharide production on formation damage, Publication SPE 19418, Presented at the SPE Formation Damage Control Symposium, Lafayette, LA, SPE, Houston, 1990, 165.
46. **Ruseska, I., Robbins, J., Costerton, J. W., and Lashen, E. S.,** Biocide testing against corrosion-causing oil-field bacteria helps control plugging, *Oil Gas J.*, 80, 253, 1982.
47. **Reis, M. A., Lemos, P. C., Almeida, J. S., and Carrondo, J. T.,** Evidence for the intrinsic toxicity of H_2S to sulphate-reducing bacteria, *Appl. Microbiol. Biotechnol.*, 36, 145, 1991.
48. **Shimada, K.,** Removal of heavy metals from mine wastewater using sulfate-reducing bacteria, *J. Water Wastewater, Jpn.*, 31(5), 52, 1987.
49. **Klemps, R., Cypionka, H., Widdel, F., and Pfenning, N.,** Growth with hydrogen, and further physiological characteristics of *Desulfotomaculum* species, *Arch. Microbiol.*, 143, 203, 1985.
50. **McCartney, D. M. and Oleszkiewicz, J. A.,** Sulfide inhibition of anaerobic degradation of lactate and acetate, *Water Res.*, 25, 203, 1991.
51. **Min, H. and Zinder, S. H.,** Isolation and characterization of a thermophilic sulfate-reducing bacterium *Desulfotomaculum thermoacetoxidans* sp. nov, *Arch. Microbiol.*, 153, 399, 1990.
52. **Speece, R. E.,** Anaerobic biotechnology for industrial wastewater treatment, *Environ. Sci. Technol.*, 17, 416A, 1983.
53. **Davies, D. G., Chakrabarty, A. M., and Geesey, G. G.,** Exopolysaccharide production in biofilms: substratum activation of alginate gene expression by *Pseudomonas aeruginosa*, *Appl. Environ. Microbiol.*, 59, 1181, 1993.
54. **Fukui, M. and Takai, S.,** Colony formation of free-living and particle-associated sulfate-reducing bacteria, *FEMS Microbiol. Ecol.*, 73, 85, 1990.

55. **Nielsen, P. H. and Hvitved-Jacobsen, T.,** Effect of sulfate and organic matter on the hydrogen sulfide formation in biofilms of filled sanitary sewers, *J. Water Pollut. Control Fed.*, 60, 627, 1988.
56. **Holder, G. A.,** Prediction of sulfide build-up in field sanitary sewers, *J. Environ. Eng.*, 112, 199, 1986.
57. **Holder, G. A., Vaughan, G., and Drew, W.,** Kinetic studies of the microbiological conversion of sulfate to hydrogen sulfide and their relevance to sulphide generation within sewers, *Water Sci. Technol.*, 17, 183, 1984.
58. **Herbert, B. N., Gilbert, P. D., Stockdale, H., and Watkinson, R. J.,** Factors controlling the activity of SRB in reservoirs during water injection, Publication SPE 13978, Soc. Petrol. Eng., Houston, 1985, 1.

Section III
BIOCORROSION

13

Microbial Corrosion of Mild Steel in a Biofilm System

WhonChee Lee, Zbigniew Lewandowski, William G. Characklis, and Per H. Nielsen

I. INTRODUCTION

Mild steel is commonly used in industrial water systems such as recirculating cooling water and water injection during oil production. Biofouling and biocorrosion are major concerns in those systems. Considerable attention has been devoted to the microbial corrosion of ferrous metals, and, in particular, the corrosion of mild steel by sulfate-reducing bacteria (SRB) has been the subject of investigation since 1934.[1]

Microbial corrosion usually results from the presence of biofilms at the metal surface. The biofilm system consists of three compartments: (1) the bulk liquid, (2) the biofilm, and (3) the substratum. Biofilms influence the corrosion processes by changing the local chemistry near the metal surface. The interfacial water chemistry between metal surface and biofilm is determined by (1) corrosion processes, (2) microbial activity and metabolic products, (3) mass transport processes within the system compartments, and (4) bulk water chemistry. As corrosion and biofilm accumulation are dynamic processes, the interfacial water chemistry changes with time. Thus, microbial corrosion is the net result of the interaction between various compartments which are in dynamic flux. To study microbial corrosion, both the conceptual models and experimental methodology must reflect and respond to these circumstances.

0-87371-928-X/94/$0.00+$.50

Our approach to simulate microbial corrosion was to grow a biofilm containing SRB on a metal substratum either in a totally anaerobic system or in a system containing aerobic bulk liquid. The microbial activities were limited to the sessile biofilm cells by conducting experiments at a high dilution rate to wash out the planktonic cells. *In situ* electrochemical measurements (DC and AC methods), *in situ* microelectrode measurements (dissolved oxygen, pH, and dissolved sulfide), sulfide compounds analysis, and surface analysis were performed periodically on coupons introduced into the systems.

This chapter reviews several mechanisms proposed to account for the SRB-enhanced corrosion of mild steel. The rate-controlling step for SRB-enhanced corrosion is suggested.

II. PROPOSED MECHANISMS FOR SRB-ENHANCED CORROSION

Based on laboratory and field studies, four mechanisms for SRB-enhanced corrosion were suggested to explain the high corrosion rate observed in the field:

1. Hydrogenase mechanism
2. Iron sulfide mechanism
3. Hydrogenase + iron sulfide mechanism
4. Aerobic/anaerobic mechanism

Von Wolzogen Kühr and van der Klugt[1] proposed the cathodic depolarization theory for SRB-enhanced corrosion in a soil environment. They suggested that SRB could remove hydrogen from a cathodic area on the iron surface by the hydrogenase enzyme. This enhances the anodic dissolution of iron.

Miller and Tiller[2] have proposed cathodic depolarization induced by microbially produced FeS to be the major factor affecting corrosion. King et al.[3] and Booth et al.[4] demonstrated that weight losses of steel were proportional to the concentration of ferrous sulfide present and were dependent on the stoichiometry of the particular ferrous sulfides. They concluded that the accelerated corrosion of mild steel in the presence of SRB was due to formation of iron sulfide.

King and Miller[5] supported the hydrogenase theory, but proposed that cathodic depolarization occurred on the iron sulfide instead of the unreacted iron surface. The role of bacteria could be either to remove the accumulated hydrogen or to produce iron sulfide to depolarize the steel.

The importance of oxygen on the SRB-related corrosion of ferrous metal and alloys has been emphasized by several authors. Accelerated corrosion rates have been observed when iron, corroded in the presence of SRB, is exposed to intermittent aerobic/anaerobic conditions.[6] Hardy and Bown[7] showed a similar effect of aeration on mild steel. They observed that sulfide film formed in a batch reactor during growth was not uniform, but consisted of an adherent base layer and loose deposits of precipitated material at a few sites. Pitting corrosion occurred beneath these loose deposits during an aeration period subsequent to growth. However, the role of dissolved oxygen in SRB-related corrosion is not well understood.

III. SRB BIOFILM

Accumulation of biofilm in a totally anaerobic bulk liquid is linked to the activity of a number of facultative anaerobic and obligately anaerobic microorganisms (SRB), which reduce sulfate to sulfide. Usually the growth rates and biofilm accumulation are slower than that for aerobic (or denitrifying) biofilms.[8] However, if high concentrations of an organic electron donor and sulfate as the only electron acceptor are present, a thick (several millimeters) sulfate-reducing biofilm will usually develop within days or weeks if SRB are present in the bulk water. The activity of SRB in the biofilm is frequently expressed as the sulfate reduction rate or sulfide production rate. In systems with low sulfate concentrations, e.g., some freshwater systems, the concentration of sulfate can be limiting for the SRB activity within the biofilm. Nielsen[8] showed an example of sulfate limitation in a totally anaerobic biofilm growing at high lactate concentration in a continuous-flow biofilm reactor. Diffusion of sulfate into the biofilm was limiting for the activity at sulfate concentrations below 0.5 to 1.0 m*M*, depending on the biofilm thickness. Once the biofilm thickness exceeded the depth of sulfate penetration into the biofilm, the sulfate removal rate was unaffected by further biofilm accumulations. The specific biofilm activity (56 to 93 μmol SO_4 cm^{-3} h^{-1}) was very high in this study due to the presence of high concentrations of lactate, which are seldom characteristic of natural or industrial water systems. At lower concentrations of metabolizable electron donors, much lower specific activities are to be expected.

SRB can be active within anaerobic microniches or in an anaerobic bottom layer when the bulk water contains oxygen. Formation and maintenance of such microniches are explained by two factors: (1) the respiration of aerobic bacteria scavenges oxygen and favors growth conditions for SRB, and (2) H_2S produced by SRB is a reductant that reacts with oxygen. Thus, if once established, colonies of SRB can protect themselves against oxygen. Depending on the physical conditions prevailing and the primary nutrients available, microbial consortia or biofilm may contain a wide range of aerobic, facultative, and anaerobic organisms, which together are capable of hydrolytic and fermentative metabolism of primary nutrient sources such as carbohydrate, protein, and other organic hydrocarbons. The initial activities of the aerobic and facultative species will tend to utilize the available oxygen at a rate faster than the diffusional rate from the bulk liquid into the biofilm with the result that the innermost parts of the consortium will become anaerobic. The sulfate-reducing activity in the anaerobic microenvironment is usually limited by availability of electron donors, but the actual activity is difficult to measure. Use of radiotracer added directly to an aerobic incubation may give some indication about the actual SRB activity,[9] but due to an internal reoxidation of the produced sulfide an underestimation can be expected.

IV. CORROSION OF MILD STEEL UNDER A TOTALLY ANAEROBIC BIOFILM

Effects of SRB on the corrosion of iron or steel have been extensively studied for at least 50 years. Most experiments were conducted in batch, semicontinuous, and continuous (chemostat) reactors containing planktonic populations. SRB biofilm development on steel surfaces and their subsequent influence on corrosion in totally anaerobic systems have only been reported recently.[10,11] The difference between a biofilm reactor and chemostat or batch reactor is that sessile SRB dominate the overall SRB activities in a biofilm reactor while planktonic SRB dominate the overall SRB activities in a chemostat or batch reactor. This section will focus on the influence of system variables — which include (1) effect of SRB activities within biofilms and (2) effect of suspended ferrous sulfide — on corrosion of mild steel in a totally anaerobic biofilm system.

A. Effect of SRB Activities within the Biofilms on Corrosion

Accelerating corrosion of steel by SRB can be attributed either directly to removal of hydrogen or indirectly to the formation of iron sulfides and dissolved sulfides. If the hydrogenase mechanism is the rate-controlling step in the overall corrosion processes, then the sessile (or biofilm) SRB should play a more significant role in determining corrosion rate than the planktonic SRB due to a decrease in diffusional distance from the metal surface. Moosavi et al.[12] have found that, based on field studies, only an indirect correlation exists between SRB activity and corrosion rate, and that the mechanisms of corrosion relate more closely to the chemical and physical nature of the corrosion products in the seabed close to an offshore platform. A radiorespirometric assay has been used to determine *in situ* SRB activity within the biofilm in those studies which only give qualitative information about SRB activities due to the limitation in simulating the real environmental conditions. McKenzie and Hamilton[13] further identified that sulfate limitation and the production of both acid-volatile and non-acid-volatile sulfides as key factors which must be incorporated into the [^{35}S]-sulfate reduction assay. Lee and Characklis[14] have demonstrated in a laboratory study that there is no correlation between biofilm (or sessile) SRB activity and corrosion rate in an iron-free medium. The biofilm SRB activity, e.g., sulfate removal rate, was controlled by the lactate loading rate. They also demonstrated that the growth rate of SRB within a biofilm was limited by nutrient transport and that the active SRB were located in the uppermost part of the biofilm. There was no detectable difference in either weight loss or electrochemical measurements of corrosion as total SRB activity increased from 21.6 to 338 mg/h during a 3-week experiment. The nature and extent of corrosion were closely related to the physical forms of iron sulfide but not to SRB activity.

B. Effect of Suspended Ferrous Sulfide

The effect of ferrous sulfide concentration on corrosion of mild steel in a suspended culture has been reported by several authors. Based on a semicontinuous culture experiment, Booth et al.[15] reported that the presence of sufficient ferrous ions in the medium to precipitate the biogenic sulfide and inhibit protective film formation led to a great increase in corrosion rate. Mara and Williams,[16] using a chemostat, established that an adherent film of sulfide on the steel surface was observed when the low corrosion rate was proportional to the bacterial growth rate in a low iron medium. After the film ruptured, the rate of corrosion was very high and independent of bacterial growth rate. Film breakdown was attributed to compressive stress established at the film-metal interface subsequent to the sulfidation of the primary corrosion product mackinawite, FeS_{1-x}, to greigite, Fe_3S_4. There was no evidence of protective film formation, and a high rate of corrosion was observed in high iron medium.

The effects of suspended ferrous sulfide on the corrosion of steel in an anaerobic biofilm reactor were studied by Lee and Characklis.[14] The biofilm was developed on steel coupons in iron-free medium, followed by weekly step increases in ferrous ion to 1, 10, and 60 mg/l in the influent. Precipitation of ferrous sulfide ($Fe^{2+} + HS^- \rightarrow FeS + H^+$) took place mainly in the liquid phase. The total dissolved sulfide concentration decreased from 50 to 40 mg/l as ferrous ion concentration in the influent increased from 1 to 10 mg/l. Very little corrosion occurred at this time, but accumulation of iron sulfide particles increased in the biofilm. However, iron-rich medium (60 mg/l) resulted in the precipitation of all the biogenic sulfide. The iron sulfide in the biofilm contacted the steel surface and the corrosion rate increased dramatically. There was no continuous and uniform protective coating of iron sulfide film formed under the biofilm, and intergranular attack was observed over the entire metal surface.

Further evidence regarding the influence of the morphology of the iron sulfide scale in a biofilm system is provided by the work of Moosavi et al.[12] They observed a clearly defined layering of corrosion products on mild steel coupons recovered from the seabed close to an offshore platform after 1 to 2 years of exposure. Next to the steel surface there was a thin black adherent layer; superimposed on this was a looser bulky black deposit, surmounted by a brown oxide layer. There was evidence of enhanced corrosion associated with break-up of this adherent film.

From the above discussion, it can be concluded that (1) the formation of adherent iron sulfide film is ferrous ion-concentration dependent and (2) the rate of corrosion is determined by the physical nature of sulfides.

V. CORROSION OF MILD STEEL UNDER AEROBIC/ ANAEROBIC BIOFILM

McKenzie and Hamilton[13] have proposed that environmental factors, most notably oxygenation, are central to determining the rate and extent of SRB-related corrosion. They established that high rates of corrosion associated with deep pitting were observed in the aerobic zone as compared to that in the anaerobic zone of a column reactor, and that corrosion products in the aerobic zone contained a higher percentage of non-acid-volatile sulfides (elemental sulfur and/or pyrite) than that in the anaerobic zone.

The initial corrosion processes of mild steel under an aerobic/anaerobic biofilm have been studied by Lee et al.[15] They demonstrated that SRB can proliferate in a biofilm even when oxygen penetrated the entire biofilm at some locations. Sulfate was not the limiting factor for SRB growth within the biofilms. The role oxygen plays in SRB-related corrosion is through the formation of different iron sulfide compounds which are corrosive toward mild steel.[10,11] In general, abiotic oxidative processes are much more rapid than biotic respiration during the early stage of biofilm accumulation. Aerobic-dominated corrosion and the corrosion rate decreased with time. SRB-enhanced corrosion occurred after the SRB had further developed and a significant amount of iron sulfide had contacted the metal. In addition to the acid-volatile sulfides (mackinawite or greigite), elemental sulfur and pyrite were also detected within the corrosion deposits. Deep pits were observed within a crevice corrosion area caused by sulfide attack. Deep pitting was attributed to a large cathode-anode ratio. The corrosion rate increased from 8 to 160 mpy as the pool of Fe–S compounds (mackinawite, pyrite, elemental sulfur) increased from 12 to 600 μg S cm^{-2}.[10,11]

VI. SUMMARY

From the various mechanisms dealing with SRB-related corrosion, it appears that corrosion caused by SRB activity is highly dependent on the structures and compositions of the corrosion deposits. Key elements in this mechanism appear to be the elemental sulfur and sulfide compounds and their location within corrosion deposits. This reinforces the importance of galvanic theory to SRB-related corrosion. Under totally anaerobic conditions, the corrosion rate of mild steel is not controlled directly by the SRB activities, but indirectly through iron sulfide formation. The corrosion rate followed first-order kinetics with respect to suspended ferrous sulfide concentration.[3,17] McNeil and Little[18] showed that mackinawite is the major corrosion product when iron alloys are corroded by SRB under totally anaerobic conditions. However, elemental sulfur and pyrite were detected in addition to mackinawite in oxic/anoxic conditions.[10,11] The effect of dissolved oxygen on SRB-related corrosion can be attributed to the formation of corrosive sulfides and/or elemental sulfur. It is difficult to determine which iron sulfide is more corrosive to mild steel, but a relationship between corrosion rate and areal sulfide concentration clearly exists.

ACKNOWLEDGMENTS

The authors gratefully acknowledge the cooperative agreement ECD-8907039 between Montana State University and the National Science Foundation. The Center for Interfacial Microbial Process Engineering Industrial Associates and Montana Science & Technology Alliance are also acknowledged for their support of this work.

REFERENCES

1. **von Wolzogen Kühr, C. A. H. and van der Klugt, L. S.,** Graphitization of cast iron as an electro-biochemical process in anaerobic soils, *Water,* 18, 147, 1934.
2. **Miller, J. D. A. and Tiller, A. K.,** Microbial corrosion of buried and immersed metal, in *Microbial Aspect of Metallurgy,* Miller, J. D. A., Ed., Elsevier, New York, chap. 3.
3. **King, R. A., Miller, J. D. A., and Walkerly, D. S.,** Corrosion of mild steel in cultures of sulfate-reducing bacteria: effect of changing soluble iron concentration during growth, *Br. Corros. J.,* 8, 89, 1973.
4. **Booth, G. H., Elford, L., and Wakerly, D. S.,** Corrosion of mild steel by sulfate-reducing bacteria: an alternative mechanism, *Br. Corros. J.,* 3, 242, 1968.
5. **King, R. A. and Miller, J. D. A.,** Corrosion by the sulfate-reducing bacteria, *Nature,* 233, 491, 1971.
6. **Starkey, R. L.,** Anaerobic corrosion — perspectives about causes, *Biologically Induced Corrosion,* Dexter, S. D., Ed., NACE, Houston, 1985, 3.
7. **Hardy, J. A. and Bown, J. L.,** The corrosion of mild steel by biogenic sulfide films exposed to air, *Corrosion,* 40, 650, 1984.
8. **Nielsen, P. H.,** Biofilm dynamics and kinetics during high-rate sulfate reduction under anaerobic conditions, *Appl. Environ. Microbiol.,* 53, 27, 1987.
9. **Hamilton, W. A. and Maxwell, S.,** Biological and corrosion activities of sulfate-reducing bacteria within natural biofilms, *Biologically Induced Corrosion,* Dexter, S. D., Ed., NACE, Houston, 1985, 131.
10. **Lee, W., Lewandowski, Z., Morrison, M., Characklis, W. G., Avci, R., and Nielsen, P. H.,** unpublished data, 1992.
11. **Nielsen, P. H., Lee, W., Lewandowski, Z., Morrison, M., Characklis, W. G.,** unpublished data, 1992.
12. **Moosavi, A., Pirrie, R., and Hamilton, W. A.,** Effect of sulfate reducing bacteria activity on performance of sacrificial anodes, *Microbially Influenced Corrosion and Biodeterioration,* Dowling, N. J., Mittelman, M. W., and Danko, J. C., Eds., The University of Tennessee, Knoxville, TN, 1990, 3.
13. **McKenzie, J. and Hamilton, W. A.,** The assay of *in situ* activities of sulfate-reducing bacteria in a laboratory marine corrosion model, *Int. Biodet.,* Elsevier Science.
14. **Lee, W. and Characklis, W. G.,** Corrosion of mild steel under an anaerobic biofilm, *Corrosion,* 49, 186, 1993.
15. **Booth, G. H., Shinn, P. M., and Wakerley, D. S.,** The influence of various strains of actively growing sulfate-reducing bacteria on the anaerobic corrosion of mild steel, *Congress International de la Corrosion Marine et des Salissures,* CREO, Paris, 1965, 363.
16. **Mara, D. D. and Williams, D. J. A.,** The mechanism of sulfide corrosion by sulfate-reducing bacteria, in *Biodeterioration of Materials,* Walter, A. H. and Van Der Plas, E. H. H., Eds., Applied Science Publishers, London, 1972, 103.
17. **King, R. A. and Wakerley, D. S.,** Corrosion of mild steel by ferrous sulfide, *Br. Corros. J.,* 8, 41, 1973.
18. **McNeil, M. B. and Little, B. J.,** Mackinawite formation during microbial corrosion, *Corrosion,* 46, 599, 1990.

Section III
BIOCORROSION

14

Indicators for Sulfate-Reducing Bacteria in Microbiologically Influenced Corrosion

Brenda Little and Patricia Wagner

I. INTRODUCTION

Microbiologically influenced corrosion (MIC) is localized corrosion and results in pitting, crevice corrosion, selective dealloying, stress corrosion cracking, or under-deposit corrosion. Since MIC does not produce unique forms of corrosion, investigators have relied on the shape, color, smell, and morphology of surface deposits in association with numbers and types of organisms to indicate MIC. Iron-oxidizing,[1–10] sulfur-oxidizing,[1,2,11,12] iron-reducing,[13] sulfate-reducing,[14–22] acid-producing,[3,4,23–32] slime-producing,[3–5,31,33] ammonium-producing,[3–5,34] and hydrogen-producing bacteria,[35] in addition to other physiological groups, have been implicated in the corrosion of metals and alloys. However, the most widely recognized and most frequently cultured bacteria in corrosion processes are the bacteria that reduce sulfate to sulfide which are collectively called sulfate-reducing bacteria (SRB).[36] SRB constitute a physiological-ecological assemblage of morphologically different types of anaerobic bacteria, which have in common the capacity to reduce sulfate to hydrogen sulfide in dissimulatory energy-conserving reactions.[37] Hydrogen sulfide can react with metals to produce metal sulfides as corrosion products. Recently, investigators demonstrated that micro-

0-87371-928-X/94/$0.00+$.50

organisms on metal surfaces produce unique environments at metal-biofilm interfaces that differ from the bulk solution in pH, dissolved oxygen, and organic and inorganic species, producing minerals whose formation cannot be predicted from strictly thermodynamic arguments.[38] Sulfide minerals that are diagnostic for MIC of iron and copper alloys have been identified. Sulfur isotope fractionation within sulfide corrosion products can also indicate microbial processing. In this chapter, traditional and innovative indicators for corrosion caused by SRB will be reviewed.

Bacteria attach to all materials exposed to natural aquatic environments and produce viscoelastic layers of immobilized cells embedded in a polymer matrix of microbial origin which is absorptive and porous. As a result, biofilms contain absorbed and entrapped materials including solutes, heavy metals, and inorganic particles in addition to cellular constituents.[39,40] The rate of biofilm formation depends on the substratum and the nature of the electrolyte.[41,42] Hydrodynamic shear stress related to flow influences transport, transfer, and reaction rates within biofilms as well as detachment. Biofilms provide protective environments for bacteria and allow different types of bacteria to flourish within different strata.[43] Bacteria within biofilms act symbiotically to produce conditions more favorable for the growth of each species. Bacteria near the fluid phase are provided with complex nutrients and oxygen from oxygenated media. Bacteria at the oxygenated fluid interface use oxygen, break down carbon sources, and produce simple polymers and fatty acids. Bacteria within the biofilm, removed from the bulk phase, further degrade organic molecules to fatty acids, carbon dioxide, and hydrogen. Successive stages of degradation depend on the chemistry of the liquid phase and the bacterial species. However, when the rate of respiration exceeds the transport rate of oxygen into the biofilm, the substratum-biofilm interface becomes anaerobic.[44] When sulfates are present, SRB predominate in anaerobic niches because hydrogenase-positive SRB can reduce hydrogen produced by acetogenic bacteria to such a low level that other bacteria cannot compete. Hydrogenase breaks hydrogen molecules down to protons and electrons and transfers energy to the organism through a series of organometallic molecules in the electron transport system.[45]

Symptoms of SRB-influenced corrosion are obvious — hydrogen sulfide odor, blackening of waters, and black sulfide corrosion products. Standard practices for evaluating the contribution of SRB to corrosion processes depend on detection and quantification of SRB using culture techniques that enumerate organisms or biochemical assays that quantify intrinsic characteristics of SRB, including enzymes and antibodies. Mineralogy of metal sulfides[38] and sulfur isotope fractionation[46] can also be used to verify the involvement of SRB in corrosion.

II. CULTURE TECHNIQUES

The American Petroleum Institute (API, New York, NY) Recommended Practice (RP-38)[47] for the enumeration of SRB in subsurface injection waters specifies sodium lactate as the carbon source. When SRB are present in the sample, they reduce sulfate in the medium to sulfide which reacts with iron in solution to produce black ferrous sulfide. Blackening of the medium over a 28-d period indicates the presence of SRB. Williams Bros. Laboratories (Tulsa, OK) and others provide API medium in 10-ml quantities in vials containing an iron nail. For tenfold dilutions, 1-ml samples are injected by syringe into media bottles. It is assumed that a single living bacterium will produce a black color. The simplest interpretation of test results is to consider that if one bottle is blackened, the sample contained at least one organism; if two bottles are blackened, the sample contained 10 organisms; three bottles, 100 organisms, and so on. This procedure is repeated as many times as necessary to produce a final dilution that does not contain organisms as judged by the blackening of the medium.

A solid medium slant or "agar deep" developed by Biosan Laboratories (Ferndale, MI) uses a modification of API medium with sodium sulfite as the reducing agent/oxygen scavenger.[48] An agar slant is inoculated by dipping a pipe cleaner into the undiluted sample and inserting it into a single vial of semisolid agar. Mineral oil and a CO_2-generating tablet are added to exclude oxygen, and the vial is capped, incubated for 5 d, and checked daily for blackening. A variation of the agar slant, developed by Nalco Chemical Co. (Chicago), uses melted agar, tryptone as nutrient, and sodium sulfite as an oxygen scavenger. The test procedure involves placing tubes of solidified agar in boiling water and cooling the tubes until the semisolid agar reaches a temperature of 40 to 45°C before adding the sample to the bottom of the tube with a pipette. Tubes are capped tightly and incubated for 3 d. Results are interpreted by multiplying the number of discrete colonies by the dilution factor. If a tube is generally black with no discrete colonies, cells are assumed to be too numerous to count and the serial dilutions were not adequate to determine an accurate estimate.

Tatnall et al.[49] evaluated API medium, agar melt, and agar deep techniques in terms of accuracy, specificity, ease of use, time required to obtain results, field vs. laboratory use, and obvious drawbacks and limitations, using known quantities of *Desulfovibrio desulfuricans* isolated from oilfield-produced water and unknown samples taken directly from oilfield-produced water without any attempt to concentrate SRB. Among the culturing techniques, RP-38 broth bottles gave accurate numbers at high cell concentrations, but overestimated numbers at low cell concentrations (Table 1). Agar deeps accurately estimated known populations

TABLE 1. Test Response vs. Known Number: API Strain

Actual count	SRB numbers obtained with:		
	RP-38	Agar deeps	Melt agar
0	NEG	$>10^5$, NEG	NEG
10^2	10	NEG	NEG
10^3	10^2	NEG	NEG
10^4	10^3–10^4	$\geq 10^4$	17, $>10^3$
10^5	10^4–10^5	$>10^4$, NEG	NEG
10^6	10^5–10^6	$>10^5$, NEG	80, 25
10^7	10^7	$>10^5$	3.5×10^2
10^8	10^7–10^8	$>10^6$	2×10^3

Double values separated by comma indicate that duplicate tests gave widely differing results.

From Tatnall, R. E., Stanton, K. M., and Ebersole, R. C., in *Corrosion 88*, Paper No. 88, National Association of Corrosion Engineers, Houston, TX, 1988. With permission.

at low cell concentrations, but greatly underestimated high cell concentrations. Melt agar tubes performed poorly at all concentrations.

The distinct advantage of culturing techniques is that they are extremely sensitive. Low numbers of SRB grow to easily detectable higher numbers in the proper culture medium. However, growth media tend to be strain specific. For example, lactate-based media like that specified by API will sustain the growth of lactate oxidizers, but not acetate-oxidizing bacteria. Attempts to prepare all-purpose SRB media have been unsuccessful. Incubating at one temperature is further selective. Culturing methods using agar media cannot distinguish between a single SRB cell and a clump of SRB cells. Although API broth bottles were the most consistently accurate of the three culturing techniques evaluated by Tatnall et al.,[49] 4 weeks were required for final results, and 28-d results were sometimes significantly different from 14- or 21-d results. Deep agar tubes were the easiest and fastest to use because no special equipment was required and results could be determined within a maximum of 5 d.

III. BIOCHEMICAL ASSAYS

Unlike culturing techniques, biochemical tests for detecting and quantifying SRB do not require SRB growth. Instead, direct methods measure constitutive properties including adenosine-5′-phosphosulfate (APS) reductase,[49] hydrogenase,[45] cell-bound antibodies,[50,51] and DNA.[52] Attempts have also been made to use adenosine triphosphate (ATP)[53] and radiorespirometric measurements for estimates of SRB activity.[54,55]

The APS reductase antibody method was developed by the Du Pont Company.[49] APS reductase is an intercellular enzyme found in all SRB. Briefly, cells are washed to remove interfering chemicals, including hydrogen sulfide, and lysed to release APS reductase. The lysed sample is washed, added to an antibody reagent, and exposed to a color-developing solution. A blue color appears within 10 min in the presence of APS reductase. The color intensity is proportional to the amount of enzyme and roughly to the number of cells from which the enzyme was extracted. Similarly, a procedure to quantify hydrogenase from SRB has been developed which requires that cells be concentrated by filtration from water samples.[45] Solids, including corrosion product and sludge, can be used without pretreatment. The sample is exposed to an enzyme-extracting solution for 15 min and placed in an anaerobic chamber from which oxygen is removed by hydrogen. Hydrogenase reacts with excess hydrogen and simultaneously reduces an indicator dye in solution. The activity of the hydrogenase is established by the development of a blue color in less than 4 h. The intensity of the blue color is proportional to the rate of hydrogen uptake by the enzyme. The technique does not attempt to estimate specific numbers of SRB. Not all SRB possess hydrogenase enzyme. Furthermore, the enzyme is not specific for SRB, i.e., other physiological types of bacteria possess hydrogenase enzyme.

Recently, Scott and Davies[36] evaluated test kits based on culture techniques and the detection of enzymes in SRB (APS reductase and hydrogenase) for detection and quantification of SRB. Liquid culture techniques included MICKIT III-C (Bioindustrial Technologies Inc., 40200 Industrial Park Circle, Georgetown, TX 78626) and SRB broth (Petrolite Co., 369 Marshall Ave., St. Louis, MO 63119). Solid or semisolid agar techniques included SANI-CHECK® SRB Test System (Biosan Laboratories, Inc., 10657 Galaxie, Ferndale, MI 48220–2133), and BUG CHECK SRB (Angus Chemical Co., Biocides Division, 2211 Sanders Rd., Northbrook, IL 60062). RapidChek SRB Detection System (Conoco Specialty Products, 600 North Dairy Ashford, Houston, TX 77079) is an APS reductase assay. The Hydrogenase test (Caproco Limited, 8741–53 Avenue, Edmonton, Alberta, Canada T6E 5E9) measures cellular hydrogenase. Test kits were received directly from manufacturers and sent to testers.[36] All kits were tested with environmental samples from the tester company and were compared with the in-house culture methods. In-house culture techniques were the media most commonly used by the organizations in their own SRB test programs and were not specifically identified except that they did include API broth and Postgate[19] media.

Results (Table 2) indicate that for produced water samples the populations estimated with culture techniques were duplicated in most cases with the APS reductase assay. The hydrogenase assay did not consistently reproduce the estimates of the other methods. There was no attempt to quantify the percentage of the natural population that was hydrogenase positive. The same general trends were observed with biofilms removed from coupons and sludge samples (Table 3).

Field and laboratory epifluorescence cell surface antibody (ECSA) methods for detecting SRB have been developed by D. H. Pope and the Materials Technology Institute of the Chemical Process Industries, Inc. (St. Louis, MO).[50,51] Both methods are based on the use and subsequent detection of specific antibodies, produced in rabbits, that react with SRB cells. A secondary antibody, produced in goats, is then reacted with the primary rabbit antibodies bound to the SRB cells. In the laboratory method, the goat antibodies are linked to a fluorochrome which enables bacterial cells marked with the secondary antibody to be viewed with an epifluorescence microscope. In the field method, the goat antibodies are conjugated with an enzyme (alkaline phosphatase) that can then be reacted with a colorless substrate to produce a visible color proportional to the quantity of SRB present. Comparisons of the laboratory fluorescent antibody (FA) method to traditional culture techniques for detecting SRB demonstrated that the laboratory FA method is more sensitive to SRB. The detection limit for the laboratory method can be as low as one SRB/ml. Laboratory and field methods gave comparable results for high numbers of SRB.[51] The detection limits for the field test are 10,000 SRB/ml. The color reagent used for the field tests is unstable at room temperature and the color reagent tends to bind nonspecifically with antibodies adsorbed directly at active sites on the filter, creating a false positive that may interfere with the detection of SRB at levels below 10,000 cells/ml.[51] Antigenic structures of marine and terrestrial strains are distinctly different; therefore, antibodies to either strain did not react with the other. Furthermore, SRB antibodies did not react with non-SRB bacteria. The developers report a poor response of rabbit antibodies developed from pure SRB cultures to mixed populations.[51] Rabbit SRB antibodies generated from fresh SRB strains from Prudhoe Bay, AK, as well as terrestrial and marine locations, were found to react better with SRB from natural sources. Tatnall et al.[49] reported that the ECSA method did not detect cells below 10^6 cells/ml in 13 samples of water from oil production in Oklahoma and Texas, representing injection waters from waterflood fields, produced water drawn directly from well heads, and separated water phase from knockout drums.[49] The ECSA method was able to detect cells in field samples better than in laboratory samples. It is believed that the antigenic properties of the cell surface change after repeated laboratory culture.

Gen-Probe Inc. (San Diego, CA) has developed a nonisotopic semiquantitative procedure for the detection of *Desulfobacterium* and *Desulfotomaculum* using DNA probes labeled with an acridinium ester and sensitive to 10^4 organisms/ml.[52] DNA probes are directed towards ribosomal RNA and may be viewed as consisting of three to four steps: (1) sample handling, (2) binding the probe to the target, (3) removal or destruction of the unbound probe, and (4) detection and quantification of the reporter group on the bound probe. The DNA probes have not been evaluated for field samples or compared with culture techniques.

ATP is a compound found in all living matter. ATP assays estimate the total number of viable organisms, not specifically SRB, by measuring the ATP in a sample. Littman[53] proposed that ATP assays could be used with oilfield water samples to estimate relative numbers of SRB. The procedure requires that a water sample be filtered to remove solids and salts and added to a reagent that releases cell ATP. An enzyme then reacts with the ATP to produce a photochemical reaction. Emitted light can be measured with a photometer and the number of bacterial cells is estimated from the total light emitted.

Because ATP is present in all living cells, one would predict that ATP methods would overestimate SRB populations. Instead, Tatnall et al.[49] found that ATP assays underestimated the number of cells by one to two orders of magnitude above 10^6 cells/ml, and the data were inconsistent.[49] ATP analysis showed reasonable correlation with API cultures only 4 of 12 times. ATP methods are subject to numerous interferences, including heavy metals found in corrosion products.

After concluding that sulfate reduction by SRB could not be accurately measured in the bulk phase, Rosser and Hamilton,[54] with subsequent modifications,[55] developed a test tube technique to perform a ^{35}S sulfate radiorespirometric assay to measure SRB metabolic activity on the surface of metal coupons after exposure to corrosive environments. The coupon is placed into anaerobic sterile medium containing ^{35}S sulfate. Oxygen-free zinc acetate is immediately injected onto an enclosed filter paper wick and the entire system is incubated. Oxygen-free hydrochloric acid is then injected past the wick into the solution. Volatile acid sulfides, including any $H_2^{35}S$ formed, are trapped during an equilibration period. The wick is removed from the tube and the radioactivity is measured using a liquid scintillation counter, after which the sulfate reduction rate is calculated. Maxwell and Hamilton[55] concluded that measurements of SRB activity on fouled engineering surfaces were a more accurate assessment of the corrosion activities of a system than similar measurements in the bulk phase. Results of radiorespirometry assays have not been compared with results from other techniques.

TABLE 2. Results of SRB Field Kit Comparisons for Waters

Sample		In-house method	MICKIT	Petrolite	BUG CHECK	SANI-CHECK®	RapidChek	Hydrogenase
Detection Limits		10^0	10^0	10^0	$<10^1$	$<10^1$	$<10^3$	NA
Produced water								
Canada		10^5	$>10^4$	10^4–10^5	BDL[a]	BDL	10^5	Weak
		10^7	$>10^4$	$>10^5$	BDL	$>10^3$	10^5	Strong
		10^4	10^3–10^4	10^2–10^3	BDL	BDL	10^3	BDL
		10^1	10^1–10^2	10^0–10^1	BDL	BDL	BDL	BDL
Texas	holding pit	10^2–10^3	10^2–10^3	10^2–10^3		$>10^4$,$>10^4$,$>10^4$	BDL	BDL
		10^3	10^2–10^3	10^1–10^2		$>10^5$,$>10^2$,BDL	10^2–10^3	Very weak
	pipeline	10^2	10^1–10^2	10^1–10^2		$>10^2$,$>10^1$,$>10^1$	10^4–10^5	Weak
		10^3–10^4	10^0–10^1	10^2–10^3		$>10^5$,$>10^5$,$>10^4$	10^3–10^4	Very weak
	FWKO	10^3	10^4–10^5	10^2–10^3	$>10^3$	$>10^3$	BDL	BDL
	Gas pipe	10^2	10^2–10^3	BDL	$>10^1$	$>10^1$	10^4–10^5	BDL
Java		10^3	10^3–10^4	10^2–10^3	BDL	BDL	10^5–10^6	Weak
		$<10^2$	BDL	BDL	BDL	BDL	BDL	Very weak
		BDL	BDL	BDL	BDL	BDL	10^4	Very weak

Power utility water							
Lake Erie	10^1–10^2	10^1–10^2	BDL	BDL,BDL,BDL	BDL,BDL,BDL		BDL
	10^2–10^3	10^2–10^3	10^1–10^2	BDL,BDL,$>10^4$	BDL,BDL,$>10^4$		BDL
Lake Huron	10^0–10^1	10^1–10^2	BDL	BDL,BDL,BDL	BDL,BDL,BDL		BDL
	10^1–10^2	10^1–10^2	BDL	BDL,BDL,BDL	BDL,BDL,BDL		BDL
Lake Ontario	10^0–10^1	10^1–10^2	10^0–10^1	BDL,BDL,BDL	BDL,BDL,BDL	BDL	BDL
	10^1–10^2	10^1–10^2	10^1–10^2	BDL,BDL,$>10^3$	BDL,BDL,BDL		BDL
Demineralized	BDL	BDL	BDL	BDL,BDL,BDL	BDL,BDL,BDL	BDL	BDL
	BDL	BDL	BDL	BDL		BDL,BDL	BDL
Process water							
Recirculated	10^1	10^4–10^5	10^2–10^3	BDL	BDL	10^5–10^6	BDL
Process filtrate	10^3	10^4–10^5	10^4–10^5	BDL	BDL	10^3	BDL
Seawater							
Java	$>10^3$	10^4–10^5	10^3–10^4	$>10^3$	BDL	10^4–10^5	Weak

[a] BDL — Below detection level.

Taken from Scott, P. J. B. and Davies, M., *Mater. Perform.*, 31(5), 64, 1992. With permission.

TABLE 3. Results of SRB Field Kit Comparisons for Coupons and Sludges

Sample		In-house method	MICKIT	Petrolite	BUG CHECK	SANI-CHECK®	RapidCheck	Hydrogenase
Detection limits		10^0	10^0	10^0	$<10^1$	$<10^1$	$<10^3$	NA
Coupons								
Java	produced water	$<10^2$	BDL[a]	10^0–10^1	BDL	BDL	BDL	Strong
	seawater	$<10^1$	BDL	10^0–10^1	BDL	BDL	BDL	BDL
		$<10^2$	10^0–10^1	BDL	BDL	BDL	BDL	Very weak
Sludges								
Canada	produced water	10^6	10^4	$>10^5$	BDL	BDL	10^5	BDL
		10^6	$>10^4$	$>10^5$	BDL	BDL	10^5	Strong
Texas	injection	$<10^1$	10^3–10^4	10^0–10^1	BDL	BDL	10^5–10^6	Very weak
	rat hole	BDL	10^1–10^2	10^0–10^1	BDL	BDL	BDL	BDL
Java	seawater	$<10^2$	BDL	BDL	BDL	BDL	10^3	Very weak
		$<10^1$	10^1–10^2	10^2–10^3	BDL	BDL	10^4	Weak

[a] BDL — below detection level.

Taken from Scott, P. J. B. and Davies, M., *Mater. Perform.*, 31(5), 64, 1992. With permission.

IV. BIOGENIC PRODUCTION OF SULFIDES

Conversion of metals to metal sulfides as a result of SRB activity has been studied since the late 1800s.[56] Baas-Becking and Moore[57] list sulfides produced by *Desulfovibrio desulfuricans* over a large pH range. Mackinawite (tetragonal FeS_{1-x}) was produced from iron and iron oxides. On continued exposure to SRB, mackinawite alters to greigite (Fe_3S_4) and smythite (Fe_9S_{11}) and finally to pyrrhotite (FeS_{1+x}).[38] SRB in thin biofilms on the surfaces of pottery[58] and silver[59] can produce pyrite films from iron-rich waters. Pyrite is not a typical iron corrosion product, but SRB can produce pyrite from mackinawite in contact with elemental sulfur.[60] Abiotic aqueous synthesis of these minerals, with the possible exception of pyrite, requires H_2S pressures higher than those found in the natural environment. The region of stability of mackinawite is wholly outside the region of normal natural environmental conditions, and its presence in corrosion products generally is proof that the corrosion was induced by SRB.

Digenite (Cu_5S_9), chalcocite (Cu_2S), and covellite (CuS) have been produced by SRB on copper surfaces.[57,61–63] Macdonald et al.[63] reported the formation of djurleite ($Cu_{1.96}S$) from SRB activity on copper alloys. Chalcocite is the most characteristic corrosion product in SRB-induced corrosion of copper. Baas-Becking and Moore[57] claim that chalcocite cannot be formed abiotically at room temperature. They further maintain that microbiological formation of chalcocite is as an alteration product of digenite, the first product of SRB-induced corrosion of copper alloys. Djurleite is important in SRB-induced corrosion of copper alloys because it reportedly forms a protective sulfide film[62] and is difficult to synthesize abiotically at room temperature.[64]

SRB in contact with silver produce acanthite (monoclinic Ag_2S).[38] SRB-induced corrosion of zinc produces a zinc sulfide reported to be sphalerite (ZnS).[57] SRB attack on lead carbonates produces galena (PbS).[57] Galena was reported more recently as a lead corrosion product in SRB-induced corrosion of lead-tin alloys.[59]

McNeil and Little[38] analyzed sulfide mineral deposits on copper alloys colonized by SRB in an attempt to identify specific mineralogies that could be used to fingerprint SRB activity. The copper sulfide mineral found in all combinations of copper-containing substrata and cultures was chalcocite (Table 4).[38] The authors concluded that the presence of chalcocite was an indicator of SRB-induced corrosion of copper. The compound was not observed in sterile controls and its presence in near-surface environments could not be explained thermodynamically.

TABLE 4. Copper Sulfide Minerals in Corrosion Products

	Bacterial Cultures						
	I	II	III	IV	V	VI	VII
99Cu	Chalcocite Digenite**	Chalcocite Digenite	Chalcocite Djurleite*	Chalcocite Digenite* Spionkopite*		Chalcocite Digenite	Chalcocite Djurleite*
90Cu 10Ni	Chalcocite Djurleite	Chalcocite Covellite* Digenite*	Chalcocite Spionkopite* **	Chalcocite Djurleite*	Chalcocite Djurleite	Chalcocite Djurleite*	Chalcocite Djurleite Digenite*
70Cu 30NI						Chalcocite Djurleite Anilite*	Chalcocite Djurleite

* Very low concentration.

** Pattern has many poorly defined peaks.

V. SULFUR ISOTOPE FRACTIONATION

The stable isotopes of sulfur (^{32}S and ^{34}S), naturally present in the sulfate source, are selectively metabolized during sulfate reduction by SRB and the resulting sulfide is enriched in ^{32}S.[65] The ^{34}S isotope accumulates in the starting sulfate as the ^{32}S isotope is selectively concentrated in the sulfide. Conventionally, the amplitude of each isotope is not reported individually, but a ratio is established and compared to the isotope ratio of a standard to yield a value $\delta^{34}S$ expressed as parts per thousand. Negative $\delta^{34}S$ values indicate a concentration of ^{32}S, and positive values indicate an accumulation of ^{34}S. Isotope fractionation by SRB has been used to interpret the origin of sulfides in modern and ancient environments, including sediments and groundwaters.[66–71] During abiotic chemical reduction of sulfate, a kinetic isotope effect results in a predominance of ^{32}S in the sulfide.[72] Bacterial sulfate reduction results in $\delta^{34}S$ greater than that predicted by the chemical kinetic isotope effect.[73] Enhanced isotope fractionation has been explained in terms of additive effects associated with two or more enzymes in the microbial sulfate reduction pathway.[74] Isotopic fractionation during sulfate reduction is inversely proportional to the rate of reduction when lactate and ethanol are used as electron donors and is directly proportional when molecular hydrogen is the electron donor.[75] Both temperature and sulfate concentration influence the rate of reduction and therefore the fractionation, but the impact of sulfate concentration is reportedly minor.[73] The most negative values for $\delta^{34}S$ have been reported for microbial reductions with slow reaction rates when the sulfide can be trapped and analyzed immediately after formation.[76]

Sulfur isotope fractionation was demonstrated by Little et al.[46] in sulfide corrosion deposits. Isotope fractionation is reported in Table 5 as $\delta^{34}S$ as parts per thousand (‰) as follows:

$$\delta^{34}S\ \text{‰} = \frac{^{34}S/^{32}S\ \text{sample} - {}^{34}S/^{32}S\ \text{standard}}{^{34}S^{32}S\ \text{standard}} \times 1000 \qquad (1)$$

Negative $\delta^{34}S$ values indicate a concentration of ^{32}S and positive numbers indicate an accumulation of the heavier ^{34}S. The standard for the sulfur isotope is meteorite troilite (FeS).

^{32}S accumulated in sulfide-rich corrosion products, and ^{34}S was concentrated in the residual sulfate in the culture medium. Accumulation of the lighter isotope was related to surface derivatization or corrosion as measured by weight loss. The demonstration of sulfur isotope fractionation in sulfide corrosion products verifies that SRB were responsible for sulfate conversion. However, the technique is destructive and complicated.

TABLE 5. Sulfur Isotope and Corrosion Data After 3 Months of Exposure

Sample	Sulfur concentration in solid ppm	$\delta^{34}S$ in solid ‰	Sulfur concentration in medium ppm	$\delta^{34}S$ in medium ‰	Weight loss mg/mm^2 × 10^{-3} for similarly treated foils
Control	747	10.2	372	0.3	15.9
	900	12.3	514	0.4	30.2
	613	16.5	580	0.2	29.9
SRB Culture I	26,627	−0.4	7	*	226.3
	36,618	−1.3	20	3.7	131.7
	36,695	−0.8	16	8.7	178.6
SRB Culture II	4,321	0.6	20	23.8	99.4
	3,371	1.5	18	25.8	227.2
	19,490	−0.7	11	*	169.5

* Insufficient sulfur.

VI. CONCLUSIONS

Despite recent advances in the detection of specific cellular constituents of SRB, mineralogical fingerprints, and isotope fractionation techniques, it appears that the most reliable methods for the detection of SRB in water, sludges, biofilms, and corrosion products involve culture techniques. It should be noted that despite the attention that has been given SRB and their role in corrosion processes, there is no direct correlation between numbers of SRB and the corrosion that one can predict from their presence.[3] For example, Pope[5] cultured high numbers of SRB from some nuclear power plant systems without observing MIC. He further observed severe corrosion in the absence of SRB, but in the presence of acid-producing bacteria.[40] Therefore, it is unreasonable to conclude that the quantification of a single physiological group will provide a predictive tool for all MIC. The use of sulfur isotope fractionation and sulfide mineralogy to verify SRB conversion of sulfate to sulfide is complicated and appropriate for research purposes.

ACKNOWLEDGMENTS

This work was supported by the Office of Naval Research, Program Element 0601153N, through the Defense Research Sciences Program, NRL contribution number BC 005:92:333.

REFERENCES

1. **Sanders, P. F. and Hamilton, W. A.,** Biological and corrosion activities of sulphate-reducing bacteria in industrial process plant, in *Biologically Induced Corrosion,* National Association of Corrosion Engineers, Houston, TX, 1986, 47.
2. **Eidsa, G. and Risberg, E.,** Sampling for the investigation of sulfate reducing bacteria and corrosion on offshore structures, in *Biologically Induced Corrosion,* National Association of Corrosion Engineers, Houston, TX, 1986, 109.
3. **Soracco, R. J., Pope, D. H., Eggars, J. M., and Effinger, T. N.,** Microbiologically influenced corrosion investigations in electric power generating stations, in *Corrosion 88,* Paper No. 83, National Association of Corrosion Engineers, Houston, TX, 1988.
4. **Pope, D. H.,** Microbial Corrosion in Fossil-Fired Power Plants — A Study of Microbiologically Influenced Corrosion and a Practical Guide for Its Treatment and Prevention, Electric Power Research Institute, Palo Alto, CA, 1987.
5. **Pope, D. H.,** A Study of MIC in Nuclear Power Plants and a Practical Guide for Countermeasures, Final Report NP-4582, Electric Power Research Institute, Palo Alto, CA, 1986.
6. **Ghiorse, W. C.,** Biology of iron- and manganese-depositing bacteria, *Annu. Rev. Microbiol.,* 38, 515, 1984.
7. **Starkey, R. L.,** Transformations of iron by bacteria in water, *J. Am. Water Works Assoc.,* 37(10), 963, 1945.
8. **Tatnall, R. E.,** Fundamentals of bacteria-induced corrosion, *Mater. Perform.,* 20(9), 32, 1981.
9. **Kobrin, G.,** Corrosion by microbiological organisms in natural waters, *Mater. Perform.,* 15(7), 38, 1976.
10. **Kobrin, G.,** Reflections on microbiologically induced corrosion of stainless steels, in *Biologically Induced Corrosion,* National Association of Corrosion Engineers, Houston, TX, 1986, 33.
11. **Eashwar, M., Subramanian, G., Chandrasekaran, P., and Balakrishnan, K.,** Probing microbiologically-influenced corrosion of steel during putrefaction of seawater, in *Corrosion 90,* Paper No. 120, National Association of Corrosion Engineers, Houston, TX, 1990.
12. **Tatnall, R. E.,** Case histories: bacteria-induced corrosion, in *Corrosion 81,* Paper No. 130, National Association of Corrosion Engineers, Houston, TX, 1981.
13. **Obuekwe, C. O., Westlake, D. W. S., Plambeck, J. A., and Cook, F. D.,** Corrosion of mild steel in cultures of ferric iron reducing bacterium isolated from crude oil. 1. Polarization characteristics, *Corrosion,* 37(8), 461, 1981.
14. **Little, B., Wagner, P., Ray, R., and McNeil, M.,** Microbiologically influenced corrosion in copper and nickel seawater piping systems, *Mar. Technol. Soc. J.,* 24, 10, 1990.
15. **Nickels, J. S., Bobbie, R. J., Martz, R. F., Smith, G. A., White, D. C., and Richards, N. L.,** Effect of silicate grain shape, structure, and location on the biomass and community structure of colonizing marine microbiota, *Appl. Environ. Microbiol.,* 41, 1262, 1981.
16. **Miller, J. D. A. and Tiller, A. K.,** *Microbial Aspects of Metallurgy,* Elsevier, New York, 1970, 61.
17. **Kasahara, K. and Kajiyama, F.,** Role of sulfate reducing bacteria in the localized corrosion of buried pipes, in *Biologically Induced Corrosion,* National Association of Corrosion Engineers, Houston, TX, 1986, 171.
18. **Iverson, W. P.,** *Microbial Iron Metabolism,* Neilands, J. B., Ed., Academic Press, New York, 1974, 475.

19. **Postgate, J. R.,** *The Sulphate Reducing Bacteria,* Cambridge University Press, Cambridge, England, 1979, 26.
20. **Bibb, M.,** Bacterial corrosion in the South African power industry, in *Biologically Induced Corrosion,* National Association of Corrosion Engineers, Houston, TX, 1986, 96.
21. **Scott, P. J. B. and Davies, M.,** Microbiologically influenced corrosion of alloy 904L, *Mater. Perform.,* 28(5), 57, 1989.
22. **Little, B., Wagner, P., Jacobus, J., and Janus, L.,** Evaluation of microbiologically-induced corrosion in an estuary, *Estuaries,* 12(3), 138, 1989.
23. **Licina, G. C.,** Ed., *Sourcebook for Microbiologically Influenced Corrosion in Nuclear Power Plants,* Electric Power Research Institute, Palo Alto, CA, 1988.
24. **Little, B., Wagner, P., Gerchakov, S. M., Walch, M., and Mitchell, R.,** The involvement of a thermophilic bacterium in corrosion processes, *Corrosion,* 42(9), 533, 1986.
25. **Little, B., Wagner, P., and Duquette, D.,** Microbiologically induced increase in corrosion current density of stainless steel under cathodic protection, *Corrosion,* 44(5), 270, 1988.
26. **Stranger-Johannessen, M.,** Fungal corrosion of the steel interior of a ship's holds, in *Biodeterioration VI,* CAB Int., Slough, England, 1986, 218.
27. **Salvarezza, R. C. and Videla, H. A.,** Microbiological corrosion in fuel storage tanks. Part 1. Anodic behaviour, *Acta Gientifica Venezalana,* 35, 244, 1984.
28. **Pope, D. H., Zintel, T. P., Kuruvilla, A. K., and Siebert, O. W.,** Organic acid corrosion of carbon steel: a mechanism of microbiologically influenced corrosion, in *Corrosion 88,* Paper No. 79, National Association of Corrosion Engineers, Houston, TX, 1988.
29. **Dias, O. C. and Bromel, M. C.,** Microbially induced organic acid underdeposit attack in a gas pipeline, *Mater. Perform.,* 29(4), 53, 1990.
30. **Honneysett, D. G., van den Bergh, W. D., and O'Brien, P. F.,** Microbiological corrosion control in a cooling water system, *Mater. Perform.,* 24, 34, 1985.
31. **Tatnall, R.,** Experiments in Microbiologically Influenced Corrosion, in *Corrosion 84,* Paper No. 95, National Association of Corrosion Engineers, Houston, TX, 1981.
32. **Gouda, V. K., Banat, M. I., Riad, T. W., and Mansour, S. I.,** Microbial-induced corrosion of Monel 400 in seawater, in *Corrosion 90,* Paper No. 107, National Association of Corrosion Engineers, Houston, TX, 1990.
33. **Nivens, D. E., Nichols, P. D., Henson, J. M., Geesey, G. G., and White, D. C.,** Reversible acceleration of the corrosion of AISI stainless steel exposed to seawater induced by growth and secretions of the marine bacterium *Vibrio natriegens, Corrosion,* 42(4), 204, 1986.
34. **Pope, D. H., Duquette, D. J., Johannes, A. H., and Wayner, P. C.,** Microbiologically influenced corrosion of industrial alloys, *Mater. Perform.,* 23(4), 4, 1984.
35. **Walch, M. and Mitchell, R.,** Role of microorganisms in hydrogen embrittlement of metals, in *Corrosion 83,* Paper No. 249, National Association of Corrosion Engineers, Houston, TX, 1983.
36. **Scott, P. J. B. and Davies, M.,** Survey of field kits for sulfate-reducing bacteria (SRB), *Mater. Perform.,* 31(5), 64, 1992.
37. **Pfennig, N., Widdel, F., and Truper, H. G.,** in *The Prokaryotes,* Balows, A. and Schlegel, H. G., Eds., Springer-Verlag, New York, 1981, 926.
38. **McNeil, M. and Little, B. J.,** Mackinawite formation during microbial corrosion, *Corrosion,* 46, 599, 1990.
39. **Pope, D. H., Duquette, D., Wayner, P., Johannes, A., and Freeman, A.,** Microbiologically influenced corrosion: a-state-of-the-art review, MTI no. 13, National Association of Corrosion Engineers, Houston, TX, 1989.
40. **Characklis, W. G. and Marshall, K. C.,** Biofilms: a basis for an interdisciplinary approach, in *Biofilms,* Wiley Science, New York, 1990, 4.

41. **Marszalek, D. S., Gerchakov, M., and Udey, L. R.,** Influence of substrate composition on marine microfouling, *Appl. Environ. Microbiol.*, 38, 987, 1979.
42. **Zachary, A., Taylor, M. E., Scott, F. E., and Colwell, R. R.,** Marine microbial colonization of material surfaces, in *Biodeterioration,* Proc. 4th International Symposium, Pitman, London, 1980, 171.
43. **Hamilton, W. A. and Maxwell, S.,** Biological and corrosion activities of sulphate-reducing bacteria within natural biofilms, in *Biologically Induced Corrosion,* National Association of Corrosion Engineers, Houston, TX, 1986, 131.
44. **Little, B., Ray, R., Wagner, P., Lewandowski, Z., Lee, W. C., Characklis, W. G., and Mansfeld, F.,** Impact of biofouling on the electrochemical behaviour of 304 stainless steel in natural seawater, *Biofouling,* 3, 45, 1991.
45. **Boivin, J., Laishley, E. J., Bryant, R. D., and Costerton, J. W.,** The influence of enzyme systems on MIC, in *Corrosion 90,* Paper No. 128, National Association of Corrosion Engineers, Houston, TX, 1990.
46. **Little, B., Wagner, P., and Jones-Meehan, J.,** Sulfur isotope fractionation by sulfate-reducing bacteria in corrosion products, *Biofouling,* 6, 279, 1993.
47. American Petroleum Institute, API Recommended Practice for Biological Analysis of Subsurface Injection Waters, *API,* New York, 1965.
48. **Rossmoore, L. A., Wireman, J. W., and Rossmore, H. W.,** Rapid field method for the detection and enumeration of sulfate reducers, in *Biodeterioration VI,* CAB Int., Slough, England, 1986.
49. **Tatnall, R. E., Stanton, K. M., and Ebersole, R. C.,** Methods of testing for the presence of sulfate-reducing bacteria, in *Corrosion 88,* Paper No. 88, National Association of Corrosion Engineers, Houston, TX, 1988.
50. **Pope, D. H.,** Discussion of methods for the detection of microorganisms involved in microbiologically influenced corrosion, in *Biologically Induced Corrosion,* National Association of Corrosion Engineers, Houston, TX, 1986, 275.
51. **Pope, D. H.,** Development of methods to detect sulfate-reducing bacteria — agents of microbiologically influenced corrosion, Materials Technology Institute of the Chemical Process Industries, Inc., MTI no. 37, National Association of Corrosion Engineers, Houston, TX, 1990.
52. **Hogan, J. J.,** A rapid, non-radioactive DNA probe for the detection of SRBs, presented at Institute of Gas Technology Symposium on Gas, Oil, Coal and Environmental Biotechnology, New Orleans, LA, 1990.
53. **Littman, E. S.,** Oilfield bactericide parameters as measured by ATP analysis, International Symposium of Oil Field Chemistry of the Society of Petroleum Engineers of AIME, Paper No. 5312, Dallas, TX, 1975.
54. **Rosser, H. R. and Hamilton, W. A.,** Simple assay for accurate determination of [^{35}S] sulfate reduction activity, *Appl. Environ. Microbiol.*, 45, 1956, 1983.
55. **Maxwell, S. and Hamilton, W. A.,** The activity of sulphate-reducing bacteria on metal surfaces in an oilfield situation, in *Biologically Induced Corrosion,* National Association of Corrosion Engineers, Houston, TX, 1986, 284.
56. **de Gouvernain, M.,** Sulfiding of copper and iron by a prolonged stay in the thermal spring at Bourbon l'Archambault, *Compt. Rend.*, 80, 1297, 1875.
57. **Baas-Becking, G. M. and Moore, D.,** Biogenic sulfides, *Econ. Geol.*, 56, 259, 1961.
58. **Duncan, S. J. and Ganiaris, H.,** Some sulphide corrosion products on copper alloys and lead alloys from London waterfront sites, in *Recent Advances in the Conservation and Analysis of Artifacts,* Black, J., Ed., University of London Summer Schools Press, London, 1987.
59. **McNeil, M. B. and Mohr, D. W.,** Copper-iron sulfide minerals during corrosion of artifacts and possible implications for pseudoguilding, *Geoarchaeology,* 8, 23, 1993.
60. **Berner, R. A.,** The synthesis of framboidal pyrite, *Econ. Geol.*, 64, 383, 1969.

61. **North, N. A. and MacLeod, I. D.,** Corrosion of metals, in *Conservation of Marine Archaeological Objects,* Pearson, C., Ed., Butterworths, London, 1986.
62. **Mor, E. D. and Beccaria, M.,** Behaviour of copper in artificial seawater containing sulphides, *Br. Corrosion J.,* 10, 33, 1975.
63. **Macdonald, D. D., Syrett, B. C., and Wing, S. S.,** Corrosion of Cu-Ni alloys 706 and 715 in flowing sea water. 2. Effect of dissolved sulfide, *Corrosion,* 35, 367, 1979.
64. **Roseboom, E. H.,** An investigation of the system Cu-S and some natural copper sulfides between 25 degrees and 700 degrees C., *Econ. Geol.,* 61, 641, 1966.
65. **Chambers, L. A. and Trudinger, P. A.,** Microbiological fractionation of stable sulfur isotopes: a review and critique, *Geomicrobiol. J.,* 1, 249, 1979.
66. **Dickman, M., Thode, H. G., Rao, S., and Anderson, R.,** Downcore sulphur isotope ratios and diatom inferred pH in artificially acidified Canadian Shield Lake, in *Environmental Pollution,* Dempster, J. P. and Manning, W. J., Eds., Elsevier Applied Science, Oxford, UK, 1988, 265.
67. **Dockins, W. S., Olson, G. J., McFeters, G. A., Turbak, S. C., and Lee, R. W.,** Sulfate reduction in ground water of southeastern Montana, USGS/WRD/WRI-80/040, 1, 1980.
68. **Kaplan, I. R., Sweeney, R. E., and Nissenbaum, A.,** Sulfur isotope studies on Red Sea geothermal brines and sediments, in *Hot Brines and Recent Heavy Metal Deposits in the Red Sea,* Degens, E. T. and Ross, D. A., Eds., Springer-Verlag, New York, 1969, 474.
69. **Lyalikova, N. N.,** Role of bacteria in oxidation of sulfide ores in copper-nickel deposits of Kola Peninsula, *Mikrobiologiya,* 30(1), 135, 1961.
70. **Sweeney, R. E. and Kaplan, I. R.,** Stable isotope composition of dissolved sulfate and hydrogen sulfide in the Black Sea, *Mar. Chem.,* 9, 145, 1980.
71. **Vaynshteyn, M. B., Tokarev, V. G., Shakola, V. A., Yu Lein, A., and Ivanov, M. V.,** The geochemical activity of sulfate-reducing bacteria in sediments in the western part of the Black Sea, *Geokhimiya,* 7, 1032, 1985.
72. **Harrison, A. G. and Thode, H. G.,** The kinetic isotope effect in the chemical reduction of sulphate, *Trans. Faraday Soc.,* 53, 1648, 1957.
73. **Kaplan, I. R. and Rittenberg, S. C.,** Microbiological fractionation of sulphur isotopes, *J. Gen. Microbiol.,* 34, 195, 1964.
74. **Rees, C. E.,** A steady state model for sulphur isotope fractionation in bacterial reduction processes, *Geochim. Cosmochim. Acta,* 37, 1141, 1973.
75. **Kaplan, I. R.,** Stable isotopes as a guide to biogeochemical processes, *Proc. R. Soc. London B.,* 189, 183, 1975.
76. **Kaplan, I. R.,** Global Geochemistry, Canoga Park, CA, personal communication, 1991.

Section III
BIOCORROSION

15

Biocorrosion of Nonferrous Metal Surfaces

H. A. Videla

I. INTRODUCTION

Biocorrosion and biofouling of metal surfaces are due to biological and electrochemical processes mediated by microorganisms attached to the metal surface or embedded in biofilms.

Biofilms lead to an important modification of the metal-solution interface, forming a barrier to the exchange of elements between the metal surface and the aqueous environment. Also, reactions between metabolites produced by bacteria and metals take place underneath biofilms or within the biofilm. Thus, the key to the alteration of conditions at a metal surface, and therefore the enhancement or retardation of corrosion, is the formation of the biofilm.[1] A biofilm is a gelatinous structure of 95% or more water made of a matrix of extracellular polymeric substances (EPS) of polysaccharidic nature, bacterial cells, and diverse inorganic detritus.

Generally, bacterial population develops rapidly, in some cases within hours of immersion, followed by the production and accumulation of EPS.

The production of EPS stabilizes and protects microorganisms on the surface. Additionally, reproduction of attached bacteria creates colonies that lead to an important concentration of physical and chemical actions at localized sites on the surface.

0-87371-928-X/94/$0.00+$.50

Aggregates of different species of bacteria, forming colonies in neighboring areas and interacting on a metabolic level, lead to the formation of consortia. Such consortia are not only composed of different species of bacteria, but can also include microfungi, microalgae, and even macroscopic organisms. Interactions within microbial consortia are generally very complex, consuming and producing substances, in many cases without any evidence of these activities in the bulk water phase.[2]

Biofilms are influenced by both the substratum and the bulk phase. The interaction of all three phases (metallic substratum, biofilm, and liquid medium) results in concentration gradients and other localized effects particularly relevant to corrosion. For instance, anaerobic conditions established beneath very thin (10 to 20 μm) biofilms,[3] together with appropriate nutrient conditions, provide an ideal environment for certain microorganisms like sulfate-reducing bacteria (SRB), intimately involved with corrosion. Consequently, most corrosion reactions in biocorrosion cases are of a localized nature, usually resulting in pitting, crevice corrosion, stress corrosion cracking, corrosion fatigue, or selective dealloying.

Essentially, biofilm effects are restricted to localized areas that constitute physical heterogeneities on the metal surface resulting in the formation of local anodes or cathodes. For example, nonuniform or patchy colonization by bacteria results in the formation of differential aeration cells, where areas under respiring colonies become anodic and metal dissolution is enhanced. Conversely, surrounding areas become cathodic, and the counterreaction of oxygen reduction occurs. This biologically modified metal-solution interface is present in the majority of biocorrosion cases, independently of the nature of the metal surface, and strikingly changes the metal behavior in the aqueous environment.

The two main influences of biofilms on corrosion are contrasting, either retarding or accelerating metal dissolution. Retardation means a reduction in the metal chemical reactivity and may be due to a "barrier effect" of the biofilm. There is evidence of this in the literature.[4,5] As biofilms are rarely uniform, the opposite effect of enhanced metal reactivity is prevalent as a consequence of the permanent separation of anodic and cathodic sites, the breakdown of inorganic passive films, or the stimulation of either the anodic or cathodic reactions.

It has been recently found[6] that the interactions between biofilms and inorganic corrosion product layers at the metal-solution interface govern the passive or active behavior of metal surfaces in biologically conditioned media. The nonferrous metal surfaces selected in this chapter — copper-nickel alloys, aluminum alloys, and titanium — exhibit marked dissimilar biocorrosion/biofilm interactions at the metal-solution interface. The different biocorrosion mechanisms acting in each case will be illustrated with practical cases to facilitate the understanding of the complex bioelectrochemical processes involved in each case.

II. DIFFERENT BIOFILMS' EFFECTS ON CORROSION

According to Characklis,[7] biofilms influence corrosion through the following simultaneous or consecutive effects:

1. Influence of microbial respiration on the metal substratum, especially if a patchy biofilm is formed
2. Influence on the substratum of corrosive metabolic products produced in the biofilm by microbial activity
3. Restriction of the passage of charged species through the biofilm matrix
4. Alteration of the medium conductivity by the EPS that forms the biofilm matrix
5. A possible chelation of metal ions by the EPS
6. Impeding an effective action of biocides
7. Altering the composition of corrosion inhibitors

The practical implications of these effects include (a) increase in heat transfer resistance; (b) reduction in fluid flow, both effects causing energy losses in the power and chemical processing industries; (c) enhanced localized corrosion of the metallic substratum; (d) deterioration of potable and natural water quality; (e) reduction of the quality of certain industrial products (e.g., pulp and paper); and (f) undesired debris production.

An important factor to consider when analyzing the different mechanisms by which biofilms affect corrosion is the electrochemical behavior of the metallic substratum. One must ask how the microbial processes important to biofilm formation interact with inorganic corrosion products at the interface, to alter the passive behavior of the metal surface.

The mechanisms by which biofilms can modify the protective action of inorganic passive films is not well understood at present, mainly because the attempts made in the literature to correlate biofilms with corrosion have been made after extended exposures of the metal samples. Under these circumstances an important accumulation of biomass has occurred on the metal surface, hindering an adequate interpretation of the interactions at the metal-solution interface.

The three nonferrous metal surfaces studied in this chapter exhibit different passive behaviors in neutral and slightly alkaline aqueous environments. Titanium is characterized by a strong passive behavior in aggressive media. The corrosion resistance of aluminum and copper-nickel alloys is strongly influenced by the different types of corrosion product layers rapidly formed at the metal surface after its immersion in the aqueous environment.

III. TITANIUM

The resistance of titanium to corrosion depends on a protective film of TiO_2 that is stable over a wide range of environmental conditions. Titanium possesses three main characteristics that account for its use in aggressive environments: (a) good resistance to seawater and other chloride solutions; (b) good performance in the presence of hypochlorites and chlorine solutions, and (c) resistance to nitric acid solutions. In addition, its alloying with palladium greatly increases its resistance to hydrochloric acid environments.[8] The most widely used grade of commercial titanium is ASTM Grade 2, which contains some oxygen.

Several reports on the titanium resistance to biocorrosion have been recently published.[9–12] Titanium is characterized by presenting very high anodic pitting and repassivation potentials that make the metal less susceptible to pitting corrosion, the most common type of attack in biocorrosion. It has been emphasized that titanium is fully resistant to reduced chemical species generally associated with anaerobic microbial activity, like ammonia, sulfides, hydrogen sulfide, nitrites, ferrous ions, and organosulfur compounds, as well as to biogenic organic acids, like acetic, succinic, lactic, and maleic, over a wide concentration range.[10]

Because of its resistance to chloride solutions, titanium has been used as heat exchanger material when seawater is used as a coolant. Recent results[12] reported bacterial adhesion on titanium surfaces exposed to bacterial cultures for only 5 h in laboratory experiments using artificial seawater. Patchy biofilms were formed under these circumstances. A decrease of the redox and corrosion potentials with time was found in these experiments, although no evidence of corrosion attack was found after 5 d exposure to bacterial cultures.

An experimental evaluation of the resistance of titanium to biocorrosion in extreme environments created by bacteria was reported by Little et al.[11] Long-term exposures to seawater, acid conditions in the presence of mesophilic and thermophilic iron-oxidizing bacteria, chloride/sulfide environments associated with mesophilic and thermophilic SRB, and titanium hydrides formation at welds in the presence of hydrogen producing bacteria, were successively evaluated. According to the authors,[11] after long-term exposures (1 year) to seawater, titanium surfaces remained macroscopically clean, but covered by a thin biofilm of diatoms and microalgae.

Titanium surfaces exposed to pure cultures of iron/sulfur-oxidizing bacteria and to facultative anaerobic cultures containing SRB for 75 d, showed very low corrosion rates based on electrochemical data. Besides, surface analysis using energy dispersive spectroscopy (EDS) coupled to an environmental scanning electron microscope (ESEM) evidenced no localized corrosion of the titanium under any of the exposure conditions.

In addition, exposures of titanium weld materials to hydrogen-producing bacteria revealed no evidence of hydrogen absorption in the exposed specimens. These results are in good agreement with another report[10] claiming the excellent resistance of titanium to hydrogen absorption, attributed to the oxide film acting as a diffusion barrier.

As a constructional material used in nuclear power plant systems, titanium appears to be one of the most resistant materials examined to date.[9] However, its susceptibility to biofouling has been pointed out, and consequently its use could be limited when the biological content of the intake water is high. In this respect, it was shown[13] that, during the early stages of biofouling, the nature of the metal surface plays a relevant role in biofilm development, facilitating or hindering microbial attachment.

Corrosion-resistant materials, like titanium, present an ideal substratum for microbial colonization, whereas nonpassive metal surfaces usually make microbial adhesion more difficult.[14] After exposing different kinds of metal surfaces to flowing seawater with high levels of pollution, the amount of biomass deposited increased in the following order: copper $<$ 70:30 copper-nickel $<$ brass $<$ aluminum $<$ AISI 304 stainless steel $<$ titanium.[13]

In summary, specific examples of titanium use in natural seawater have demonstrated its immunity to biocorrosion.

IV. COPPER-NICKEL ALLOYS

Copper based alloys such as 90:10 or 70:30 copper-nickels, admiralty brass, and aluminum bronze are commonly used as heat exchanger materials because of their good resistance to both macro- and microfouling in marine environments. In spite of these well-known antifouling properties, it was reported[15] that after several months of exposure, bacteria can be found entrapped between corrosion product layers and EPS. Biofilms seem to alter the arrangement of passive layers, creating sandwiched structures. Under these circumstances, the toxic effect of cupric ions is counteracted by metal ions binding by the EPS. In this layer-over-layer structure, the partial detachment of biomass may influence the removal of inorganic corrosion products, contributing to the increase in biofilm heterogeneity and to the increase in the corrosion rate through a differential aeration.[16]

In sterile seawater, corrosion hazard can be related to the presence of chlorides and to the effects derived from poorly passivating inorganic films. However, in polluted seawater, several other factors may influence the corrosion resistance of copper-nickel alloys, such as the presence of micro- and macroorganisms capable of attaching to the metal surface and sulfur anions (mainly sulfides, bisulfides, and H_2S) related to the degree of pollution or to the presence of SRB.[17,18] Generally, a complex electrochemical behavior has been found in polluted saline media where the simultaneous presence of chlorides and sulfides alters the distribution and the structure of biofilms and passive layers.[19]

Samples of 90:10 copper-nickel alloy immersed in artificial chloride + sulfide solutions, and kept in the vicinity of the open circuit potential, showed a complex passive film formed by two different layers: (1) a thin, adherent inner layer of cuprous sulfide and (2) a porous outer layer of cuprous hydroxychloride.[20] Conversely, in polluted seawater, several layers of different composition are formed.[21] These layers are porous, thick, and poorly adherent, whereas the inner layer is thinner with grooves at the grain boundaries.[22] Generally, corrosion product layers formed in artificial seawater are more uniform and compact than those formed in natural seawater. This could be attributed to the adsorption of organics and biological material, and the precipitation of copper salts and sulfur compounds. Biofouling and corrosion studies of stainless steel and 70:30 copper-nickel samples exposed for several weeks to natural seawater,[23] showed that biofilms affect copper-nickel passivity by reducing the adhesion of precipitated corrosion products and facilitating the removal of passive layers. The final result is an enhancement of differential aeration effects.

Organisms like the protozoans *Zoothamnium* sp. are frequently found in polluted seawater and can facilitate detachment of passive layers through adhesion effects developed at the fixation points.[24]

Sulfide anions present in polluted seawater play a relevant role in corrosion by leading to the formation of an imperfect oxide layer of poorly protective characteristics.[25] In this way, copper-nickel dissolution is higher than that observed in the presence of chloride anions alone. Similar results have been recently reported for aluminum brass samples immersed in sulfide-contaminated media.[12] Energy dispersive X-ray analysis (EDXA) of corrosion product layers on copper-nickel alloys validate the relevant role of metabolic products of SRB bacteria and the alteration of passive layer composition through the combined effects of differential aeration and chemical characteristics at restricted areas underneath microbial colonies.[12]

Besides, differences in the substratum composition can influence colonization by marine organisms. In this regard, the effect of different iron contents on colonization of 90:10 copper-nickel alloy has been reported.[26] It was found that alloys without iron were rapidly corroded and scarcely colonized. Moreover, it was established that for iron concentrations higher than 2.5%, macrofouling settlement was favored.

V. ALUMINUM ALLOYS

Aluminum is a reactive metal, although it rapidly develops an oxide film — stable in many neutral and acid media — that protects the metal surface from corrosion in aggressive environments. This film can be artificially produced through an electrochemical process (anodizing). Notwithstanding that pure aluminum is soft and mechanically weak, it can be alloyed and heat treated to cover a broad range of mechanical requirements. Aluminum alloys 2014 and 2024 are Al-Cu materials with substantial Mg and Zn added to improve aging kinetics at low temperatures. These alloys have been widely used in the aircraft industry as well as alloy 2219, which is essentially an Al-Cu alloy with minor elements (Mn, Zr, Ti, V), characterized by an improved high-temperature strength and weldability.

Aluminum and aluminum alloys are prone to biocorrosion effects. The major part of the literature has been addressed to the biocorrosion of aluminum alloys for aircraft fuel tanks by microbial contaminants of jet fuels.[27–29] If fuels are completely free of water, microbial utilization of the hydrocarbon components is nil. However, when water becomes available, growth of microorganisms is possible, and the hydrocarbons are used as a main carbon source. Chemical contaminants and water provide nitrogen and the necessary trace elements for growth. Microbial activity generates small ecosystems such as tubercles, slime deposits, or biofilms which retain water on the metal surface.

Microorganisms isolated from contaminated fuels include different species of fungi, bacteria, and yeast. The fungus *Hormoconis resinae* (formerly *Cladosporium resinae*) and some species of *Aspergillus, Penicillium,* and *Fusarium* have been reported as frequent contaminants in jet fuels along with the bacterium *Pseudomonas aeruginosa*.[30]

Aluminum passivity is warranted by the stability of the oxide film formed on its surface. The breakdown of passive oxide films by aggressive anions such as chloride frequently causes the failures of aluminum alloys in aqueous environments, leading to severe pitting of the metal.[31] It has been reported that the water phase present in contaminated aircraft fuel contains chloride concentration level in the order of 10^{-2} to 10^{-4} *M*.[32] An important decrease of pitting corrosion resistance for 2024 aluminum alloy has been found in this chloride range by using electrochemical polarization methods.[33] Conversely, nitrate and phosphate anions usually present in bilge water have an opposite effect. Phosphate ions by their buffering action could prevent the localized acidification necessary to initiate the pitting process.[34] Nitrates are considered the most efficient pitting inhibitors of the corrosion of aluminum in chloride media.[35] Microbial growth and hydrocarbon biodegradation is limited mainly by the available concentrations of nitrogen and phosphorous (generally present as nitrates and phosphates) in the medium. Previous information concerning nitrogen metabolism of *H. resinae*[36] indicates that this fungus can preferentially select nitrate if there are ammonium and nitrate ions in the medium. Chromatographic analysis of the fuel phase that has been in contact with *H. resinae* cultures containing nitrates as nitrogen source show a high degree of degradation of the hydrocarbon chains.[37] This result would account for the enhanced corrosion attack of 2024 aluminum alloy in these cultures. Thus, the chloride-nitrate ratio (corrosive ion-passivating ion ratio) is very important in establishing the feasibility of mechanisms for biocorrosion of aluminum alloys in chloride-containing media.

Two other relevant factors must be taken into account: (1) a local drop in pH due to organic acid production by fungal hydrocarbon degradation; localized acidification results in a dissolution of the protective oxide film with the subsequent pitting of the bare metal; and (2) an enhanced corrosiveness due to the sum of acidic metabolites, differential aeration effects, and nitrate consumption in the regions underneath microbial colonies. In this regard, it has been reported[38] that fungal adhesion processes enhance metal dissolution at the fixation points of the mycelium to the metal, thus acting as nucleation centers for pitting. The surfactant activity of fatty acids, occasionally assisted by mechanical effects of the liquid phase movement, could account for the breakdown of passive films at certain points of the aluminum alloy surface. Once this breakdown is produced, the initiation and progression of pitting is established.

A recent article on the corrosion resistance of 2219-T87 aluminum alloy exposed to dilute biologically active solutions,[39] presents evidence in agreement with the results discussed above. Bacterial consortia were associated with anodic sites on the metal surface, and pitting occurred adjacent to the biofilm. Metallurgical features like welding or inclusions increase biocorrosion hazard.

VI. SUMMARY

Biocorrosion of different nonferrous metal surfaces analyzed in this review clearly demonstrates the relevant role of microbial adhesion and interactions of corrosion products with metabolites at the metal-solution interface. Titanium appears to be the most biocorrosion-resistant metal tested to date. Titanium passivity is enhanced in the presence of the oxidizing or reducing agents and by-products associated either with aerobic or anaerobic microbial environments.

Microbially assimilable materials (like high levels of organics in seawater) counteract the toxicity of copper-nickel alloys, facilitating bacterial adherence. This bacterial attachment enhances corrosion through several mechanisms: (1) generating diffusional barriers at the metal-solution interface, (2) changing the structural characteristics of the corrosion product layers, (3) producing corrosive metabolites, (4) facilitating the removal of inorganic passive layers, and (5) creating differential aeration conditions due to a patchy distribution of biofilms and corrosion products.

Aluminum alloys are prone to localized corrosion (mainly pitting) in biologically contaminated media due to the breakdown of either passive oxide or hydroxide layers in restricted areas of the surface where microbial attachment occurs. This can be due to (1) a local increase in the proton concentration derived from organic acid production, (2) an increase in the redox potential of the local environment, (3) a surfactant activity of certain metabolites (fatty acids) that cause a loss of stability of passive films, (4) a decrease of inhibitor concentration by microbial consumption, or (5) an increase of the chloride-inhibitor ratio.

REFERENCES

1. **Edyvean, R. G. J. and Videla, H. A.,** Biological corrosion, *Interdiscip. Sci. Rev.,* 16, 267, 1991.
2. **Costerton, J. W. and Geesey, G. G.,** The microbial ecology of surface colonization and of consequent corrosion, in *Biologically Induced Corrosion,* Dexter, S. C., Ed., NACE, Houston, 223, 1986.
3. **Hamilton, W. A.,** Sulphate-reducing bacteria and anaerobic corrosion, *Annu. Rev. Microbiol.,* 39, 195, 1985.
4. **Bultman, J. D., Southwell, C. R., and Hummer, C. W.,** Biocorrosion of structural steels in seawater, *Rev. Coat. Corros.,* 2, 187, 1977.
5. **Videla, H. A. and Guiamet, P. S.,** Protective action of Serratia marcescens in relation to the corrosion of aluminum and its alloys, in *Biodeterioration Research 1,* Llewellyn, G. C. and O'Rear, C. E., Eds., Plenum Press, New York, 1987, 275.
6. **Videla, H. A.,** Metal dissolution/redox in biofilms, in *Structure and Function of Biofilms,* Characklis, W. G. and Wilderer, P. A., Eds., John Wiley & Sons, Chichester, UK, 1989, 301.
7. **Characklis, W. G.,** Fouling biofilms and corrosion, *Microbial Corrosion I,* Sequeira, C. A. C. and Tiller, A. K., Eds., Elsevier Applied Science, London, 1988, 95.
8. **Abkowitz, S., Bannon, B., Broadwell, R., Forney, C. E., Jr., Harvey, B., Herman, W., Kane, R., Minkler, W., Newman, J. R., Schley, J. R., Schultz, R., and Soltow, K.,** in *Titanium, the Choice,* Herman, W. E., Broadwell, R., Kessler, H. D., Monses, J., and Hockaday, G. M., Eds., Titanium Development Association, Dayton, OH, 1990, 4.
9. **Licina, G. J.,** An overview of microbiologically influenced corrosion in nuclear power plant systems, *Mater. Perform.,* 28, 55, 1989.
10. **Schutz, R. W.,** A case for titanium's resistance to microbiologically influenced corrosion, *Mater. Perform.,* 30, 58, 1991.

11. **Little, B. J., Wagner, P. A., and Ray, R. I.,** An experimental evaluation of titanium resistance to microbiologically influenced corrosion, *Corrosion 92,* Paper No. 173, NACE, Houston, TX, 1992, 12.
12. **Videla, H. A., Gomez de Saravia, S. G., and Mele, M. L. F. de,** MIC of heat exchanger materials in marine media contaminated with sulphate-reducing bacteria, *Corrosion 92,* Paper No. 189, NACE, Houston, TX, 1992, 9.
13. **Videla, H. A., Mele, M. F. L. de, and Brankevich, G.,** Microfouling of several metal surfaces in polluted sea water and its relation with corrosion, *Corrosion 87,* Paper No. 365, NACE, Houston, TX, 1987, 16.
14. **Videla, H. A.,** Biological corrosion and biofilm effects on metal biodeterioration, in *Biodeterioration Research 2,* O'Rear, C. E. and Llewellyn, G. C., Eds., Plenum Press, New York, 1989, 39.
15. **Blunn, G.,** Biological fouling of copper and copper alloys, in *Biodeterioration 6,* Barry, S., Houghton, D. R., Llewellyn, G. C., and O'Rear, C. E., Eds., CAB International, London, 1986, 567.
16. **Videla, H. A.,** Mechanisms of MIC, in *Proceedings Argentine-USA Workshop on Biodeterioration,* (CONICET-NSF), Videla, H. A., Ed., Aquatec Quimica, Sao Paulo, 1986, 43.
17. **Videla, H. A., Mele, M. F. L. de, and Gómez de Saravia, S. G.,** Relationship between biofilms and inorganic passive layers in the corrosion of copper-nickel alloys in chloride environments, *Corrosion 89,* Paper No. 185, NACE, Houston, TX, 1989, 7.
18. **Chamberlain, A. H. L., Simmonds, S. E., and Garner, B. J.,** Marine copper-tolerant sulphate-reducing bacteria and their effects on 90/10 copper-nickel (CA 706), *Int. Biodeter.,* 24, 213, 1988.
19. **Gómez de Saravia, S. G., Mele, M. F. L. de, and Videla, H. A.,** The interaction of corrosion products and biofouling on 70/30 cupronickel in polluted seawater, *Biofouling,* 7, 141, 1993.
20. **Kato, C. and Pickering, H. W.,** A rotation disk study of the corrosion behavior of Cu-9.4 Ni 1.7 alloy in aqueous NaCl solution, *J. Electrochem. Soc.,* 131, 1219, 1984.
21. **Mukhopadhayay, N. and Baskaran, S.,** Characterization of corrosion products on cupronickel 70:30 alloy in sulfide-polluted seawater, *Corrosion,* 42, 113, 1984.
22. **Syrett, B.,** The mechanism of accelerated corrosion of copper-nickel alloys in sulfide polluted seawater, *Corros. Sci.,* 21, 187, 1981.
23. **Videla, H. A., Mele, M. F. L. de, and Brankevich, G. J.,** Assessment of corrosion and microfouling of several metals in polluted seawater, *Corrosion,* 44, 423, 1988.
24. **Videla, H. A., Mele, M. F. L. de, and Brankevich, G.,** Biofouling and corrosion of stainless steel and 70/30 copper-nickel samples after several weeks of immersion in seawater, *Corrosion 89,* Paper No. 291, NACE, Houston, TX, 1989, 10.
25. **Mele, M. F. L. de, Brankevich, G., and Videla, H. A.,** Corrosion of CuNi30Fe in artificial solutions and natural seawater. Influence of biofouling, *Br. Corros. J.,* 24, 211, 1989.
26. **Chamberlain, A. H. L. and Garner, B. J.,** The influence of iron content on the biofouling resistance of 90/10 copper/nickel alloys, *Biofouling,* 1, 79, 1988.
27. **Iverson, W. P.,** A possible role for sulfate-reducers in the corrosion of aluminum alloys, *Electrochem. Technol.,* 5, 77, 1967.
28. **Hedrick, H. G.,** Microbiological corrosion of aluminum, *Mater. Prot.,* 9, 27, 1970.
29. **Neihof, R. and May, M.,** Microbial and particulate contamination in fuel tanks on naval ships, *Int. Biodeter. Bull.,* 19, 59, 1983.
30. **Neihof, R.,** Microbes in fuel: an overview with a naval perspective, in *Proc. 2nd Int. Conf. on Long-Term Storage Stabilities of Liquid Fuels,* Southwest Research Institute, San Antonio, TX, 2, 215, 1986.
31. **Foroulis, Z. A. and Thubrikar, M. J.,** On the correspondence between critical pitting potentials and pitting of aluminum under conditions of natural immersion, *J. Electrochem. Soc.,* 122, 1296, 1975.

32. **Scott, J. A. and Hill, E. C.,** Microbial aspects of subsonic and supersonic aircraft, *Microbiology Symposium,* The Institute of Petroleum, London, 1971, 27.
33. **Salvarezza, R. C, Mele, M. F. L. de, and Videla, H. A.,** Mechanisms of the microbial corrosion of aluminum alloys, *Corrosion,* 39, 26, 1983.
34. **Galvele, J. R.,** Present state of understanding of the breakdown of passivity and repassivation, *Proceedings 4th International Symposium on Passivity,* Airlie, VA, 1987, 285.
35. **Szklarska-Smialowska, Z.,** Pitting corrosion inhibitors, in *Pitting Corrosion of Metals,* NACE, Houston, TX, 1986, 290.
36. **Parbery, D. G.,** Biological problems in jet-aviation fuel and the biology of Amorphoteca resinae, *Mater. Org.,* 6, 161, 1971.
37. **Videla, H. A., Guiamet, P. S., Dovalle, S. and Reinoso, E. H.,** Effects of fungal and bacterial contaminants of kerosene fuels on the corrosion of storage and distribution systems, *Corrosion 88,* Paper No. 91, NACE, Houston, TX, 1988, 20.
38. **Videla, H. A.,** The action of Cladosporium resinae growth on the electrochemical behavior of aluminum, in *Biologically Induced Corrosion,* Dexter, S. C., Ed., NACE, Houston, TX, 1986, 215.
39. **Walsh, D., Danford, M., and Qiong, Q.,** The corrosion resistance of aluminum 2219-T87 to dilute biologically active solutions, *Corrosion 92,* Paper No. 166, NACE, Houston, TX, 1992, 14.

Section III
BIOCORROSION

16

Unusual Types of Pitting Corrosion of Copper Tubes in Potable Water Systems

G. G. Geesey, P. J. Bremer, W. R. Fischer, D. Wagner, C. W. Keevil, J. Walker, A. H. L. Chamberlain, and P. Angell

I. INTRODUCTION

Copper tube has long been the standard for many plumbing applications throughout the world. Of the total copper consumption in Europe and the U.S. in 1989, copper tube represents 11 and 14%, respectively.[1] This translates to approximately 250,000 tons of copper tube in each of these regions annually. The longstanding success of copper as a conduit for a wide variety of water delivery systems has been attributed to numerous features including durability, corrosion resistance, and antimicrobial activity.

Unusual pitting corrosion failures have, however, occurred in a small number of institutional buildings in different regions of the world in recent years. Incidences of an abnormal form of pitting corrosion were primarily in hot and cold water distribution systems of hospitals. While water provided from the mains to the facilities was initially potable, subsequent storage at some of the facilities may have resulted in changes in water quality before introduction to the water distribution system within the buildings. The pitting appeared to be unlike any previously reported failures. After extensive analysis, it was discovered that a film

0-87371-928-X/94/$0.00+$.50

of microbial origin was always present in areas where pitting was observed. The occurrence of a biofilm was believed to contribute in some way to this new form of pitting corrosion of copper plumbing material used in these institutional buildings.

Most corrosion engineers and scientists are not familiar with the potential roles that microbes may have in corrosion reactions. What are the criteria that can be used to implicate microorganisms in pitting corrosion of copper or other metals? It is usually the slimy texture of the corrosion deposit that leads most investigators to suggest a microbiological role. The slimy deposit is actually the result of the biofilm mode of growth. Ironically, biofilms are ubiquitous in nature and are present to varying degrees on virtually every surface in contact with nonsterile water. Microbial biofilms are present on corroding and noncorroding surfaces.

The reason that biofilms had not been associated with pitting corrosion of copper plumbing tube prior to 1985, when it was first reported in a German hospital, may be that no one specifically looked for or suspected their involvement. Although sulfate-reducing bacteria (SRB) had been implicated in certain types of corrosion as far back as the 1930s where anaerobic conditions and high sulfate concentrations occurred,[2] microbially influenced corrosion (MIC) had not been considered to be important in the corrosion of copper tubing through which oxygen-containing water with low sulfate concentrations was circulated. Although polysaccharide-containing biofilms are present on most pipe surfaces, the detection of this and other microbial products led Fischer et al.[3] to establish a connection between the microbiological slime layer and underlying pits in the corroding copper tubing in the hospitals where the problems have occurred. Considerable research has been conducted in the past 5 years to gain a better understanding of the role that biofilm microorganisms can play in pitting corrosion of copper tube. Through this research, we have obtained some understanding of the factors such as water chemistry, temperature and system design, installation, operation, and maintenance that predisposes a system to corrosive microbial attack. This chapter summarizes what we have learned on this subject.

II. TYPES OF PITTING IN COPPER TUBE

A. Classical Types of Pitting

Several different types of pitting corrosion are known to occur in copper tubing used for water service in institutional buildings.[4] Type 1 pitting occurs in cold water lines that contain water supplied from deep bore holes.[5] This water contains low concentrations of microorganisms and organic carbon and high concentrations of dissolved inorganic ions. Pits develop in the inner wall of copper tube. The pits contain soft crystalline

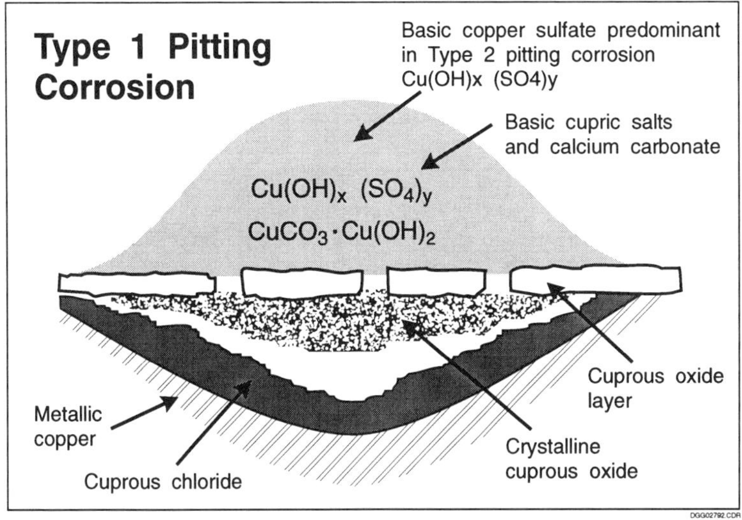

FIGURE 1. Type 1 pitting corrosion.

cuprous oxide under a membrane of cuprous oxide crystals (Figure 1). Cuprous chloride also occurs in the pits. Basic copper carbonate mounds occur over the pits. Type 1 pitting occurs in the presence of certain hard well waters. Luce[6] has shown that Type 1 pitting propensity of such waters depends on pH, dissolved oxygen, chloride, sulfate, sodium, and nitrate. There are no indications to date which suggest that microorganisms are involved in this type of pitting corrosion.

Copper tubing is fabricated to different hardness: hard, half-hard, and soft. In the past, a film derived from the drawing lubricant that becomes carbonized during the softening (bright annealing) process has been related in a rather complicated way to Type 1 pitting corrosion of annealed copper tube.[7] Abrasive cleaning or controlled oxidation of the tube bore is now carried out as a final step in the manufacturing process to remove the film. Hard-drawn tubes, not being bright annealed in the course of fabrication, are much less likely to contain carbon films. However, carbon films are sometimes found in hard-drawn tubes, presumably as a result of exposure to unusually high temperatures during the drawing process. No influence of tube hardness has been detected with respect to the unusual type of pitting corrosion.[3,8–10]

Type 2 pitting occurs in systems which circulate hot, soft water. The pits contain hard, crystalline cuprous oxide and are covered by small nodules composed of copper oxides and basic copper sulfate (Figure 1). An adherent cupric oxide-cuprous oxide layer exists over the unpitted surface. Type 2 pitting occurs primarily at temperatures above 60°C or where the water is below pH 7.4 or where the bicarbonate-sulfate ratio

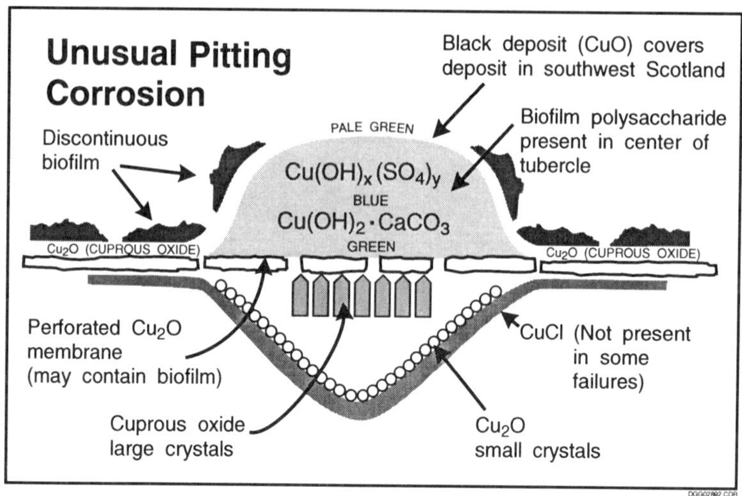

FIGURE 2. Unusual type of pitting corrosion.

in the water is less than 1. Like Type 1 pitting, no evidence currently exists that suggests microbiological involvement in this form of pitting corrosion.

Type 3 pitting is characterized by groups of small hemispherical pits under a common covering of basic copper sulfate. An oxide membrane extends across all the pits in the group with a perforation above the center of each pit. Up to 1% sulfide may be found in the pits. This type of pitting occurs in pipes carrying cold water with high pH, low hardness, and low mineral and low organic content. This type of pitting has been restricted to two areas in Sweden. No involvement of microorganisms has been demonstrated in this type of pitting corrosion.

B. Unusual Types of Pitting

Two unusual types of pitting corrosion have been observed which are thought to involve microbial biofilms.[11] One resembles Type 3 in that numerous pits occur beneath a common basic copper sulfate crust with the oxide membrane across the pits being randomly perforated. This type of pitting has been reported in certain hospitals in southwest Scotland which use surface waters that contain relatively high concentrations of dissolved organic compounds, suspended particulates including microorganisms, and low buffering capacity (pH 7.4 to 9.3). The other new type of pitting occurs in cold, warm, and hot water systems and exhibits features of both Type 1 and Type 2 pitting (Figure 2).[9] It resembles Type 1 pitting in that the pits are hemispherical and contain soft crystalline cuprous oxide with varying amounts of cuprous chloride under a cuprous oxide membrane. It resembles Type 2 pitting in that the oxide on the surface between the pits is largely cupric. The mounds above the pits are principally basic copper sulfate, often with a deposit of powdery cupric oxide around the periphery and on parts of the deposit itself.

Immersion of the tubing in 25% nitric acid indicated the absence of a carbon film and hence the pitting was not Type 1. A dried gelatinous film was present that stained positive with periodic acid Schiffs base (PAS) reagent or Gentiana violet, indicating the presence of polysaccharide.[9,11] This unusual type of pitting has been referred to as "Type 1 ½" pitting and has been in Saudi Arabia, Germany, Scotland, and more recently in southwest England. A description of the corrosion products from copper tubing collected from the various institutional facilities has been reviewed.[9,11,12]

With one exception, cases in which these unusual types of pitting occurred were from soft water supplies. The waters in which the unusual type of pitting corrosion have been observed have total hardness of between 25 to 40 mg/l $CaCO_3$, alkalinity (carbonate hardness) between 10 and 20 mg/l $CaCO_3$, chloride between 15 and 20 mg/l, and sulfate between 10 and 30 mg/l. In most cases, the sulfate content is approximately twice the carbonate content. Since the water is so poorly buffered, pH is quite variable in water associated with this type of pitting. A more complete description of water characteristics where the unusual types of pitting corrosion have been observed is presented by Fischer et al.[9] and by Wagner et al.[13]

In addition to those cases described above, there have been other reports of pitting corrosion in copper tube exposed to soft supply waters containing chloride and sulfate ions in Kuwait.[14] Although their description of the pits resembled that of Type 1½ pitting corrosion, they did not look for or observe the presence of a biofilm. Instead, they proposed yet another mechanism that involved attack of the protective cuprous oxide film along grain boundaries by chloride ions.

III. DESCRIPTION OF PITTING CORROSION OF COPPER IN INSTITUTIONAL WATER SYSTEMS

A. Saudi Arabia Hospital Study

An unusual type of pitting corrosion of copper tubing was reported in a hospital in Saudi Arabia. There were significant differences in the occurrence of pitting in sections of the hospital piping system that received water from different sources.[14] The unusual type of pitting was most prevalent in pipe sections of the hospital that contained warm water. Although no microbiological studies were performed on failed tubing from this system, PAS-positive material was recovered from the pipe wall.[11]

B. Corrosion Characteristics of German Hospital Copper Water Lines

Pitting corrosion similar to that observed in Saudi Arabia and from a hospital in eastern Scotland was reported in a hospital in Germany shortly after it had opened in 1986.[9] Approximately one third of the water system exhibited failures. Coincident with the pitting phenomena was a temporary and locally high concentration of dissolved copper ions and solid copper corrosion products in the water. Problems occurred with highest frequency in deadleg sections of cold and warm water lines. The tubes contained a deposit of adherent cupric oxide and showed pitting under mounds of basic cupric sulfate with loose powdery cuprous oxide inside, on top of, and around the perimeter of the pit.

Chemical analysis of the corrosion products revealed copper complexes with organic compounds of microbial origin such as pyruvate, lactate, and histidine.[15] A biofilm was detected and found to contain polysaccharides, oligopeptides, and *n*-acetylated derivatives of glucose, mannose, and galactose. Extended absorption fine structure spectroscopy detected Cu^{1+} complexes with imidazole residues of histidine in the biofilm.[16]

Microbiological evaluation of the corrosion deposits recovered from a county hospital in Germany was carried out by Wagner et al.[10,13] The presence of high numbers of surface-associated bacteria was not always correlated with pitting. The range of culturable bacterial species was quite variable even among the pitted pipe samples, indicating that one species was not responsible for all the observed pitting corrosion. Nevertheless, three isolates were consistently recovered: two strains of *Pseudomonas paucimobilis,* each with different pigmentation, and *P. solanacearum.*[11] Whereas all three isolates were capable of growth as suspended cultures at temperatures as high as 40°C, one strain of *P. paucimobilis,* when grown in a mixed-population biofilm, tolerated temperatures of 50°C.[11] The isolate of *P. solanacearum* was found to be a facultative anaerobe capable of nitrate respiration to dinitrogen and capable of producing large quantities of uronic acid-containing exopolysaccharide which was enhanced at elevated temperatures. The three isolates displayed copper tolerance, particularly at elevated temperatures which coincided with enhanced exopolysaccharide production.[16] Thus, there may be a relationship between temperature, exopolysaccharide production, copper tolerance, and pitting corrosion. Additional research needs to be carried out to establish any correlation between these factors.

Water that had passed through the water distribution system of the hospital was run through test loops of copper tube.[10,13] After 2 years, the manifestations of pitting and generalized corrosion were reproduced, although at a slower rate than was observed previously in the hospital piping system. A black layer of cupric oxide developed on the surface of the tubing, tubercles of malachite or posniakite were observed, and both uniform attack and pitting attack occurred at the same time. Unlike the tubing in the hospital water distribution system, no continuous biofilm was recovered from the tubing in the test loop. Thus, simulation of the unusual corrosion reaction has been achieved, but at a slower rate and without evidence of a biofilm.

C. Corrosion Characteristics of Facilities in Southwest England

Detailed examination of copper tubing that suffered from the unusual types of pitting corrosion in two hospitals in southwest England revealed failures with characteristics similar to those in Germany, Saudi Arabia, and Scotland. Pitting was associated almost exclusively with pipes bearing black cupric oxide films containing a microbial polymer biofilm.[11] The pitted pipes were invariably from locations where ambient temperatures were 30 to 40°C and where water flow rates were low or intermittent, or in some instances stagnant. The water sources for the two hospitals in southwest England which experienced the unusual type of pitting corrosion were obtained from different rivers. The water was soft with high humic acid content.

A study was conducted to relate bacterial densities with pitting using 14 pipe samples collected from the two hospitals that experienced the unusual pitting corrosion, as well as a third hospital in the region that did not experience the unusual type of corrosion.[11] No SRB were detected in any of the samples based on anaerobic incubation in Postgate C medium at 30°C for 3 to 5 d. Twelve isolates were identified by the API 20NE system as falling into one of the three following genera: *Methylobacterium, Pseudomonas,* and *Aeromonas.* All samples examined had some bacteria present. The unusual type of pitting corrosion, as determined by the presence of perforations, well-developed pits, and black cupric oxide over the majority of the surface, occurred only where bacterial densities exceeded 100-colony forming units (cfu)/ml. In some cases, pitting was observed where bacterial densities were low, but polysaccharide abundance was high, based on PAS base staining.[18]

D. Hospital Survey in Southwest Scotland

Severe pitting corrosion has been reported in several hospitals in southwest Scotland. In one hospital alone, 64 failures were reported between 1982 to 1988. Angell et al.[11] reported a pepper pot pitting on the tube surface that exhibited some of the features of the unusual type of pitting corrosion reported in the other hospitals described above. Tubes contained a film that stained PAS- and alcian-positive, suggesting the presence of substituted and unsubstituted polysaccharides. A superficial deposit of organic material was also detected that was thought to be composed of humic substances.

Examination of the pitted areas from which the corrosion product mound had largely detached revealed a copper (I) oxide layer with a number of pepper pot holes in it.[11] The perforations in the cuprous oxide layer of these pits were larger than those associated with the pits in tubing from the hospitals in Germany, Saudi Arabia, and southwest England (Figure 2). A short distance from the pit, a deposit of cuprous oxide was observed beneath a layer containing spherical cupric oxide nodules and biofilm. McEvoy and Colbourne[19] found that more severely corroded tubes contained a more fully developed biofilm. Microbiological analysis of the black tubercles covering the perforated areas indicated the presence of SRB and a variety of aerobes, including *Pseudomonas* and *Alcaligenes* spp., pink-pigmented facultatively methylotrophic bacteria, and fungi.[20] By contrast, fewer and much smaller black nodules were observed in the tubing taken from hospitals that had not experienced pitting corrosion. No pitting was observed in tubing with thin biofilms containing fewer rod- and cocci-shaped bacteria.[20]

IV. FACTORS CONTRIBUTING TO MICROBIALLY INFLUENCED PITTING CORROSION OF COPPER TUBING

A. Temperature

Substandard water maintenance was thought to contribute to the corrosion problems. The institutional buildings experiencing pitting corrosion had a lower temperature in their hot water systems than those without corrosion. Practical experience has shown that MIC has not been a problem in water systems maintained above 60°C. Studies by Walker et al.[34] showed that the hot water in the hospitals which experienced the pitting corrosion was not maintained at a sufficiently high temperature to inhibit microbial growth. Most of the time the temperature was below 50°C, which enabled the bacteria to proliferate. A two-stage continuous culture model was used to mimic the corrosive environment of one of the hospitals.[34] Using a microbial inoculum from the corroded copper tubing, these authors showed that the biofilm could be established at temperatures up to 55°C. However, at temperatures above 55°C the biofilm was greatly reduced. That pitting corrosion and biofilm accumulation were much more pronounced at temperatures below 55°C than above that temperature suggests the involvement of a biological rather than a nonbiological reaction. It was concluded that the temperature of the hot water supply has a significant role in the control of MIC in these institutional buildings.[34]

B. Water Chemistry

Although the quality of the supply water was considered to be good, the level of assimilable organic carbon (AOC) was found to be higher than in other areas where corrosion problems were absent. The water in hospitals which did not maintain their hot water above 50°C exhibited higher AOC, presumably due to sloughing of biofilm from the walls of the tubing. In this respect, the growth of biofilm microorganisms on the walls of the piping can lead to the degradation of water quality. Besides serving as nutrients for the bacteria, some of the organic carbon (i.e., humic substances in the supply water) are strong chelators of copper and counteract the bactericidal properties of copper ions in solution. Aluminum-chelating phenolic material was recovered from the biofilm/corrosion deposits.[35]

Dissolved oxygen concentration in the water of hospitals experiencing corrosion dropped to 0.1 mg/l during the 11-h period that hospital activities were minimal (between 7 p.m. and 6 a.m.). This drop was believed to be due to the respiratory activities of the biofilm bacteria, since the planktonic bacterial densities were not high enough to account for the oxygen demand. The establishment of anaerobic conditions was thought to promote the growth of SRB which were found in the biofilm and may have contributed to the corrosion reactions.

The supply water also contained fine particulates which accumulated in the storage tanks and promoted microbial growth. A combination of the factors identified above were thought to promote the extensive biofilm growth on the inner pipe surface and contribute to the pitting of the underlying copper surface. Although a specific mechanism was not identified for the microbial role in the corrosion process, it was felt that removal of the humic substances from the water, increasing the water temperature to 55 to 60°C, and a general cleanup of the system was needed to control the corrosion problem. Unfortunately, once this type of corrosion is allowed to start, it is difficult to control.

C. Design and Installation

In addition to water characteristics and temperature, plumbing design and installation seem to play a significant role in system susceptibility to MIC.[8] Stagnation of water can occur through the existence of deadlegs built into the system. These regions are particularly vulnerable to microbial accumulation and are difficult to treat. Stagnation also occurs when water is left in the system for extended periods of time following pressure testing before the system goes into operation. Systems designed with long horizontal runs of tubing are susceptible to MIC. This feature promotes the accumulation of sediment on the bottom of the tubing, which increases the surface area for microbial colonization and growth and promotes the development of anaerobic conditions in these areas which are conducive for the growth of SRB and other potentially corrosive anaerobic species. Inadequate filtration of source water can also lead to sediment accumulation in various parts of the system. Poor soldering practices can lead to the accumulation of a bead on the inside of the tubing at joints. This irregularity in surface contour has long been known to produce eddies in the water flow pattern and promote erosion corrosion that is free of deposits of any kind. The protruding bead also acts to entrain bacterial cells that accumulate as a biofilm in the eddies.

V. LABORATORY SIMULATIONS OF MICROBIALLY INFLUENCED CORROSION OF COPPER

A. Attempts to Reproduce Unusual Type of Pitting Corrosion Observed in the Field

Controlled laboratory studies were carried out to determine whether the three bacterial isolates recovered from the corroded tubing in the German hospital, when grown as a mixed population in a synthetic water, could reproduce the pitting corrosion of copper observed in the field.[11] Flow-through reactors containing copper rings exposed to synthetic water with a chemistry similar to that used by several of the hospitals described above that experienced the unusual type of pitting corrosion were operated in the presence and absence of bacteria and the corrosion rates compared after 5 months. The copper surfaces exposed to these conditions exhibited none of the features of the pitted copper tubes recovered from the hospitals, although there were important differences between the materials recovered from the sterile and inoculated reactors. The sterile copper surfaces exhibited slight superficial corrosion with a basic cupric carbonate deposit. The remainder of the surface contained a cuprous oxide film. On the outer edge of the copper ring a pit was observed which contained crystalline cuprous oxide.

The rings from the inoculated reactor contained a biofilm with polysaccharide. Corrosion was observed over a broad area, producing an orange layer of cuprous oxide and in some places a more coherent cuprous oxide film surrounded by deposits of basic copper carbonate and sulfate crystals. In other areas of the ring pits containing loose copper oxide were observed under basic copper carbonate deposits. A film of black cupric oxide occupied areas surrounding the pits. Although this laboratory study failed to reproduce the spherical nodules of cupric oxide characteristic of the unusual type of pitting observed in the hospitals described above, pitting was observed in the presence of polysaccharide-producing bacterial biofilms.

B. FTIR Studies

Sensitive analytical techniques have been developed to study MIC of copper in real time, in a nondestructive manner, and without disturbance to the system. Bremer and Geesey[21] demonstrated that bacteria isolated from pits in corroded copper tubing removed from a heat exchanger could be grown either as a batch culture or as a continuous culture on a germanium internal reflection element (IRE) and monitored by attenuated total reflectance Fourier transform infrared spectroscopy (ATR/FTIR) for up to 417 h (Figure 3). The spectra revealed the accumulation of microbial products over time at the surface of the IRE. These products consisted largely of protein and polysaccharide as well as other unidentified components. Operation in the double beam mode using a sterile IRE exposed to a sterile liquid bacterial culture medium allowed accurate subtraction of components from the bulk aqueous phase that had adsorbed to the surface of the inoculated IRE (Figure 4).

ATR/FTIR was adapted for MIC studies of copper by depositing ultrathin films of copper on the IRE (Figure 5).[22] Copper films of 6 to 7 nm nominal thickness, when deposited by vapor deposition, appeared to be continuous based on X-ray photoelectron spectroscopic analysis and contained a cuprous oxide layer at the surface. Atomic force microscopic evaluation of the copper thin films revealed them to be coalescing aggregates of copper atoms (Figure 6). When IREs coated with 6- to 7-nm-thick films were submerged in an aqueous medium, there was sufficient transmission of IR radiation through the Cu to obtain significant water absorbance at 1640 cm^{-1}. The intensity of the water absorption band was found to be very sensitive to changes in thickness of the copper film.[22–24] This feature can be used to study the effect of microbial biofilms and their products on the integrity of the oxidized copper surfaces in aqueous environments.

One strain of *P. paucimobilis*, recovered from the corroded copper tubing in the German hospital, when grown as a biofilm on thin films of copper on Ge IREs, promoted higher rates of copper corrosion than sterile controls subjected to otherwise similar conditions.[17] That attached bacteria and not suspended bacterial cells promoted the observed deterioration of the copper film was demonstrated by flow-through studies. After an initial inoculation to initiate biofilm development on the copper film, no additional bacteria were added to the aqueous medium flowing over the surface. Corrosion occurred only after biofilm had developed.[17] These results suggest that pure cultures of bacteria isolated from pitted copper tubing can destabilize copper under highly controlled laboratory conditions.

Several bacteria were isolated from corrosion deposits on copper tubing exposed to tap water from laboratory faucets at California State University, Long Beach, and used to study the MIC of copper. One

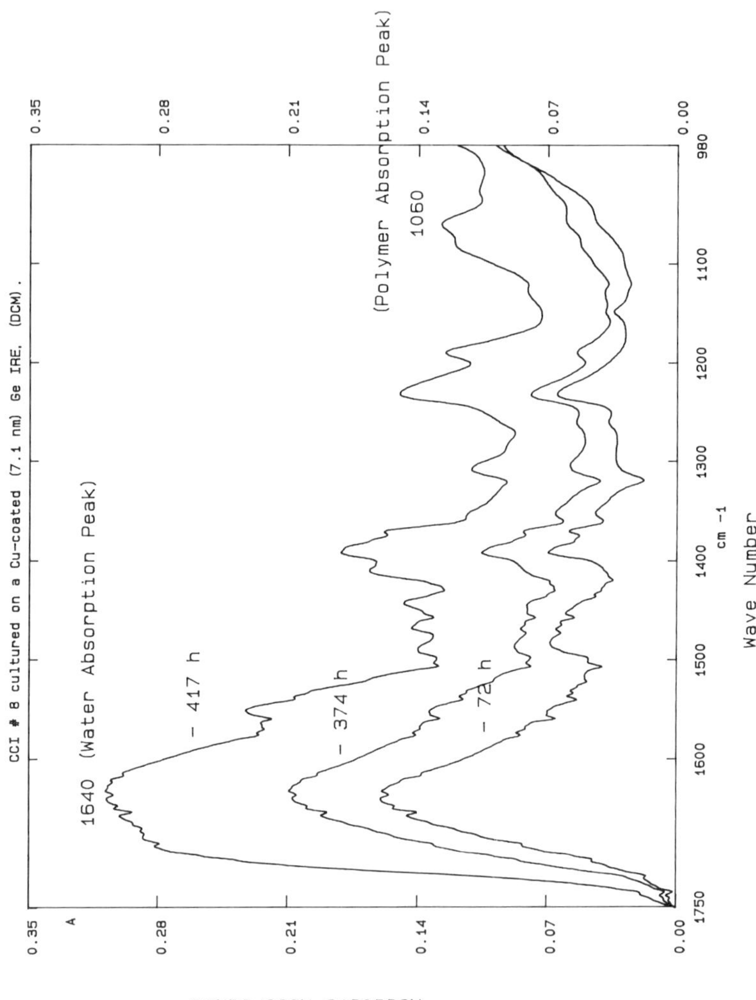

FIGURE 3. Infrared spectra collected at various times after inoculation of bacterial isolate CCI#8 in an ATR cell containing a Cu-coated germanium IRE.

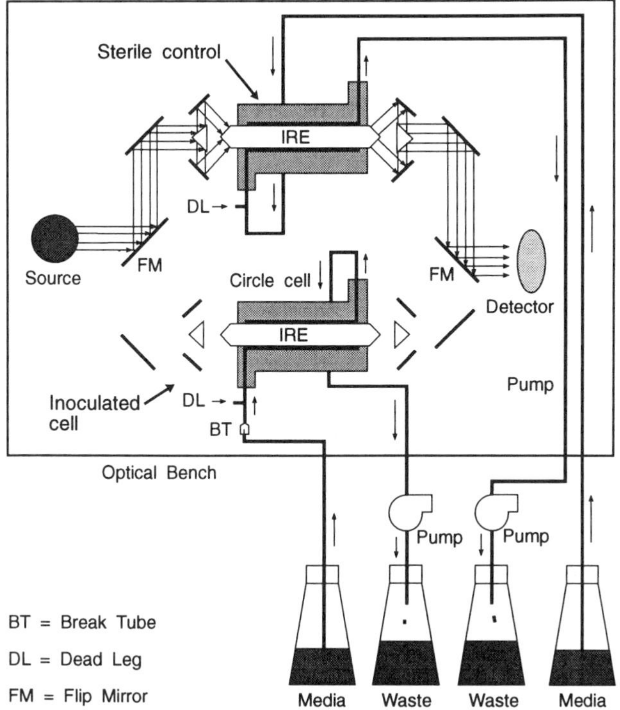

FIGURE 4. Schematic diagram of the optical bench of an FTIR spectrometer containing two cylindrical ATR cells operated in the dual beam mode.

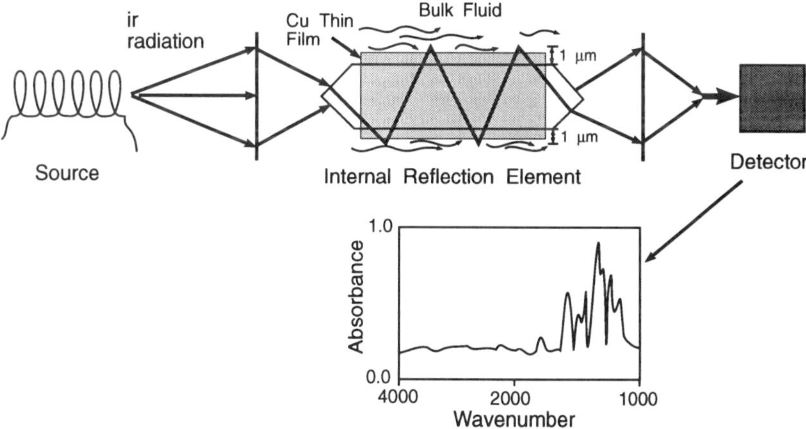

FIGURE 5. Schematic diagram of a copper-coated germanium IRE exposed to flowing medium inoculated with film-forming bacteria.

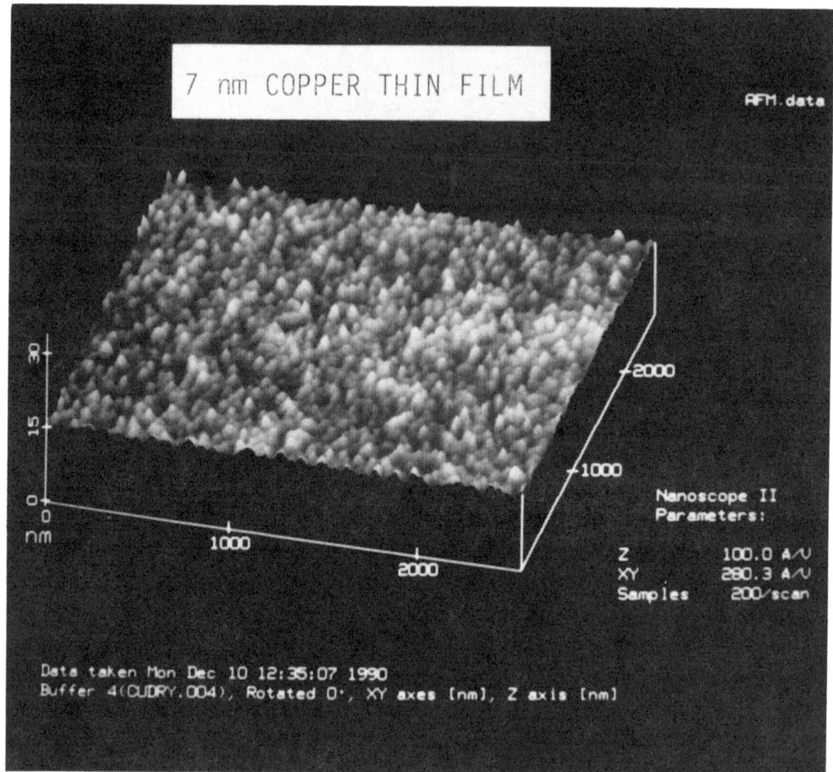

FIGURE 6. Atomic force micrograph of a 7-nm-thick copper film deposited on a germanium coupon by physical vapor deposition.

isolate, referred to as CCI#8, when exposed to a copper-coated IRE in stagnant culture media, produced an immediate decrease in the stability of the copper film based on changes in the intensity of the water absorbance band.[25] When this experiment was repeated with another bacterial isolate (CCI#11), the copper film remained stable in the presence of this bacterium over a 77-h period. These studies suggest that some, but not all, bacteria promote deterioration of copper thin films.

A similar study was performed in a flow-controlled sampling cell.[25] When CCI#8 was inoculated into the flowing medium that passed over the copper film, very little deterioration of the metal was observed over a 2.5-h period. Unlike the static culture experiment described above, no change in the stability of the copper film was observed after flow was suspended and the sampling cell was maintained as a static culture. However, when flow was resumed after 64 h as a static culture, there was an immediate loss of film stability. There was visual evidence of a biofilm on the inoculated IRE and localized deterioration of the copper film under the biofilm. No bacterial growth or deterioration of the copper film was observed in the sampling cell that was maintained as a sterile control under the same flow regime. The results suggest that flow conditions influence biofilm-induced deterioration of the copper film and that in the absence of bacteria, flow has little or no effect over the range of flow conditions tested in these experiments. It should be noted that flow conditions have been shown to affect copper corrosion without the assistance of microorganisms.

The experiments described above were rerun over a longer time period under different flow regimes to obtain a better understanding of the relationship between flow, biofilm growth, and deterioration of the copper film.[26] Exposure of the copper thin film to CCI#8 under flowing conditions for 330 h resulted in only slight deterioration of the metal. This corresponded to a corrosion rate of 8.4×10^{-4} mpy.[27] Shortly after flow was stopped, the corrosion rate increased to 8.8×10^{-3} mpy.[27] These rates represent the average corrosion rate over the entire IRE and not where focused attack occurred under portions of the biofilm.

At the same time that the copper film deteriorated there was an increase in the amount of exopolysaccharide that accumulated on the surface.[26] No increase in accumulation of other cell components such as protein was observed during this time. The accumulation of biofilm on the copper-coated IRE was confirmed by visual examination and plating of homogenized biofilm at the termination of the experiment. Again, the area under the biofilm was discolored, verifying the copper deterioration detected by the FTIR. The data confirm earlier studies that changes in flow rate and allowing the system to become stagnant promote biofilm-induced copper corrosion.

Some bacterial biofilms protect copper from the destructive action of other bacteria. Biofilms of CCI#11, grown under flowing or quiescent conditions, caused no detectable deterioration of the copper film, even over exposure periods of up to 500 h.[26] Once established, biofilms of CCI#11 protected the copper film from attack by CCI#8 over a 300-h period under flowing or quiescent conditions. Protection was not due to exclusion of CCI#8 from the surface by the biofilm of CCI#11, since changes in the infrared spectrum of the biofilm were observed after inoculation of CCI#8 to the system and approximately equal numbers of both bacterial types were recovered after plating samples of the biofilm on solid culture medium after termination of the experiment. These results demonstrate for the first time that a particular type of bacterium can protect the surface of copper from the aggressive localized attack of another type of bacterium. Since biofilms are present on most surfaces submerged in aqueous environments and only a very small fraction of these surfaces are subject to pitting corrosion, it is not unreasonable to suggest that some bacteria play a protective role in MIC of copper tubing.

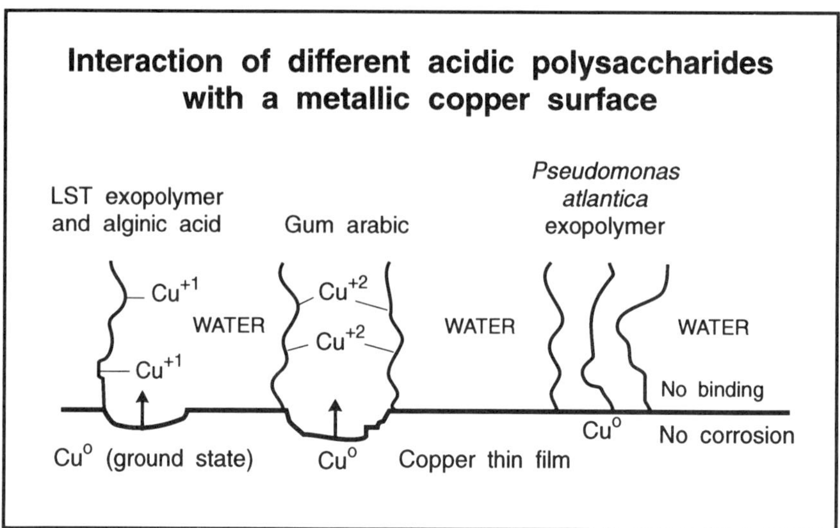

FIGURE 7. Interactions of different acidic polysaccharides with a metallic copper surface. Different polymer chemistries stabilize the copper in different valence states.

C. Role of Microbial Exopolysaccharides in Localized MIC of Copper

Studies by Fischer et al.[3] and Paradies et al.[15] indicated that polysaccharides and oligopeptides were associated with corrosive biofilms. Exopolysaccharides isolated from biofilm bacteria have been shown to destabilize copper thin films. When a copper-coated IRE was exposed to a 1% solution of a crude acidic exopolymer from a sediment bacterium (FRI), the copper film corroded immediately.[28] A similar phenomenon was observed when the copper-coated IRE was exposed to a suspension of other acidic polysaccharides such as alginic acid and gum arabic.[24] The results suggest that metallic copper is sensitive to a wide range of acidic polysaccharides, including those produced by some biofilm-forming bacteria that colonize copper tube.

Results from several studies suggest that the metallic copper is oxidized by acidic polysaccharides. X-ray photoelectron spectroscopy has shown that some of the copper deposited on the IRE after exposure to gum arabic and alginic acid was oxidized to Cu^{2+}, that which was exposed to other bacterial polysaccharides was oxidized to Cu^{1+}, and that which was exposed to yet other types of bacterial exopolysaccharides underwent little oxidation and remained as Cu^0 (Figure 7).[29–30] These results are consistent with a corrosion mechanism based on a copper concentration cell formed by the excretion of exopolysaccharides with different affinities for copper ions by different bacterial species within a biofilm.[31–32]

Paradies et al.[8] presented evidence that the imidazole or L-histidine groups of the oligopeptide chains in biofilms participate in complex formation with Cu^{1+}. These authors also have detected a Cu^{3+} complex in corrosive biofilms which was formed through a reaction with peroxide. A corrosion mechanism involving these complexes has been proposed.

The chemistry and possibly the structure of exopolysaccharides of copper-corroding bacteria are influenced by the presence of copper in the surrounding environment. Bremer and Geesey[33] demonstrated that the copper-complexing carboxyl groups contributed by uronic acid subunits are significantly more abundant in exopolysaccharides produced by cells exposed to copper than those produced in environments with little or no copper present. The exopolymer produced in the presence of copper bound approximately 16% of its weight in copper. It is likely that some microorganisms produce large amounts of copper-complexing exopolymers to protect themselves from the toxic effects that Cu^{2+} has on essential reactions inside the cell. These polymers accumulate during biofilm growth in a heterogeneous manner over the surface. This heterogeneity may promote localized dissolution of metallic copper, which takes the form of pitting. Whether pitting propensity is simply related to the quantity of exopolymer produced or a reflection of subtle differences in exopolymer chemistry or structure remains to be determined. Further research will be required to gain a more complete understanding of the role of specific bacterial exopolymers in pitting corrosion of copper. Nevertheless, significant progress has been made in demonstrating the participation of microbial biofilms in this phenomenon.

VI. PREVENTION OF MIC OF COPPER TUBING USED IN POTABLE WATER SYSTEMS

Based on the information presented above, it appears that a combination of factors contribute to the unusual forms of pitting corrosion of copper tube in institutional buildings such as hospitals. These factors include the use of soft water with low pH, high suspended solids, and AOC content, long-term or periodic stagnation of water in the pipeworks which produces widely fluctuating oxygen concentrations, maintaining water temperatures that promote rapid growth and activity of naturally occuring bacteria that form biofilms on the pipe wall, and the lack of an adequate monitoring program to periodically evaluate water quality and pipe wall condition. Thus, it should be possible to avoid the undesirable corrosion by identifying and specifying limits for these parameters.

Nuttall and Rich[36] described the key factors for design, installation, and commissioning of copper plumbing systems in large public buildings. From a microbiological point of view, they were concerned more with control of *Legionella pneumophilla* than of corrosion-causing microorganisms, since there was little information available on MIC of copper tube at that time. More recently, Fischer et al.[9] have summarized many of the precautions that should be followed when water chemistry, temperature, and system operation predispose the system to MIC. Based on these previous reports and the studies described above, a number of recommendations can be made that reduce the likelihood of microbially induced pitting corrosion of copper tubing. The recommendations fall into three categories: system design, installation, and operation.

The system should be designed to minimize the possibility of water stagnation. Deadlegs (sections of tubing which contain stagnant water for long periods of time without being refreshed in any way) should be eliminated. In large installations, pumps should be installed that maintain water flow velocities between 0.7 to 1.5 m/s at all times. Filtration equipment may be desirable for removal of suspended particulates in the source water that could accumulate along horizontal runs of plumbing.

Although little attention has been given to monitoring programs, periodic examination of the interior pipe wall surface for deposit accumulation, tubercle formation, and biofouling permits recognition of a potential MIC problem and initiation of corrective action before control over the problem is lost. Monitoring of this nature requires installation of removable, replaceable pipe sections at different locations in the system, periodic inspections, and a plan of corrective action when MIC is observed. This alternative could be less costly than replacement of all failed tubing in a system that is particularly prone to MIC.

Installation of the plumbing system should be carried out according to copper tube manufacturers' recommended procedures. Soldered joints should be smooth and free of solder beads on the inside pipe surface. Debris introduced into the tubing during installation should be removed before pressure testing. No water should be introduced into the system for pressure testing or for other reasons until it is certain that the system can be operated under the conditions for which it was designed soon after testing has been completed. The system should never be allowed to go out of operation without draining all of the water. Only high-quality water should be used during pressure testing. If there is a delay between the time the system is pressure tested and put into commission, water should be circulated through all wetted tubing at regular intervals to prevent stagnation. Immediately prior to commissioning, the system should be sterilized to kill any harmful microorganisms that inadvertently entered the system during previous operations.

Operating conditions should be based on health-related, microbiological standards applicable to potable water systems. The use of high-quality water is strongly recommended. Water hardness should be adjusted to establish a protective oxide film on the inner surface of the plumbing material. Of all the recommendations made, maintaining proper water temperature in the lines is the most important for minimizing MIC. Water temperature should be maintained below 25°C for cold water and above 55°C for hot water at all times. Maintenance of these temperatures may require water flow through the pipes at regular intervals.

In summary, certain water chemistry, system design, and installation and operation practices predispose copper tubing to microbially induced pitting corrosion which results in through-wall perforations that are costly to repair. The phenomenon appears to be related to a combination of chemical and biological factors which act in concert to cause the observed failures in institutional buildings. By altering several of these factors at the same time, it should be possible to reduce or eliminate the problem.

ACKNOWLEDGMENTS

Preparation of this review was supported by the International Copper Association grant #413, U.S. National Science Foundation grant DMR-9196070, and Cooperative Agreement ECD 8907039 between the U.S. National Science Foundation and Montana State University. We wish to thank Hector Campbell, Otto von Franque, Brian Moreton, and William Dresher for their helpful comments and suggestions in the preparation of this paper.

REFERENCES

1. **Voutilainen, P.,** Overview of developments in the copper industry, *Copper '90. Refining, Fabrication, Markets,* The Institute of Metals, The Bourne Press, Bournemouth, Dorset, England, 1990, 1.
2. **von Wolzogen Kuhr, C. A. H. and van der Vlugt, I. S.,** The graphitization of cast iron as an electrochemical process in anaerobic soils, *Water,* 18, 147, 1934.
3. **Fischer, W., Haenssel, I., and Paradies, H. H.,** First results of microbial induced corrosion on copper pipes, *Microbial Corrosion-1,* Sequeira, C. A. C. and Tiller, A. K., Eds., Elsevier Applied Science, London, 1989, 300.
4. **Campbell, H.,** Electrochemical Factors: An Overview, Report to International Copper Association Workshop, Frankfurt, Germany, December, 1990.
5. **Lucey, V. F.,** Mechanism of pitting corrosion of copper in supply waters, *Br. Corros. J.,* 2, 175, 1967.

6. **Lucey, V. F.,** Assessment of type 1 pitting corrosion characteristics of potable waters, *Corrosion of Copper and Copper Alloys in Buildings,* Japan Copper Development Association, Tokyo, 1982, 86.
7. **Campbell, H. S.,** Pitting corrosion in copper water pipes caused by films of carbonaceous material produced during manufacture, *J. Inst. Metals,* 77, 345, 1950.
8. **Paradies, H. H., Haenssel, I., Fischer, W., and Wagner, D.,** *Microbiologically Induced Corrosion on Copper Pipes,* INCRA Final Report No. 404, 1990.
9. **Fischer, W., Paradies, H. H., Wagner, D., and Haenssel, I.,** Copper deterioration in a water distribution system of a county hospital in Germany caused by microbially induced corrosion. 1. Description of the problem, *Werkst. Korros.,* 43, 56, 1992.
10. **Wagner, D., Fischer, W., and Paradies, H. H.,** Copper deterioration in a water distribution system of a county hospital in Germany caused by microbially induced corrosion. 2. Simulation of the corrosion process in two test rigs installed in this hospital, *Werkst. Korros.,* 43, 496, 1992.
11. **Angell, P., Campbell, H. S., and Chamberlain, A. H. L.,** *Microbial Involvement in Corrosion of Copper in Fresh Water,* Interim Report, International Copper Association, New York, 1990.
12. **Fischer, W., Paradies, H. H., Haenssel, I., and Wagner, D.,** Copper deterioration in a water distribution system of a county hospital in Germany caused by microbial induced corrosion, *Microbially Influenced Corrosion and Biodeterioration,* Dowling, N. J., Mittelman, M. W., and Danko, J. C., Eds., University of Tennessee Press, Knoxville, 1990, 8–47.
13. **Wagner, D., Fischer, W. R., and Paradies, H. H.,** Test methods on microbial induced corrosion in different loops, *Proceedings of the 12th Scandinavian Corrosion Congress and Eurocorr,* Dipoli, Finland, 92, 1992, 651.
14. **Shalaby, H. M., Al-Kharafi, F. M., and Gouda, V. K.,** A morphological study of pitting corrosion of copper in soft tap water, *Corrosion,* 45(7), 536, 1989.
15. **Paradies, H. H., Fischer, W. R., Haenssel, I., and Wagner, D.,** Characterization of metal biofilm interactions by extended absorption fine structure spectroscopy, *Microbial Corrosion-2,* Sequeira, C. A. C. and Tiller, A. K., Eds, European Federation of Corrosion, Publication No. 8, The Institute of Materials, 1992, pp168–188.
16. **Chamberlain, A. H. L. and Angell, P.,** Influences of microorganisms on pitting of copper tube, *Microbially Influenced Corrosion and Biodeterioration,* Dowling, N. J., Mittelman, M. W., and Danko, J. C., Eds., University of Tennessee Press, Knoxville, 1990, 3–65.
17. **Geesey, G. G. and Bremer, P. J.,** Evaluation of corrosive action of biofilms of *Pseudomonas paucimobilis* on copper, *Proceedings of the 1st Pan-American Corrosion and Protection Congress,* Vol. 1, Mar del Plata, Argentina, 1992, 349.
18. **Chamberlain, A. H. L., Angell, P., and Campbell, H. S.,** Staining procedures for characterizing biofilms in corrosion investigations, *Br. Corros. J.,* 23, 197, 1988.
19. **McEvoy, J. and Colbourne, J. S.,** *Glasgow Hospital Survey Pitting Corrosion of Copper Tube,* Report to International Copper Research Association, New York, 1988.
20. **Keevil, C. W., Walker, J. T., McEvoy, J., and Colbourne, J. S.,** Detection of biofilms associated with pitting corrosion of copper pipework in Scottish hospitals, *Biocorrosion,* Gaylarde, L. C. and Morton, L. H. E., Eds., Biodeterioration Society, Kew, England, 1989, 99.
21. **Bremer, P. J. and Geesey, G. G.,** An evaluation of biofilm development utilizing non-destructive attenuated total reflectance fourier transform infrared spectroscopy, *Biofouling,* 3, 89, 1991.
22. **Bremer, P. J., Geesey, G. G., Drake, B., Jolley, J. G., and Hankins, M. R.,** Characterization of a thin copper film to investigate microbial biofilm formation, *Surf. Interface Anal.,* 17, 767, 1991.

23. **Iwaoka, T., Griffiths, P. R., Kitasako, J. T., and Geesey, G. G.,** Copper coated cylindrical internal reflectance elements for investigating interfacial phenomena, *Appl. Spectrosc.*, 40, 1062, 1986.
24. **Jolley, J. G., Geesey, G. G., Hankins, M. R., Wright, R. B., and Wichlacz, P. L.,** In situ, real time FT-IR/CIR/ATR study of the biocorrosion of copper by gum arabic, alginic acid, bacterial culture supernatant and *Pseudomonas atlantica* exopolymer, *Appl. Spectrosc.*, 43, 1062, 1989.
25. **Geesey, G. G. and Bremer, P. J.,** Applications of fourier transform infrared spectrometry to studies of copper corrosion under bacterial biofilms, *Mar. Technol. Soc. J.*, 24(3), 36, 1990.
26. **Bremer, P. J. and Geesey, G. G.,** Laboratory-based model of microbiologically induced corrosion of copper, *Appl. Environ. Microbiol.*, 57, 1956, 1991.
27. **Bremer, P. J. and Geesey, G. G.,** Nondestructive, real-time studies on the interaction of bacterial biofilms on submerged copper surfaces, *Microbially Influenced Corrosion and Biodeterioration*, Dowling, N. J., Mittelman, M. W., and Danko, J. C., Eds., University of Tennessee Press, Knoxville, 1990, 8–37.
28. **Geesey, G. G., Iwaoka, T., and Griffiths, P. R.,** Characterization of interfacial phenomena occurring during exposure of a thin copper film to an aqueous suspension of an acidic polysaccharide, *J. Colloid Interface Sci.*, 120, 370, 1987.
29. **Jolley, J. G., Geesey, G. G., Hankins, M. R., Wright, R. B., and Wichlacz, P. L.,** Auger electron spectroscopy and X-ray photoelectron spectroscopy of biocorrosion of copper by gum arabic, BCS and *Pseudomonas atlantica* exopolymer, *J. Surface Interface Anal.*, 11, 371, 1988.
30. **Jolley, J. G., Geesey, G. G., Hankins, M. R., Wright, R. B., and Wichlacz, P. L.,** Auger electron and X-ray photoelectron spectroscopic study of the biocorrosion of copper by alginic acid polysaccharide, *Appl. Surface Sci.*, 37, 469, 1989.
31. **Geesey, G. G.,** What is biocorrosion?, *Biofouling and Biocorrosion in Industrial Water Systems*, Flemming, H.-C. and Geesey, G. G., Eds., Springer-Verlag, Heidelberg, 1991, 155.
32. **Geesey, G. G., Mittelman, M. W., Iwaoka, T., and Griffiths, P. R.,** Role of bacterial exopolymers in the deterioration of metallic copper surfaces, *Mater. Perform.*, 25, 37, 1986.
33. **Bremer, P. J. and Geesey, G. G.,** Properties of the exopolymer of a bacterium that causes pitting corrosion of copper, *Short Communications of the 1991 International Marine Biotechnology Conference*, W. C. Brown, Dubuque, IA, 1993.
34. **Walker, J. T., Dowsett, A. B., Dennis, P. J. L., and Keevil, C. W.,** Continuous culture of biofilm associated with copper corrosion, *Int. Biodeter.*, 27, 121, 1991.
35. **Angell, P. and Chamberlain, A. H. L.,** *Status Report*, International Copper Research Association Project 405, New York, 1989.
36. **Nuttall, J. L. and Rich, T.,** *Key Factors in the Design, Installation and Commissioning of Copper Plumbing Systems in Large Public Buildings*, Final report to International Tube Association, New York, 1988.

Section III
BIOCORROSION

17

Development of Electrochemical Test Methods for the Study of Localized Corrosion Phenomena in Biocorrosion

F. Mansfeld and H. Xiao

I. INTRODUCTION

Considering the fact that the vast majority — if not all — reactions involved in corrosion processes are of electrochemical nature, it is not surprising that the use of electrochemical techniques has been very successful in basic and applied studies of corrosion. This applies to both mechanistic studies in laboratory investigations and corrosion monitoring in field applications. In evaluating the use of electrochemical monitoring techniques for the detection and quantification of microbiologically influenced corrosion (MIC), a few limitations of such techniques have to be considered. A weakness of electrochemical techniques is the failure of most techniques to give quantitative results in cases of localized corrosion. These techniques give average readings for the surface of a test electrode, and it is not clear whether a measured corrosion current corresponds to uniform corrosion of the entire surface or to localized corrosion of just a few sites on this surface. In the latter case, corrosion rates will be severely underestimated if the measured corrosion loss is not normalized to the area at which localized corrosion occurs — a

0-87371-928-X/94/$0.00+$.50

quantity which is usually unknown. This general disadvantage of electrochemical techniques is especially bothersome in the case of MIC, where most corrosion processes are of a localized nature. Most recently, techniques such as electrochemical impedance spectroscopy (EIS) have been shown to contain information which is specifically related to localized corrosion processes.[1–3] However, so far EIS has been used mainly as a laboratory technique for mechanistic studies. Another new technique is electrochemical noise analysis (ENA), which is considered by some as the ideal tool for the study of localized corrosion. Unfortunately, very few studies have demonstrated that localized corrosion can be detected for a system for which no previous information concerning its corrosion mechanism is available.

In a joint project with the Center for Interfacial Microbial Process Engineering at Montana State University, a number of electrochemical techniques are being evaluated for the study of localized corrosion phenomena in MIC. These techniques include EIS, ENA, and the thin foil technique for monitoring of pit growth rates. EIS has been used successfully in a number of areas of corrosion research such as the evaluation of inhibitors, polymer coatings, anodic films, and general studies of the kinetics of corrosion reactions.[4]

ENA has been used in investigations of various types of corrosion for a variety of materials. Most frequently, this technique is applied in studies of localized corrosion phenomena. Different research groups have been working in three different directions of data analysis, namely spectral analysis, statistical analysis, and theoretical approach. Several models of electrochemical noise generation during corrosion processes have been established,[5–9] but most of the theories lack support from experimental investigations.

For the purpose of applying the electrochemical noise technique to MIC, this chapter is focusing on the spectral analysis of pitting corrosion of pure Al and Al alloys, with some description of the thin foil technique for detecting pit penetration. Electrochemical noise measurements follow either fluctuations of potential, often E_{corr}, or current as a function of time or experimental conditions. In this chapter, most measurements are conducted on electrode couples of the same material which eliminates the noise from other materials such as reference or counter electrodes. Many published papers concentrate on potential noise alone,[10–24] while only a small number of papers studied current noise.[25–29] In the present investigation, current noise as well as potential noise have been evaluated, and the current noise appears to be more related to localized corrosion than potential noise. The experimental noise data sets have been analyzed by an analysis program based on the maximum entropy method (MEM) through which the frequency domain information of noise data can be presented as power spectral density (PSD).

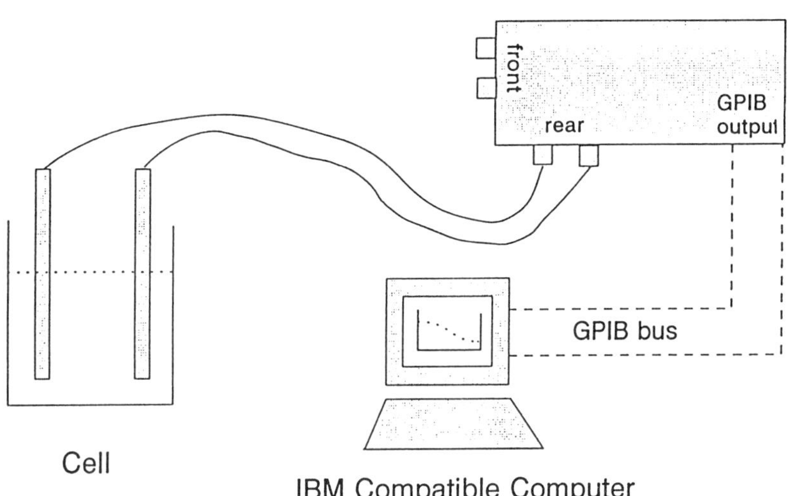

FIGURE 1. Experimental arrangement for potential noise measurements.

Another concern of this investigation is to detect pit penetration by the thin foil technique which was introduced by Hunkeler and Boehni.[30] Combined with the dual-cell design pioneered by Little and co-workers[31] for MIC studies, an electrochemical cell was designed to monitor pit propagation. This cell has two electrochemically separated compartments, making it possible to use two different solution environments in a given experiment. This approach is very useful for studies of MIC, in which corrosion is often of a localized nature.

II. EXPERIMENTAL APPROACH

A. Potential Noise Measurements

The experimental arrangement for potential noise measurements is shown in Figure 1. Two identical samples immersed in the test solution are connected to a HP3457A digital multimeter which is controlled by an IBM-compatible computer. The potential fluctuations between the two samples are determined at a constant rate of 2 points per second for up to 2048 s and stored in the computer for further analysis. Pt wires were tested first in 0.5 *N* NaCl solution as background material which does not corrode in this environment. Al 2009/SiC-T8 metal matrix composite (MMC) samples containing 20% SiC particles were tested in 0.5 *N* NaCl, 0.1 *M* Na_2SO_4, and deionized water for up to 2 weeks.

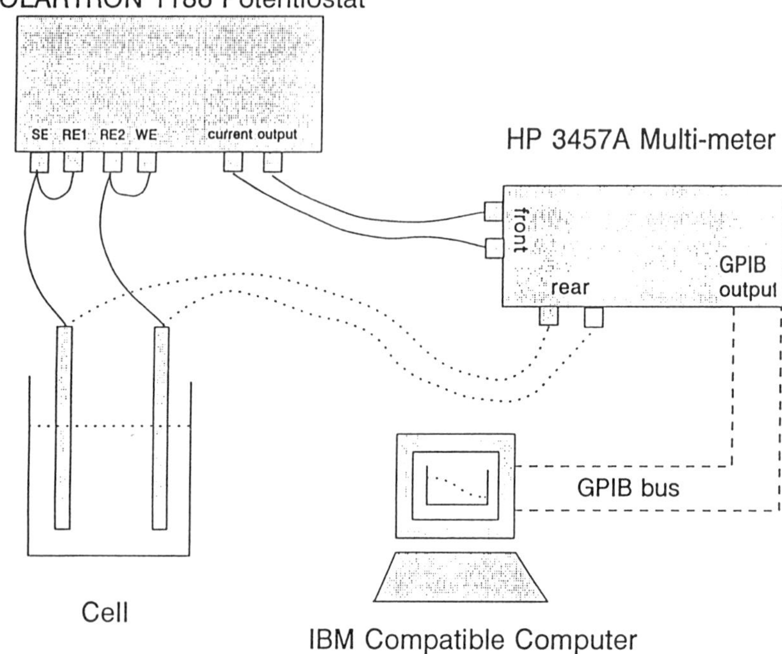

FIGURE 2. Experimental arrangement for current noise measurements.

B. Current Noise Measurements

The experimental arrangement for the recording of current noise is shown in Figure 2. In addition to the equipment used in potential noise measurements, a Schlumberger model 1186 potentiostat is employed as a zero resistance ammeter (ZRA). The current fluctuations are sampled at the current output of the potentiostat by the computer-controlled multimeter. The sampling rate was 2 points per second and the collected data were stored in the computer for further processing. Pt and the MMC Al 2009/SiC were tested as in the potential noise measurements.

C. Thin Foil Tests

The experimental arrangement for measuring pit propagation rates is shown in Figure 3. In a newly designed cell, a piece of metal foil with known thickness has one side exposed to the solution in the cell, while the other side is covered by a thin sheet of filter paper. Next to the filter paper, two closely spaced detector electrodes are pressed tightly to the filter paper by the outer cell wall. A small voltage is applied between the two parallel detector electrodes. When the paper is dry, no current

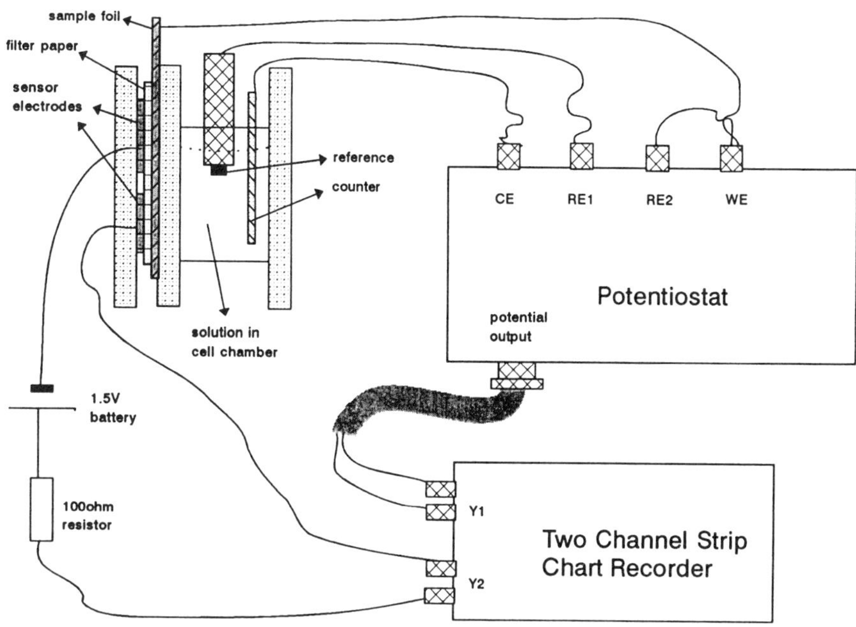

FIGURE 3. Experimental arrangement for thin foil measurements.

flows across the two detector electrodes; however, if a pit has penetrated the test sample, the paper will be wetted and a current will be detected. Using this test, the time for the fastest growing pit to penetrate the foil can be recorded for foils of different thicknesses and a pit-propagation rate law can be determined. In order to test the experimental arrangement shown in Figure 3, Al and Ni foils with varying thickness were exposed in 0.5 *N* NaCl solution. In order to reduce the time for foil penetration, the foils were polarized by a constant anodic current which exceeded the passivation current. The design used in the present study is slightly different from that of Hunkeler and Boehni,[30] in which a voltage was applied between the test sample and the detector electrode, which may polarize the test electrode in an unpredictable and undesirable way.

In the thin foil tests reported here, only one half of the special test cell was used. The filter paper and the detector electrodes could be replaced by another half cell, so that both sides of the test sample would be exposed to different test solutions. For example, one solution could contain microorganisms, while the other would be an abiotic solution of the same composition similar to the dual-cell approach pioneered by Little et al.[31] In further stages of this project, this PITCELL will be explored for its application in studies of MIC phenomena. For these tests a modification of the detection system will be necessary.

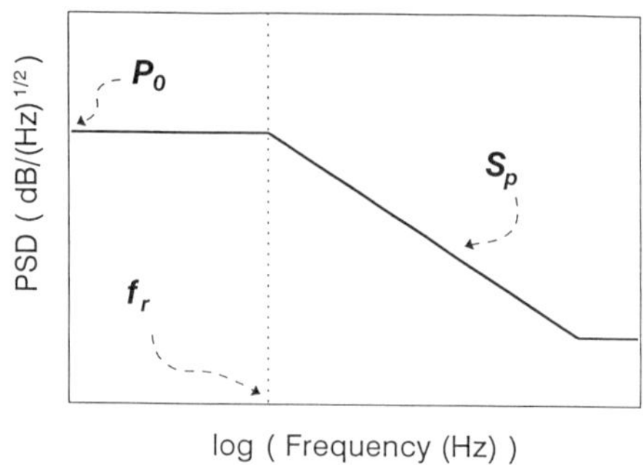

FIGURE 4. Schematic PSD plot.

D. Electrochemical Impedance Spectroscopy (EIS)

For a comparison and as confirmation of the results obtained with the techniques described above, EIS measurements have been conducted for all materials and solutions. A significant amount of information has been collected previously with EIS in this laboratory for localized corrosion of Al alloys, and models for the analysis of EIS data have been prepared.[32] EIS data have been collected as a function of exposure time immediately after each measurement of electrochemical noise data.

III. DATA ANALYSIS AND EXPERIMENTAL RESULTS

A. Spectral Analysis of Electrochemical Noise Data

For the analysis of electrochemical noise data, the fluctuations of potential or current are often difficult to interpret and are probably less meaningful than the frequency response behavior. The PSD is defined as the average power of a signal in a certain frequency range.[33] A typical PSD plot of electrochemical noise data is given in Figure 4, where the three most important parameters are the low-frequency limit P_o, the roll-off frequency f_r, and the slope S_p. The correlation of these parameters with the mechanism of the corrosion process is still uncertain in most cases. However, it can be expected that higher values of P_o correspond to larger fluctuations of the measured signal and therefore to the higher intensity of localized attack and that f_r and S_p are related to the type of corrosion of a given material (e.g., uniform vs. localized).

In order to obtain the frequency domain information, the time series of noise data has to be transformed by certain functions. A computer program using the MEM has been developed to obtain the PSD of the noise data collected in the present study. The MEM estimation of PSD is described by the following equations:[34,35]

$$PSD(f) = \frac{a_o}{|1 + \sum_{k=1}^{M} a_k z^k|^2} = \sum_{j=-M}^{M} \Phi_j z^j \tag{1}$$

where Φ_j is the autocorrelation function of the time series, a_k is the MEM coefficient array, and M is called the order or the number of poles of the estimation approximation. Once the MEM coefficient a_k is obtained by Equation 2,

$$\begin{bmatrix} \Phi_0 & \Phi_1 & \Phi_2 & \cdots & \Phi_M \\ \Phi_1 & \Phi_0 & \Phi_1 & \cdots & \Phi_{M-1} \\ \Phi_2 & \Phi_1 & \Phi_0 & \cdots & \Phi_{M-2} \\ \cdots & \cdots & & & \\ \Phi_M & \Phi_{M-1} & \Phi_{M-2} & \cdots & \Phi_0 \end{bmatrix} \cdot \begin{bmatrix} 1 \\ a_1 \\ a_2 \\ \cdots \\ a_M \end{bmatrix} = \begin{bmatrix} a_0 \\ 0 \\ 0 \\ \cdots \\ 0 \end{bmatrix} \tag{2}$$

the PSD can be calculated through Equation 1.

Compared to the Fast Fourier Transformation (FFT) method, the MEM estimation has the advantage of sharper spectral features. Reducing M within certain accuracy limits can increase the calculation speed and smooth the spectra.

B. Potential and Current PSD Data, EIS Data

Both potential noise and current noise data have been analyzed by the MEM program. Figure 5 shows the potential fluctuation-time records for Pt and Al 2009/SiC in 0.5 *N* NaCl solution for the first day of exposure; Figure 6 shows the corresponding PSD plots. It is interesting that the potential noise for Pt, which does not corrode in this solution, is at the same level as that for Al 2009/SiC, which pitted severely in the first day of immersion. In Figure 6, very similar P_0 values are observed for both systems. This very reproducible result gave the motivation for studying the current noise of these materials. Figure 7 shows the current fluctuation-time records for Pt and Al 2009/SiC in 0.5 *N* NaCl solution during the first day of immersion, while Figure 8 shows the corresponding PSD plots. On the scale used in Figure 7 the current fluctuations for Pt cannot be detected. The current noise PSD for Pt (Figure 8) is at the same level as the background noise from the potentiostat. However, the current noise PSD for Al 2009/SiC is much higher at all frequencies and shows the typical shape observed in our studies for pitting of Al-based materials. This phenomenon indicates that localized corrosion is not the only cause of the observed potential fluctuations. The large potential fluctuations for Pt are likely due to fluctuations of the O_2 mass transport

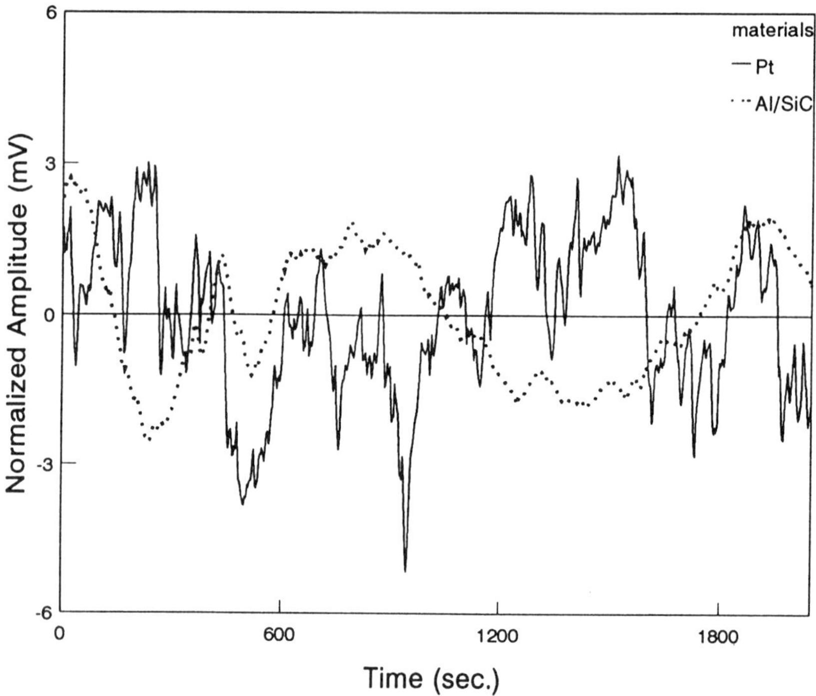

FIGURE 5. Normalized potential fluctuations for Pt and Al 2009/SiC for the first day of exposure in 0.5 *N* NaCl.

process, which lead to large changes of the open-circuit potential of the passive Pt surface. This process does not involve corrosion reactions and the corresponding current fluctuations are very small. On the other hand, the Al 2009/SiC MMC experiences passive film-breakdown processes as pits initiate during the first day of immersion and the current fluctuations are very large. Apparently the current fluctuations are more indicative of pit initiation and propagation than the potential fluctuations and therefore more applicable to the study of MIC phenomena. Recording of potential noise alone could lead to erroneous conclusions concerning the type and extent of corrosion.

For Al alloys, previous results obtained with EIS in this laboratory have shown that pits initiate within the first day of exposure to 0.5 *N* NaCl and propagate according to a time law which can be expressed as[36]

$$r = at^b \tag{3}$$

where r is the pit propagation rate, a is an experimental parameter which depends on the metal-solution system, and b is another experimental

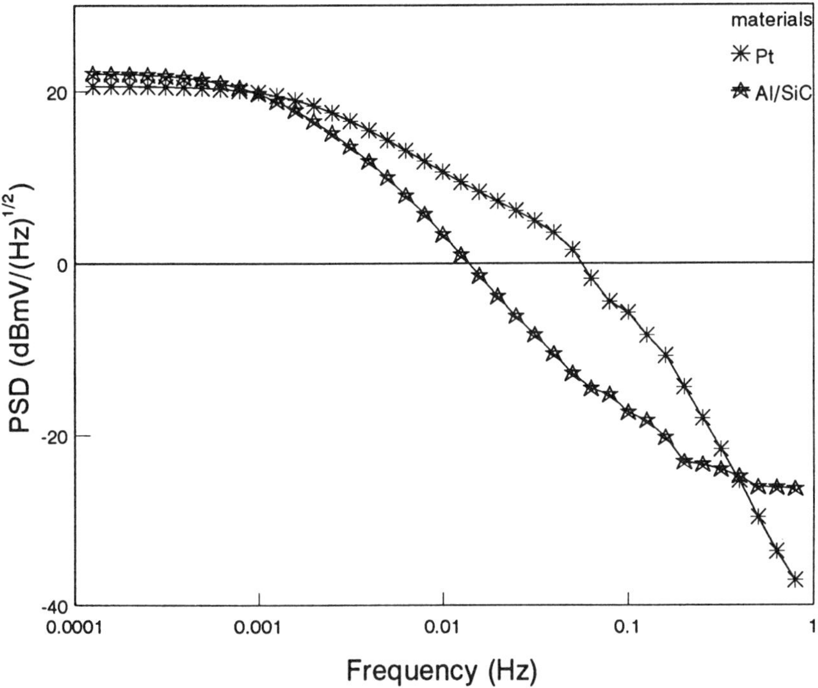

FIGURE 6. PSD plots for Pt and Al 2009/SiC for the first day of exposure in 0.5 *N* NaCl; data of Figure 5.

parameter which is indicative of the mechanism of pit propagation. Figure 9 shows the impedance curves for Al 2009/SiC in 0.5 *N* NaCl solution as a function of immersion time. All curves show the typical pitting behavior which is described by the model discussed elsewhere.[1,2] The total capacitance observed in the frequency region between 5 and 500 Hz shows a big increase from the first to the second day, indicating a large increase of the pitted area during this time. The polarization resistance of the pits which can be extracted from the spectra shown in Figure 9 follows the time dependence given in Equation 3 for $b = 1$.

To illustrate the noise behavior of Al 2009/SiC in 0.5 *N* NaCl as a function of immersion time, Figure 10 gives the potential fluctuation PSD curves for 2 weeks of immersion. No distinguishable features can be determined from these curves. However, in Figure 11, which presents the current fluctuation PSD curves for the same experimental conditions, the curve obtained during the first day of immersion is much higher than those for the remainder of the exposure period which are almost identical (except for the 7th day, for which an obvious artifact was found). All curves have a similar roll-off frequency f_r and similar linear range slope S_p (except for the 7th day). The three parameters which are

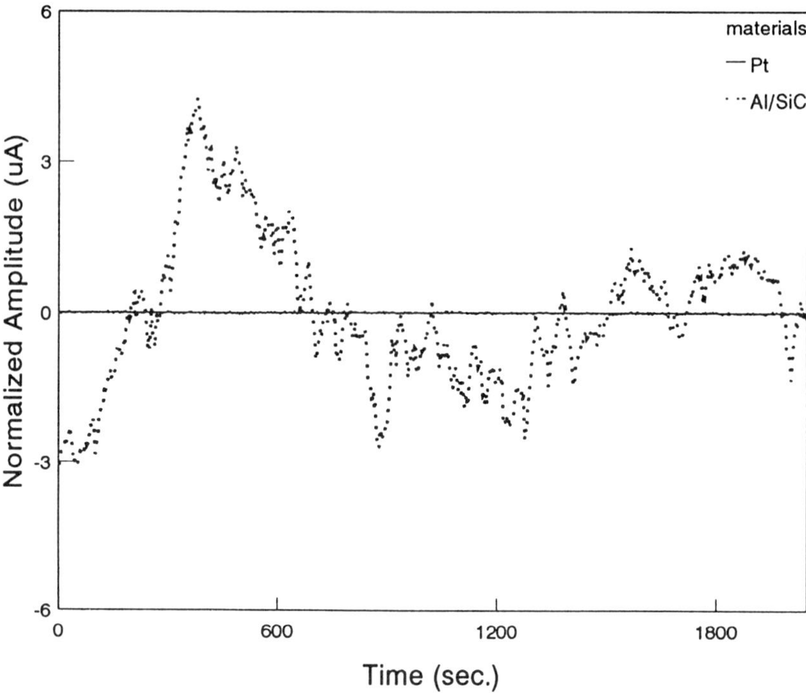

FIGURE 7. Normalized current fluctuations for Pt and Al 2009/SiC for the first day of exposure in 0.5 *N* NaCl.

considered characteristic for both the potential and current PSD curves are tabulated in Table 1 as a function of exposure time. A comparison of the P_0 values in Table 1 shows that P_0 does not change with exposure time for the potential noise data, but decreases significantly after the first day for the current noise data. The average value of f_r is about 0.5 Hz for the current noise and 0.9 Hz for the potential noise without significant changes with exposure time. The slope S_p also does not seem to depend on exposure time, but is significantly larger for the potential noise than for the current noise.

In order to compare the noise data for solutions of different corrosivity, the Al 2009/SiC MMC was also exposed to 0.1 *M* Na_2SO_4 and deionized water. The PSD curves for both potential and current fluctuations in both solutions are presented in Figures 12 and 13. The parameters P_0, f_r, and S_p are tabulated in Tables 2 and 3. The data analysis is made difficult by the peculiar shape of some of the curves, and no obvious trends could be detected. This lack of an obvious difference in the noise data characteristics in the three solutions may be due to the poor corrosion resistance of the Al 2009/SiC MMC, which corrodes sig-

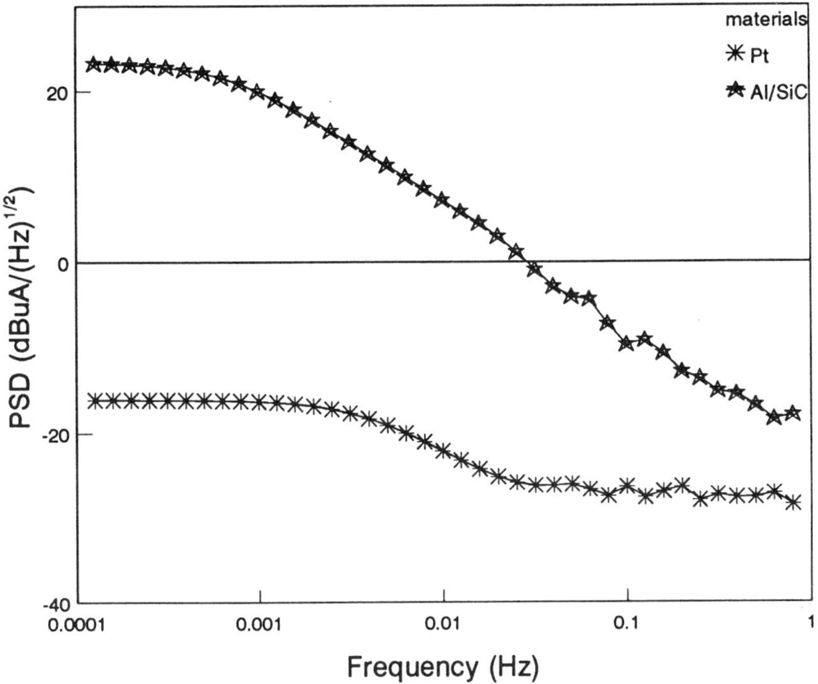

FIGURE 8. PSD plots for Pt and Al 2009/SiC for the first day of exposure in 0.5 *N* NaCl; data of Figure 7.

nificantly even in pure water. However, the current PSD curves in Figures 12 and 13 are low compared to those for the NaCl solution, and the P_0 values in Tables 2 and 3 are much lower than those for NaCl. It is interesting that the difference in the slope S_p between potential and current noise data is the same in all three solutions (Tables 1 to 3). No pits could be found on the samples exposed to the Cl^--free solutions by visual examination. The impedance plots for the Al 2009/SiC MMC in 0.1 *M* Na_2SO_4 are shown in Figure 14 as a function of immersion time. The absence of pitting is indicated by the absence of the second time constant at low frequencies (see Figure 9).

C. Thin Foil Test Results

Figure 15 shows the detector current pattern for a pure Al foil (0.0254 mm) in 0.5 *N* NaCl for 20 μA/cm² applied anodic current density. Penetration of the foil was detected after exposure for 25 min, which leads to a very high penetration rate of about 1 μm/min. By visual examination, two pits were found to have penetrated the foil. Figure 16 shows the

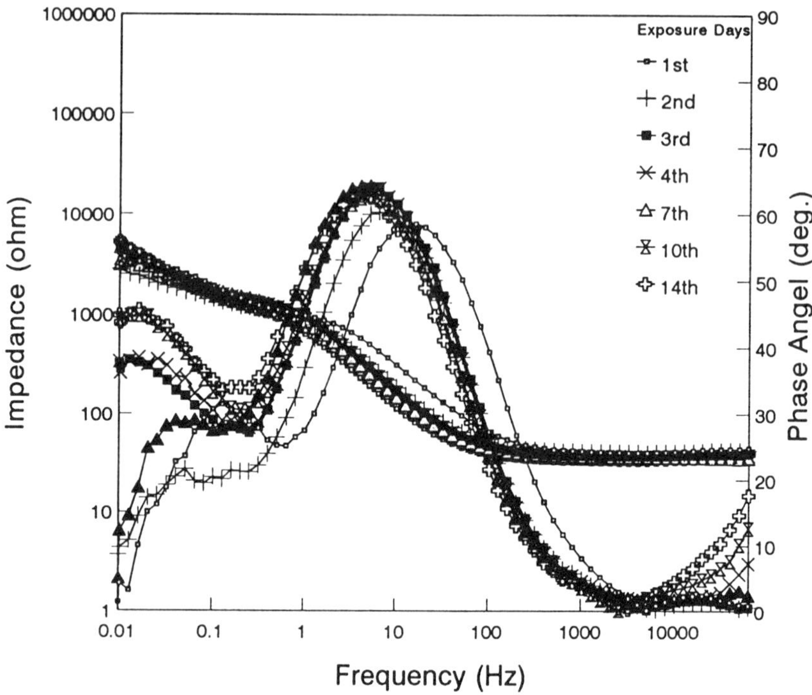

FIGURE 9. Impedance spectra for Al 2009/SiC as a function of exposure time to 0.5 *N* NaCl.

detector current pattern for a Ni foil (0.124 mm) in 0.5 *N* NaCl for an anodic polarization current density of 200 $\mu A/cm^2$. It is interesting that after the first penetration a second current burst can be seen. By microscopic observation, one pit was found on the surface. The pit shape from both sides of the foil is shown schematically in Figure 17. Apparently the growing pit produced undercutting of the exposed surface and the knife-like feature on the backside may be the cause of the second current burst triggered by an additional flux of electrolyte.

IV. SUMMARY AND CONCLUSIONS

Electrochemical impedance spectroscopy (EIS) and electrochemical noise analysis (ENA) have been evaluated for their usefulness in studies of localized corrosion phenomena. An Al-based metal matrix composite has been used as a model system because of its known susceptibility to localized corrosion. A thin foil technique has also been evaluated because of its promise as a tool for detection of pit penetration during the early stages of the corrosion process.

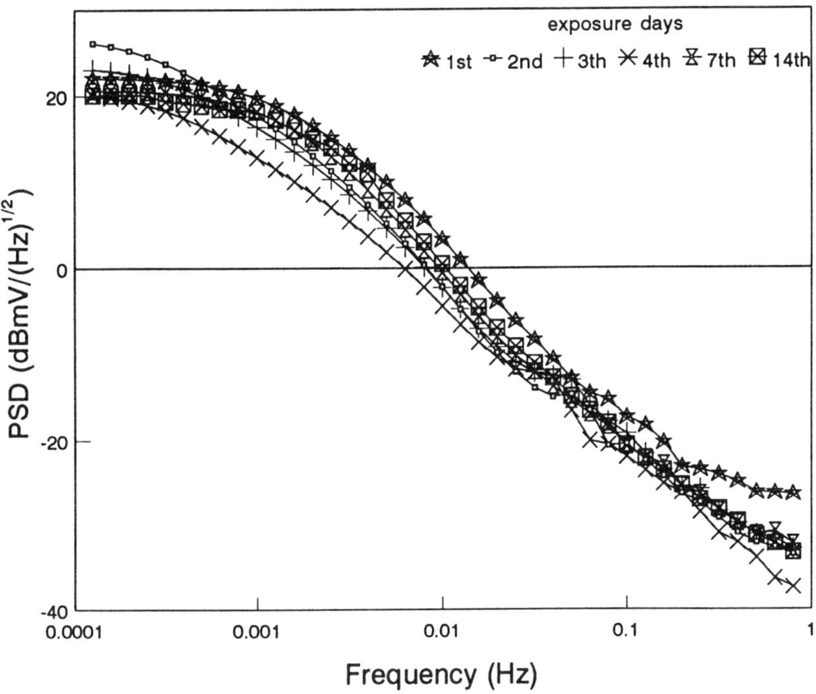

FIGURE 10. PSD plots for potential fluctuations for Al 2009/SiC as a function of exposure time to 0.5 *N* NaCl.

Emphasis has been placed on the evaluation of ENA, which is a relatively new technique with the advantage that no external signal has to be applied in order to determine the corrosion characteristics of a system. For this reason ENA is very promising as a tool, not only for laboratory studies, but perhaps even more for field studies and corrosion monitoring. The effort in the initial phases of this project has been concentrated on the development of software for the recording and analysis of electrochemical noise data. The maximum entropy method has been chosen for data analysis and the results of this analysis have been displayed as power spectrum density (PSD) plots. An important result of the initial studies is the realization that potential noise can be misleading for systems in which mass transport plays an important role, such as Pt or Al in aerated, neutral NaCl. Since the corrosion potential of Pt in this environment is poorly biased, small fluctuations of the transport of O_2 to the surface can lead to large fluctuations of E_{corr} which could be mistakenly identified as being due to localized corrosion. On the other hand, current fluctuations seem to be mainly due to localized corrosion effects. While the PSD plots give a convenient overview of the

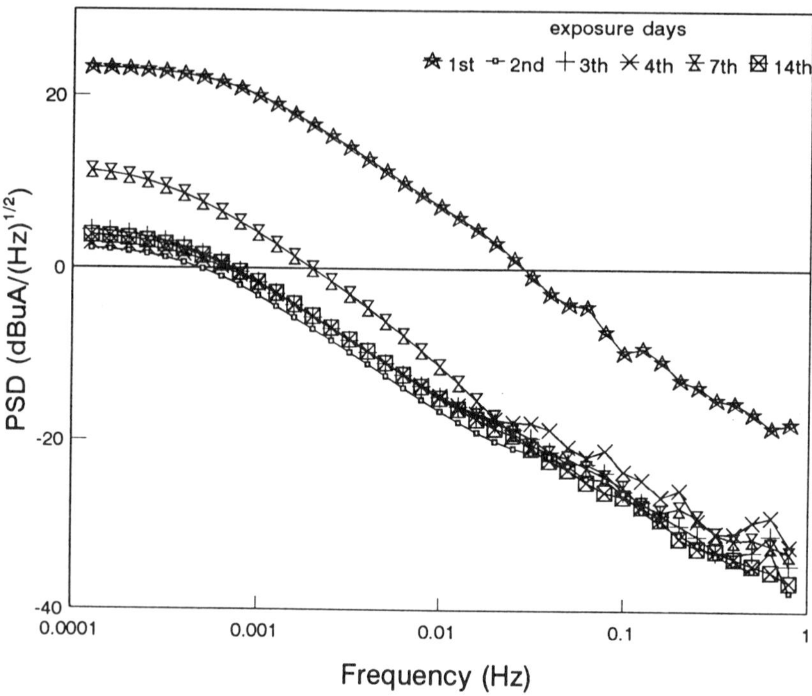

FIGURE 11. PSD plots for current fluctuations for Al 2009/SiC as a function of exposure time to 0.5 *N* NaCl.

TABLE 1. PSD Parameters for Al2009/SiC(20p)/T8 in 0.5 *N* NaCl

		Days					
		1st	2nd	3rd	4th	7th[a]	14th
Current	P_0 (dB/$\sqrt{}$Hz)	22.5	2.4	4.3	3.3	11.4	4.0
	f_r(mHz)	0.7	0.5	0.5	0.6	0.3	0.5
	Sp (dB/dec)	−13.1	−12.1	−12.4	−12.2	−14.1	−12.3
Potential	P_0 (dB/$\sqrt{}$Hz)	22.0	26.1	22.7	20.0	21.0	20.0
	f_r(mHz)	1.2	0.7	0.7	0.7	1.0	1.2
	Sp (dB/dec)	−21.8	−22.7	−20.4	−19.1	−21.6	−22.6

[a] Some artifacts are observed in the 7th day current fluctuation data.

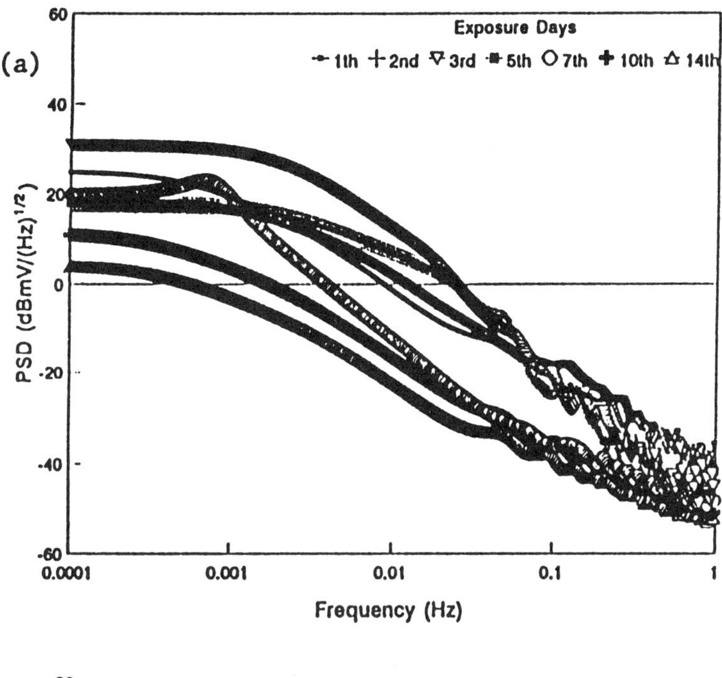

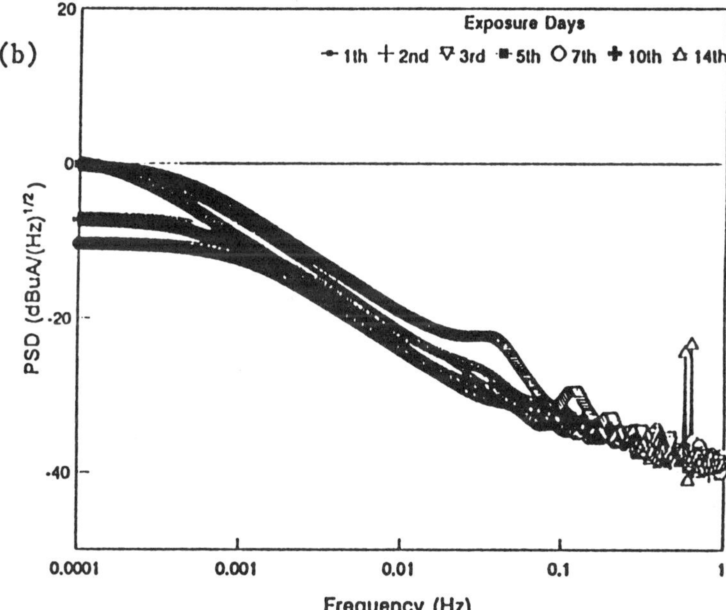

FIGURE 12. PSD plots for potential (a) and current (b) fluctuations for Al 2009/SiC as a function of exposure time to 0.5 *M* Na_2SO_4.

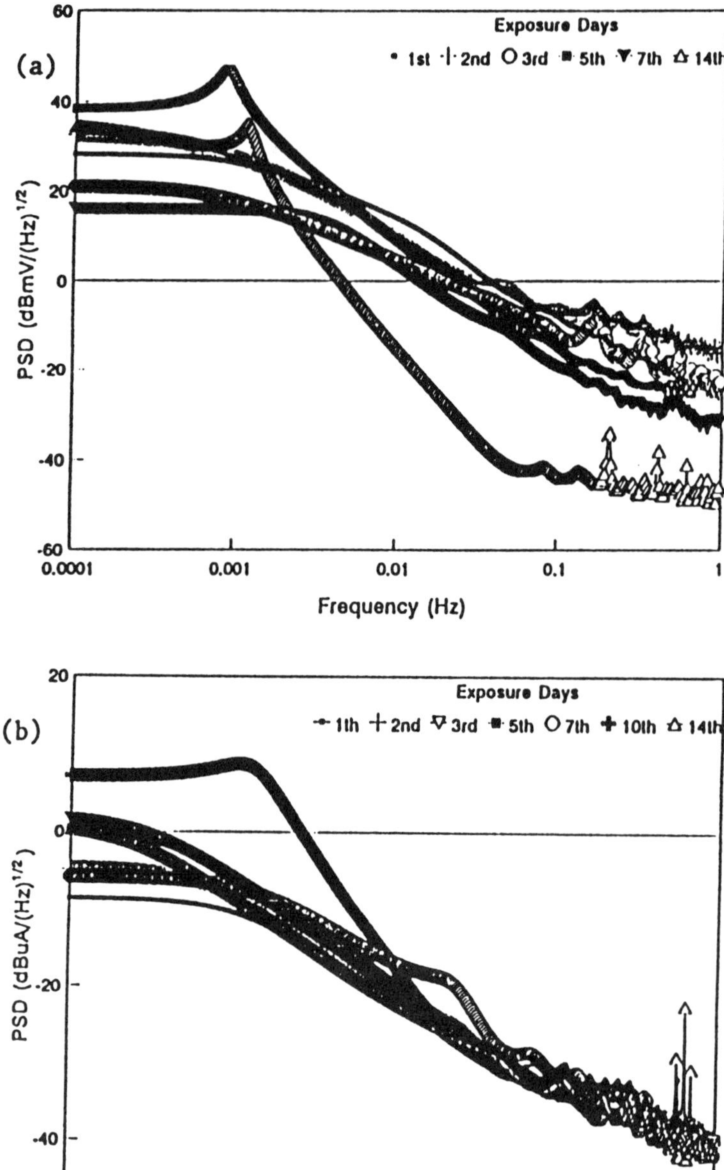

FIGURE 13. PSD plots for potential (a) and current (b) fluctuations for Al 2009/SiC as a function of exposure time to deionized water.

TABLE 2. PSD Parameters for Al2009/SiC(20p)/T8 in Na_2SO_4

		Days						
		1st	2nd	3rd	5th	7th	10th	14th
Current	P_0 (dB/√Hz)	−1	−7	0	−11	−10	−7	0
	f_r(mHz)	0.3	0.8	0.3	0.9	0.9	0.6	0.2
	Sp (dB/dec)	−13.6	−13.1	−13.1	−12.6	−11.1	−11.6	−12.2
Potential	P_0 (dB/√Hz)	25	18	31	18	20	11	4
	f_r(mHz)	0.8	—	2.0	1.9	0.8	0.5	0.8
	Sp (dB/dec)	−23.7	—	−28.9	−21.6	−28.9	−19.6	−17.6

TABLE 3. PSD Parameters for Al2009/SiC(20p)/T8 in H_2O

		Days						
		1st	2nd	3rd	5th	7th	10th	14th
Current	P_0 (dB/$\sqrt{}$Hz)	−8	−5	2	0	−6	7	0
	f_r(mHz)	1.1	0.6	0.3	0.4	1.1	1.5	0.2
	Sp (dB/dec)	−13.4	−12.9	−15.1	−14.4	−11.4[a]	−28.7	−12.4
Potential	P_0 (dB/$\sqrt{}$Hz)	29	33	21	38	16		34
	f_r(mHz)	1.9	1.1	0.7	—[b]	2.8		—[b]
	Sp (dB/dec)	−19.5	−21.8	−14.3	—[b]	−21.0		—[b]

[a] PSD plot has at least two obvious different slopes; one presented corresponds to the lower frequency one.

[b] Parameter not available because of the irregular shape of the plot.

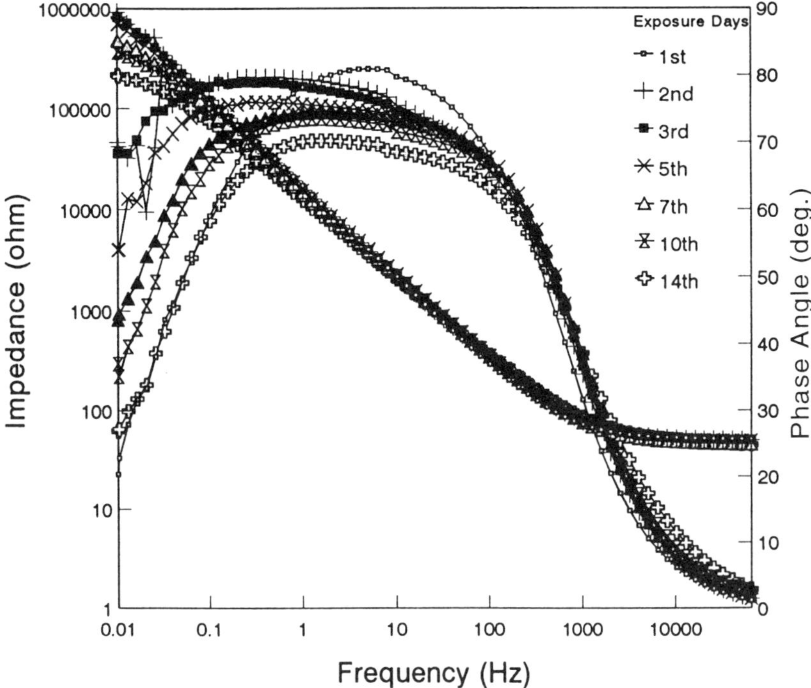

FIGURE 14. Impedance spectra for Al 2009/SiC as a function of exposure time to 0.1 *M* Na_2SO_4.

frequency dependence of the electrochemical noise, the correlation of the characteristic parameters P_o, f_r, and S_p with the extent of corrosion and the type of corrosion is not clear at present. Therefore, these parameters have been used in this study for a qualitative description of the noise characteristics of different materials and exposure to different solutions.

EIS, which is a more mature technique, has been used mainly to determine the extent of pitting of the Al-based MMC at certain time intervals, for establishment of a pit growth law according to Equation 3 and for comparison with the noise data.

A new experimental design has been completed for the thin foil technique which eliminates potential problems with previous detection systems for the penetration of the metal foil. This technique will be used at a later stage to determine pit growth rates with a nonelectrochemical technique by measuring the time for foil penetration as a function of foil thickness. The application of this technique lies in the monitoring area for MIC similar to the "pitting fuse" described by Kendig et al.[37] A metal foil can be exposed as part of a system to be monitored and an alarm

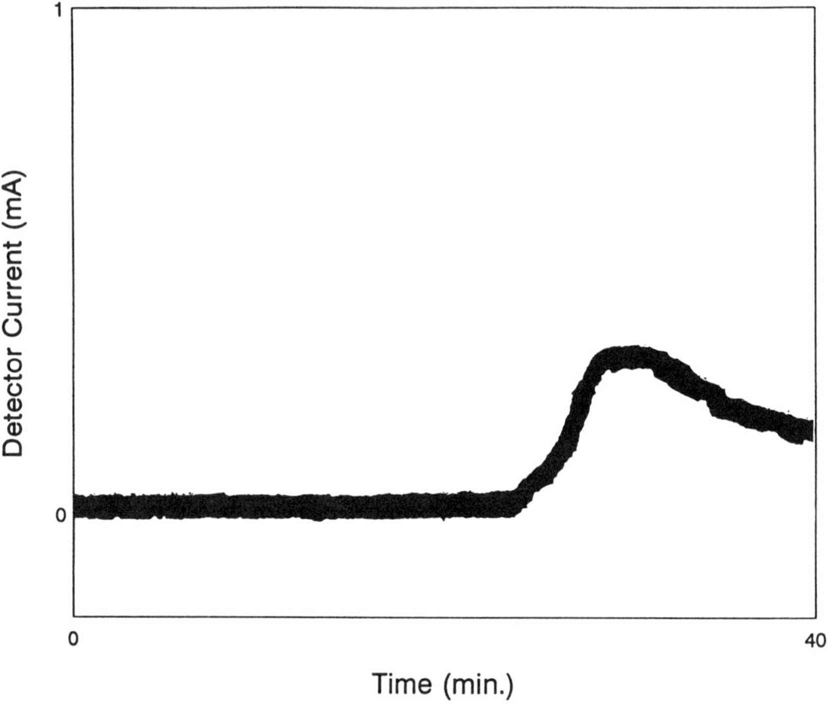

FIGURE 15. Detector current-time trace for Al foil exposed to 0.5 *N* NaCl at applied anodic current.

can be detected when this foil has been penetrated by the fastest growing pit. The foil thickness can be adjusted, considering Equation 3, to cause such an alarm in a reasonable exposure time.

An important part of this project is the transfer of the software and hardware to the Center for Interfacial Microbial Process Engineering at Montana State University, where the use of the candidate electrochemical techniques for the study of localized corrosion phenomena in MIC will be further evaluated.

ACKNOWLEDGMENT

This project has been funded by Montana State University under Subcontract No. 291435–01.

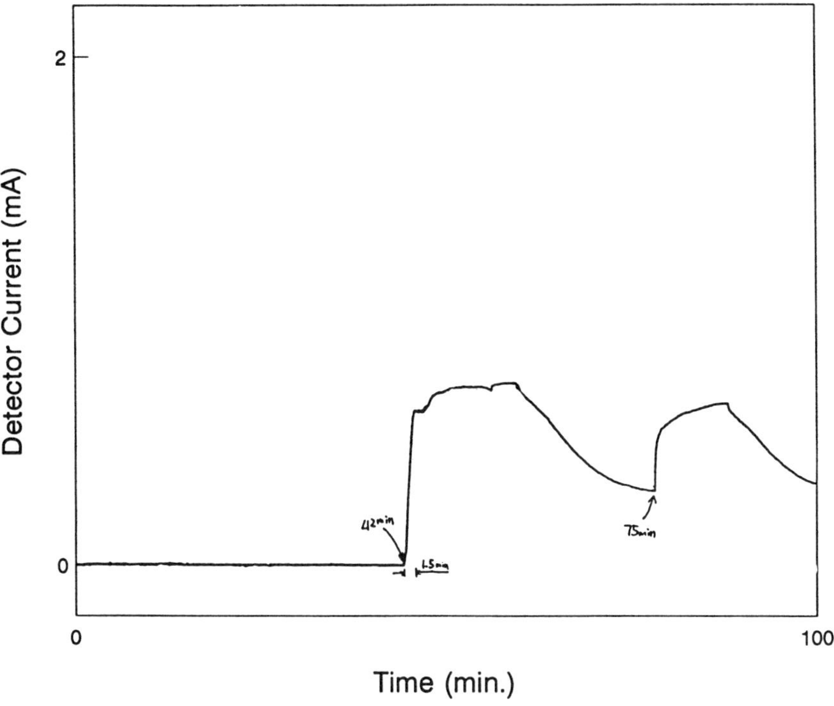

FIGURE 16. Detector current-time trace for Ni foil exposed to 0.5 *N* NaCl at applied anodic current.

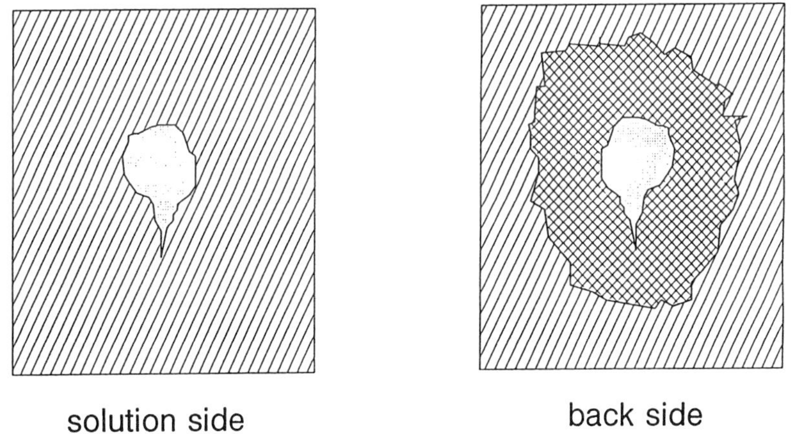

FIGURE 17. Appearance of Ni foil after pitting test (schematic).

REFERENCES

1. **Mansfeld, F., Lin, S., Kim, S., and Shih, H.,** Pitting and passivation of Al alloys and Al-based metal matrix composites, *J. Electrochem. Soc. Vol.,* 137, 78, 1990.
2. **Mansfeld, F., Lin, S., Kim, S., and Shih, H.,** Surface modification of Al alloys and Al-based metal matrix composites by chemical passivation, *Electrochem. Acta,* 34, 1123, 1989.
3. **Mansfeld, F. and Little, B.,** A technical review of electrochemical techniques applied to microbiologically influenced corrosion, *Corros. Sci.,* 32, 247, 1991.
4. **Mansfeld, F. and Lorenz, W. J.,** Electrochemical impedance spectroscopy (EIS): application in corrosion science, in *Techniques for Characterization of Electrodes and Electrode Processes,* John Wiley & Sons, New York, 1991, 581.
5. **Blanc, G., Epelboin, I., Gabrielli, C., and Keddam, M.,** Electrochemical noise generated by anodic dissolution or diffusion processes, *J. Electroanal. Chem.,* 75, 94, 1977.
6. **Okata, T.,** A theoretical analysis of the electrochemical noise during the induction period of pitting corrosion in passive metals, *J. Electroanal. Chem.,* 297, 361, 1991.
7. **Okata, T.,** Electrochemical noise considered by two-step initiation hypothesis of pitting corrosion, *J. Electrochem. Soc.,* 92, 9, 1991.
8. **Hashimoto, M., Miyajima, S., and Murata, T.,** A stochastic analysis of potential fluctuation during passive film breakdown and repair on iron, *Corros. Sci.,* 33, 885, 1992.
9. **Kendig, M. and Hagen, G.,** Monte Carlo model of electrochemical noise, *Corrosion/88,* Paper No. 383, NACE, Houston, TX, 1988.
10. **Searson, P. and Dawson, J.,** Analysis of electrochemical noise generated by corroding electrodes under open-circuit conditions, *J. Electrochem. Soc.,* 135, 1908, 1988.
11. **Uruchurtu, J. and Dawson, J.,** Noise analysis of pure aluminum under different pitting conditions, *Mater. Sci. Forum,* 8, 113, 1986.
12. **Uruchurtu, J.,** Electrochemical investigations of the activation mechanism of aluminum, *Corrosion,* 47, 472, 1991.
13. **Hladky, K. and Dawson, J.,** The measurement of localized corrosion using electrochemical noise, *Corros. Sci.,* 21, 317, 1981.
14. **Hladky, K. and Dawson, J.,** The measurement of corrosion using electrochemical l/f noise, *Corros. Sci.,* 22, 231, 1982.
15. **Loto, C. and Cottis, R.,** Electrochemical noise generation during stress corrosion cracking of the high-strength aluminum AA7075-T6 alloy, *Corrosion,* 45, 136, 1989.
16. **Loto, C. and Cottis, R.,** Electrochemical noise generation during stress corrosion cracking of alpha-brass, *Corrosion,* 43, 499, 1987.
17. **Cottis, R. and Loto, C.,** Electrochemical noise generation during SCC of a high-strength carbon steel, *Corrosion,* 46, 12, 1990.
18. **Metikos-Hukovic, M., Loncar, M., and Zevnik, C.,** Monitoring the electrochemical potential noise produced by coated metal electrodes, *Werkst. Korros.,* 40, 494, 1989.
19. **Nachstedt, K. and Heusler, K.,** Electrochemical noise at passive iron, *Electrochem. Acta,* 33, 311, 1988.
20. **Bertocci, U. and Kruger, J.,** Studies of passive film breakdown by detection and analysis of electochemical noise, *Surface Sci.,* 101, 608, 1980.
21. **Simoes, A. and Ferreira, M.,** Crevice corrosion studies on stainless steel using electrochemical noise measurements, *Br. Corros. J.,* 22, 21, 1987.
22. **Monticelli, C., Brunoro, G., Frignani, A., and Trabanelli, G.,** Evaluation of corrosion inhibitors by electrochemical noise analysis, *J. Electrochem. Soc.,* 139, 706, 1992.
23. **Hashimoto, M., Miyajima, S., and Murata, T.,** An experimental study of potential fluctuation during passive film breakdown and repair on iron, *Corros. Sci.,* 33, 905, 1992.

24. **Hashimoto, M., Miyajima, S., and Murata, T.,** A spectrum analysis of potential fluctuation during passive film breakdown and repair on iron, *Corros. Sci.*, 33, 917, 1992.
25. **Tachibana, K., Miya, K., Furuya, K., and Okamoto, G.,** Changes in the power spectral density of noise current on type 304 stainless steels during the long time passivation in sulfuric acid solutions, *Corros. Sci.*, 31, 27, 1990.
26. **Smith, S. and Francis, R.,** Use of electrochemical current noise to detect initiation of pitting conditions on copper tubes, *Br. Corros. J.*, 25, 285, 1990.
27. **Bertocci, U. and Yang-Xiang, Y.,** An examination of current fluctuations during pit initiation in Fe-Cr alloys, *J. Electrochem. Soc.*, 131, 1011, 1984.
28. **Miyata, Y., Handa, T., and Takazawa, H.,** An analysis of current fluctuations during passive film breakdown and repassivation in stainless alloys, *Corros. Sci.*, 31, 465, 1990.
29. **Tsuru, T. and Sakairi, M.,** Spectrum analysis of current fluctuations upon pit generation, *Corros. Eng.*, 39, 447, 1990.
30. **Hunkeler, F. and Boehni, H.,** Determination of pit growth rates on aluminum using a metal foil technique, *Corrosion*, 37, 645, 1981.
31. **Little, B. J., Wagner, P., Gerchakov, S. M., Walch, M., and Mitchell. R.,** The involvement of a thermophilic bacterium in corrosion processes, *Corrosion*, 42, 533, 1986.
32. **Mansfeld, F., Shih, H., and Tsai, C. H.,** Software for simulation and analysis of electrochemical impedance spectroscopy (EIS) data, *ASTM STP*, p. 174, 1154, 1992.
33. **Oppenheim, A. and Schafer, R.,** *Discrete-Time Signal Processing*, Prentice-Hall, Englewood Cliffs, NJ, 1989.
34. **Fougere, P.,** *Maximum Entropy and Bayesian Methods*, Kluwer Academic, Boston, MA, 1990.
35. **Press, W., Flannery, B., Teukolsky, A., and Vetterling, W.,** *Numerical Recipes*, Cambridge University Press, 1986.
36. **Mansfeld, F., Wang, Y., Xiao, H., and Shih, H.,** Monitoring in localized corrosion of aluminum alloys with electrochemical impedance spectroscopy (EIS), in Proc. Symp., Critical factors in localized corrosion, *Electrochem. Soc. Proc.*, 92–9, 469, 1992.
37. **Kendig, M. W., Mansfeld, F., and Jeanjaquet, S.,** The evaluation of techniques for corrosion monitoring in flat plate collectors, *Electrochem. Soc. Proc.*, 83–1, 319, 1983.

Index

C

D

N

O

T

U

V

W

X

Z